Biology of Butterflyfishes

Biology of Butterflyfishes

Editors

Morgan S. Pratchett
Professor, Marine Ecology
ARC Centre of Excellence for Coral Reef Studies
James Cook University
Townsville
Australia

Michael L. Berumen
Assistant Professor of Marine Science and Engineering
Red Sea Research Center
King Abdullah University of Science and Technology (KAUST)
Thuwal
Kingdom of Saudi Arabia

B.G. Kapoor
Formerly Professor of Zoology
Jodhpur University
Rajasthan
India

CRC Press
Taylor & Francis Group
Boca Raton London New York

CRC Press is an imprint of the
Taylor & Francis Group, an **informa** business
A SCIENCE PUBLISHERS BOOK

CRC Press
Taylor & Francis Group
6000 Broken Sound Parkway NW, Suite 300
Boca Raton, FL 33487-2742

Printed in the United States of America on acid-free paper

International Standard Book Number: 978-1-4665-8289-7 (Hardback)

Library of Congress Cataloging-in-Publication Data

Biology of butterflyfishes / editors: Morgan S. Pratchett, Michael L. Berumen, B.G. Kapoor.
p. cm.
Includes bibliographical references and index.
ISBN 978-1-4665-8289-7 (hardcover : alk. paper) 1. Chaetodontidae. I. Pratchett, Morgan S. II. Berumen, Michael L. III. Kapoor, B. G.

QL638.C48B56 2014
597'.72--dc23

2013023192

Visit the Taylor & Francis Web site at
http://www.taylorandfrancis.com

CRC Press Web site at
http://www.crcpress.com

Science Publishers Web site at
http://www.scipub.net

Foreword

The chaetodontids, popularly known as butterflyfishes, are a group that to both the public and the scientific community are highly representative of coral reef ecosystems. This reflects not only the diversity of highly visible and strikingly coloured species but their strong association with living corals both as a source of nutrition and as a habitat for newly recruited individuals. Indeed a number of earlier publications visualised this not as a passive linkage but as a functionally important association that provided an insight into the prevailing and future health of coral reefs. The role of chaetodontids as "indicator" species in the context of reef health ensured this group maintained a high profile in the coral reef literature and focused attention on obligate associations with living corals and trends in abundance linked to disturbances in reef ecosystems. However recent research has identified demographic and physiological properties of corals that provide direct metrics of their health and fitness, but more importantly have revealed a high diversity of behavioural, ecological and evolutionary processes in the chaetodontids themselves. The latter findings have enhanced our understanding of this group not just as obligate corallivores tied closely to living corals, but as a speciose group of perciforms with a complex biogeography and evolutionary history, a wealth of behavioural and life history features and a diversity of feeding interactions with the reef biota. In this context chaetodontids represent an ecological and evolutionary prism that allow us to probe the historical nature of associations of reef fishes with the changing profiles of Cenozoic reefs, the circumstances under which species may limit or expand consumption of food resources and the patterns of response to disturbance at both ecological and evolutionary time scales. And although the now-clarified fossil record of this group is sparse, the increasing focus on phylogenetic reconstruction and phylogeography has provided a platform for rigorous biogeograpical analysis, investigation of patterns in the evolution of trophic and behavioural traits and geographical vs. ecological processes of speciation.

The significance of chaetodontids and especially the genus *Chaetodon* in the context of ecological and evolutionary studies can be summarized by three observations; i) They occur within a range of reef ecosystems manifesting very different geological histories and prevailing oceanographic conditions; ii) They show a pattern of feeding behaviour that provides an unparalleled opportunity to identify food resources, establish the degree of trophic specialization and show how nutritional and foraging profiles vary with temporal and spatial change; iii) Their distinctive colour patterns and behaviours allow close and detailed observation of reproductive and feeding patterns in undisturbed groups. Dietary patterns and foraging especially in the genus *Chaetodon,* have provided a critical focus in evaluating how both evolutionary history and current ecological circumstances have shaped these traits. The fact that members of this genus occur in all of the world's great tropical ocean basins, each with a unique history and prevailing oceanographic conditions, provides a key to evaluating the hierarchy of events from evolutionary to ecological that have shaped feeding behaviours. The large number of species (many endemic) that occur in regions with turbulent oceanographic histories or currently marginal conditions is especially useful in this context. The endemic species of the Red Sea and Persian Gulf, reef systems that are subject to highly variable conditions over evolutionary and ecological time scales, provide a counterpoint for comparative studies over the broader Indo-Pacific. However although the taxonomic structure of chaetodontids is better known than many other groups of reef fishes for individual species, the match between genetic structure and phenotypes, and the propensity to hybridise in specific locations, reminds us that reconstructing patterns of species divergence will be a challenge.

An intriguing aspect of current studies concerns the history and biogeographic distribution of feeding specialisation, specifically the emergence and frequency of obligate feeding on hard corals. Although corals appear to represent an abundant food resource for benthic feeding reef fishes, it is noteworthy that relatively few species have evolved the capacity to use them. These species are dominated by members of the genus *Chaetodon*. Feeding limited to specific taxa of hard corals appears to be the ultimate in resource specialisation. However it is worth remembering that individual corals represent a diverse repository of food sources ranging from animal tissue, microscopic plant assemblages, mucous and tissues enriched to varying degrees by bacterial content. Although the external manifestations of feeding behaviour indicate a high level of specialisation it is unclear what components of the corals are targeted in a nutritional context, something that may vary between species and locations. Probing the nexus between feeding and nutritional targets in species which specialise on hard corals represents an area that will repay increased research investment.

One of the most important findings in an evolutionary context is the strong support for the hypothesis that critical episodes of chaetodontid diversification occurred during the Miocene, at a period when the distribution, structure and composition of coral reefs achieved a "modern" configuration. This pattern which seems such an obvious consequence of an integrated evolutionary development of corals and reef fishes urges a caveat. The temptation to develop one-size-fits-all approaches to the ecology and evolution of reef fishes should be tempered with detailed examination of other groups of reef fishes. The emerging picture suggests unique clade and taxon-specific histories in reef fishes in the context of the geological evolution of Cenozoic reefs and their associated oceanographic conditions. There is a need to identify those episodes in the evolution of reef fishes that can be legitimately generalised to a wide range of taxa. The most appropriate pathway for accomplishing this is to provide integrated perspectives on the evolutionary history and ecology of critical groups of reef fishes, perspectives that allow rigorous comparisons between major clades. The group of studies encapsulated in this volume represent a major advance in achieving this goal.

John H. Choat

Contents

CHAPTER 1

The Origins and Diversification of Coral Reef Butterflyfishes

David R. Bellwood[1,*] and *Morgan S. Pratchett*[2]

INTRODUCTION

The Chaetodontidae is a diverse family of percomorph fishes represented by 122 extant species, characterised by deep compressed bodies, small protruded mouths and bristle-like teeth (Allen et al., 1998). The family is dominated by fishes of the genus *Chaetodon*, which are among the most conspicuous inhabitants of coral reef environments. Two thirds of all butterflyfishes are found living within coral reef habitats, and many of these species feed mainly, if not exclusively, on reef-building corals (Cole et al., 2008; Cole and Pratchett, Chapter 5). Because of their reliance on corals for food, *Chaetodon* butterflyfishes are regarded among the most specialised and highly evolved coral reef fishes (e.g., Gosline, 1985). These fishes are inextricably linked to the corals on which they feed (Reese, 1977, 1981), but did the family originate within coral reef environments?

Reefs as we know them today are essentially a feature of the last 65 million years. The K/T (Cretaceous/Tertiary) boundary at 65 million years ago (Ma) marked the end of the Mesozoic era. In the seas, this was

[1]ARC Centre of Excellence for Coral Reef Studies and School of Marine and Tropical Biology, James Cook University, Townsville, Australia.
E-mail: david.bellwood@jcu.edu.au
[2]ARC Centre of Excellence for Coral Reef Studies, James Cook University, Townsville, Australia.
E-mail: morgan.pratchett@jcu.edu.au
*Corresponding author

characterised by the loss of the ammonites and the end of the period in which rudist bivalves were the most significant feature of carbonate reefs (Wood, 1998). Thereafter, scleractinian corals became the dominant feature of carbonate reefs and these corals formed a distinctive component of carbonate reefs for the next 65 million years. However, it appears that scleractinian-dominated coral reefs and modern coral reef fish families first appeared and then diversified at approximately the same time, in the early Cenozoic (Bellwood, 1996; Bellwood and Wainwright, 2002; Wallace and Rosen, 2006). This suggests that the origins of modern coral reefs and their associated fish families are closely linked, and it is within this system that butterflyfishes (family Chaetodontidae) have proliferated, if not originated.

A fundamental step in understanding the origins of the family Chaetodontidae, is to establish the timing of their evolution relative to key points in the evolution of coral reefs. However, dating the earliest origins of the butterflyfishes, like any group of fishes, is difficult. There are basically two types of evidence that may be used: i) establishing the maximum age of fossils that unequivocally belong to the family Chaetodontidae, and ii) reconstruction of phylogenetic lineages based on either be morphological or molecular relationships, with dating calibrated using either biogeographic events, fossils or an inferred molecular clock. Each of these approaches has its advantages and limitations, but taken in concert, should provide a reasonably robust picture of the evolutionary history of the family. The two approaches are reciprocally illuminating but for ease of discussion each will be examined in turn.

RECORDS IN ROCK: DATING THE ORIGINS OF BUTTERFLYFISHES

For most coral reef fishes, the Eocene fossils from Monte Bolca in northern Italy mark the earliest record, dated to approximately 50 million years old. The fossils of Monte Bolca were laid down in deep waters in the vicinity of shallow tropical coastal ecosystems which contained both sea grasses and hard substrata, probably including some coral outcrops. These deposits mark the first documented record of numerous reef fish families such as the Acanthuridae, Labridae, Pomacentridae, Ephippidae and many others (Bellwood, 1996). For many years this also included the Chaetodontidae. Numerous Monte Bolca specimens were originally identified as belonging to the family Chaetodontidae. Most species are in the genus *Pygaeus* (8 species) (Fig. 1.1) but other chaetodontid genera include *Parapygaeus*, *Malacopygaeus* and *Acanthopygaeus* (Blot, 1980; Bannikov, 2004). In addition, Frickhinger (1991) inadvertently illustrated the Bolca enoplosid *Enoplosus pygopterus* as *Pygaeus gazolai* and placed it within Chaetodontidae.

Fig. 1.1 For many years the earliest fossil record of the Chaetodontidae was associated with a number of *Pygaeus* species from the Eocene deposits of Monte Bolca in northern Italy. Many, like *Pygaeus agassizi* Eastman (top panel) were relatively large predatory forms, others like *Pygaeus coleanus* Agassiz (bottom panel) were smaller but with no morphological links to the family (Bannikov, 2004). *P. coleanus* is now regarded as a generalized percoid in the *incertae sedis* genus *Blotichthys*.

Unfortunately, none of these Bolca specimens were identified based on synapomorphies (derived anatomical features) that are characteristic of the family Chaetodontidae, which is a critical step in the accurate placement of species within higher taxonomic groupings. As with the Pomacentridae (Bellwood and Sorbini, 1996), parrotfishes (formerly Scaridae) (Bellwood and Schultz, 1991) and others (Blot, 1980), the earliest records attributed to the family Chaetodontidae were erroneous. In a recent re-evaluation, Bannikov (2004) rejected all Eocene material previously placed within the family. One is probably best placed in the Enoplosidae, the remainder remain *incertae sedis* (of uncertain affinity).

The recently described *Frigoichthys margaritae* has some chaetodontid affinities but in many ways it more closely resembles extant species in the family Microcanthidae. Microcanthids have broad superficial similarity to the Chaetodontidae (which is why they are included in the modern identification books as relatives (cf. Kuiter, 2002)), although microcanthids probably have greater affinity with the Kyphosidae than the Chaetodontidae (Smith and Wheeler, 2006). Several other Bolca specimens have chaetodontid-like characteristics (e.g., *A. gazolai*), but none are clearly placed within the family or can be readily identified as immediate sister taxa.

Discounting the Monte Bolca material, the earliest recorded fossil from the family Chaetodontidae is *Chaetodon penniger* (Bogachev, 1964) from the Rupelian (29–35 Ma) of Transcaucasia. This specimen is accepted by both Patterson (1993) and Bannikov (2004) as a chaetodontid. However, the basis for this decision is not clear and its status does not appear to have been critically examined in a phylogenetic context. Its status therefore, remains equivocal. A somewhat later specimen, *Chaetodon fisheuri*, was described from the Miocene of Algeria, in northern Africa. Originally described in 1927 by Arambourg, it has recently been re-examined (Carnevale, 2006) and can be placed with some confidence within the family. This fully articulated specimen bears a strong resemblance to extant butterflyfishes (Fig. 1.2).

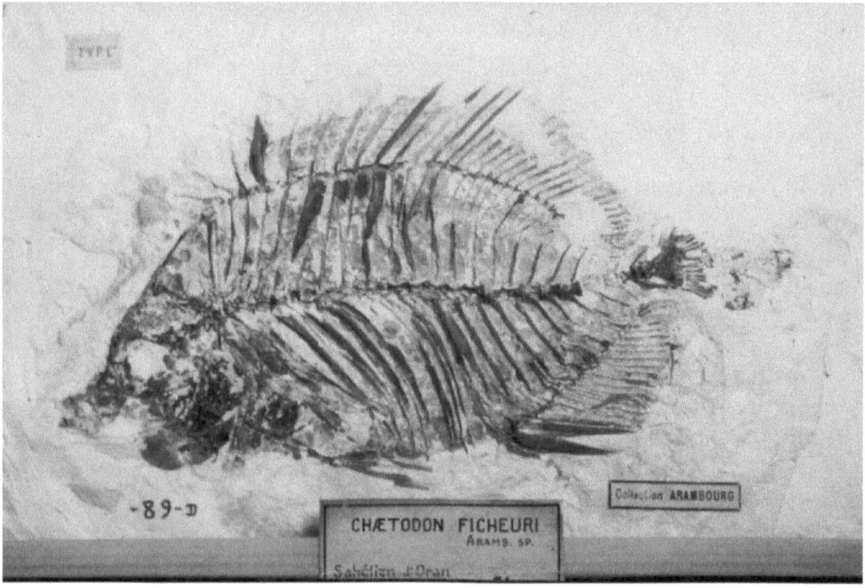

Fig. 1.2 *Chaetodon ficheuri* Arambourg from the deposits of Oran in northern Algeria is of late Miocene age. Exceptionally well preserved and clearly belonging to the family, this species has recently been re-examined (Carnevale, 2006).

Based on the fossil record, therefore, we can confidently say that the Chaetodontidae were present in the central-western Tethys, in the region now occupied by the Mediterranean, in the Miocene (5–25 Ma). There is also a strong possibility that butterflyfishes were present in the waters covering Western Europe during the Oligocene (29–35 Ma).

Phylogenetic Evidence

Robust phylogenies convey information on relationships and the relative ages of taxa. With appropriate calibrations, phylogenies can also yield valuable information on the likely ages of major lineages and the evolution of key character traits (e.g., Bellwood, 1994; Bellwood et al., 2004; Cowman and Bellwood, 2011). In recent years, the development of phylogenies based on molecular data has profoundly changed our view of chaetodontids and their relationships, and opened a new chapter in our understanding of the evolution of the group.

The Chaetodontidae were for a long time grouped with other fish families that share a similar body morphology. These deep bodied fishes were collectively termed the higher Squamipinnes (a group characterised by lateral compression and having scales (*squami*) on the bases of the fins (*pinnes*)). The higher Squamipinnes tentatively included the Ephippidae, Drepanidae, Scatophagidae, Pomacanthidae, Chaetodontidae, and the Acanthuroidei, namely the Zanclidae, Luvaridae, Siganidae and Acanthuridae (Tyler et al., 1989). However, this grouping (i.e., the monophyly of the higher squamipinnes) has been questioned and it almost certainly represents a polyphyletic group (Near et al., 2012).

Evolutionary relationships among the families formally ascribed to the higher Squamipinnes are yet to be fully resolved, with several recent analyses producing different topologies (e.g., Smith and Wheeler, 2006; Yagishita et al., 2009; Near et al., 2012). What is clear, however, is that the traditional relationship between butterflyfishes (Chaetodontidae) and angelfishes (Pomacanthidae) is well supported by molecular data (Bellwood et al., 2004; Smith and Wheeler, 2006; Yagishita et al., 2009). This suggests that numerous shared attributes between these families, including body shape, colour patterns, larval morphology, and tooth morphology, have a common evolutionary basis.

Based on a chronogram, constructed using Bayesian inference with multiple parametric priors, Bellwood et al. (2010), and later Cowman and Bellwood (2011), estimated that the family Chaetodontidae originated in the early Eocene (41.5–60.7 Ma) when it split from the Pomacanthidae. Ancestral range reconstructions suggesting that the origins and earliest diversification of the family were in the vicinity of the Indo-Australian Archipelago (Cowman and Bellwood, 2013).

Intra-relationships within the Chaetodontidae

There have been a number of phylogenies constructed for the family Chaetodontidae. The first cladogram, using morphological features, was constructed by Blum (1989) and expanded by Smith et al. (2003). These phylogenies were quickly followed by analyses using molecular data (e.g., Fessler and Westneat, 2007; Bellwood et al., 2010). The early morphological and later molecular phylogenies, however, produced discordant topologies. Although the morphological and molecular phylogenies agree in many respects, they differed in the placement of key taxa, especially *Amphichaetodon* (cf. Smith et al., 2003; Bellwood et al., 2010). In recent years, two large molecular phylogenies of the family have been produced (Fessler and Westneat, 2007; Bellwood et al., 2009). They used different representative taxa from each genus or subgenus and different molecular markers (although both included nuclear and mitochondrial DNA). The resultant phylogenetic trees were almost identical, giving considerable confidence in the inferred relationships and topologies of the trees (the two trees were subsequently combined by Cowman and Bellwood, 2011). It is these trees that will be used to explore intra-relationships with the family Chaetodontidae.

The taxonomy of generic and subgeneric groups within the Chaetodontidae is confusing. As extremely colourful fishes, they have attracted a lot of attention (Allen et al., 1998; Kuiter, 2002). This, along with a limited range of character states, has led to a multitude of generic and subgeneric names. The Chaetodontidae contains two distinct clades: the bannerfishes and the butterflyfishes (Fig. 1.3). The bannerfishes are the smaller of the two clades with just 25 species. Although small, this is a morphologically diverse clade with eight distinct genera. All eight existing genera are strongly supported by the molecular phylogenies (Fessler and Westneat, 2007; Bellwood et al., 2010; Cowman and Bellwood, 2011). These genera are not very species rich but the generic divisions are stable and distinct. The butterflyfishes, in marked contrast, are problematical.

The other major division of the family Chaetodontidae, the butterflyfishes, have just two genera but contain at least 103 species (Table 1.1). Numerous species, a lack of reliable morphological characters, and a series of revisions have resulted in a plethora of generic or subgeneric groupings. This taxonomic confusion is discussed in detail by Fessler and Westneat (2007). Basically, there is a great deal of inconsistency and the existing taxonomic classification is of little ecological, evolutionary or systematic utility. Only two genera are supported by the latest molecular phylogenies, *Prognathodes* (10 spp.) and its sister taxon *Chaetodon* (93 spp.; including the monotypic *Parachaetodon*). There are four well supported subgeneric clades within the genus *Chaetodon*, but these do not agree with any existing subgeneric

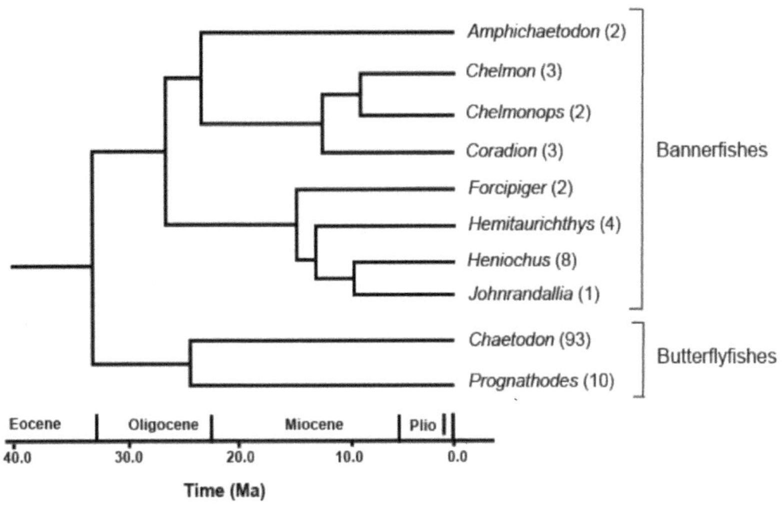

Fig. 1.3 A phylogeny of genera within the Chaetodontidae (modified after Bellwood et al., 2010). The number of species is given in parentheses.

classifications (e.g., *Megaprotodon*; taxonomic details are given in Fessler and Westneat, 2007). Fessler and Westneat (2007), Bellwood et al. (2010) and Cowman and Bellwood (2011) refer to these clades by number, 1 to 4. These clades are of considerable ecological utility; indeed the phylogenetic structure of the family underpins significant ecological and evolutionary divisions in terms of morphological innovation, biogeographic history, habitat associations, the evolution of trophic modes and susceptibility to climate change (Bellwood et al., 2010).

The bannerfishes and butterflyfishes are of comparable age (as sister taxa) but differ markedly in species numbers. This may in part reflect their biogeographic history and patterns of habitat association. The bannerfishes initially diverged from the butterflyfishes at about 26.1 Ma with divisions of the remaining bannerfish lineages between 8 and 23 Ma: *Amphichaetodon* (22.9 Ma), *Coradion* (11.9 Ma), *Chelmonops* and *Chelmon* (8.4 Ma). The bannerfish clade reflects similar patterns of Miocene diversification, at a generic level, as seen in other reef fish families such as the Labridae and Pomacanthidae (e.g., Bellwood et al., 2004; Bernardi et al., 2004; Barber and Bellwood, 2005; Cowman et al., 2009). Likewise, separation of the two *Chelmonops* species at 0.9 Ma (*curiosus* in West Australia; *truncatus* in East Australia) and possible parapatric speciation in the three species in the *Chelmon rostratus* lineage at 3.8 (*C. muelleri*—NE Australia) and 1.5 Ma (*C. marginalis*—NW Australia) reflect Pliocene division and origination of species as seen in other reef fish families (cf. Fauvelot et al., 2003; Bernardi et al., 2004). In the bannerfishes these species divisions appear to be associated with Pliocene vicariance around the Australian continent.

Table 1.1 A species list of bannerfishes and butterflyfishes. Generic and clade affiliation, diet and habitat follow Bellwood et al. (2010). Corallivores were defined as those species with more than 80% of corals comprising the diet. Reef habitat codes: coral (C), rocky (R), deep (D), sedimentary (S), rubble (RU), oceanic (O), algal (A), coastal (CO), coastal weedy (CW), inshore (I). **Note there are at least 4 more species to be named (Allen et al., 1998; Kuiter 2002).

Major clades	Genus/ minor clade	Species	Diet	Habitat
Bannerfishes	*Amphichaetodon*	*howensis*	?	R
		melbae	?	R
	Chelmon	*marginalis*	Omnivore	CO, S
		muelleri	Omnivore	CO, S, A
		rostratus	Omnivore	D, CO, S
	Chelmonops	*curiosus*	Omnivore	R
		truncatus	Omnivore	D, R
	Coradion	*altivelis*	Omnivore	C, S
		chrysozonus	Omnivore	R
		melanopus	Omnivore	D, C
	Forcipiger	*flavissimus*	Omnivore	C
		longirostris	Omnivore	C
	Hemitaurichthys	*multispinosus*	Omnivore	D
		polylepis	Omnivore	C
		thompsoni	Omnivore	D
		zoster	Omnivore	C
	Heniochus	*acuminatus*	Omnivore	CO, D
		chrysostomus	Omnivore	C
		diphreutes	Omnivore	D
		intermedius	Omnivore	C
		monoceros	Omnivore	C
		pleurotaenia	Omnivore	C
		singularis	Corallivore	D, CO
		varius	Omnivore	C
	Johnrandallia	*nigrirostris*	Omnivore	C, R
Butterflyfishes	*Chaetodon* (Clade 1)	*hoeferi*	Omnivore	R, S
		robustus	Omnivore	R, C
		marleyi	Omnivore	R, C
	Chaetodon (Clade 2)	*argentatus*	Omnivore	C, RU
		assarius	Omnivore	C
		blackburnii	Omnivore	R, C
		burgessi	Omnivore	D, C
		cintrinellus	Omnivore	R
		daedalma	Omnivore	R
		declivis	Omnivore	R, D
		dialeucos	Omnivore	

Table 1.1 contd....

Table 1.1 contd.

Major clades	Genus/ minor clade	Species	Diet	Habitat
		dolosus	Omnivore	D
		excelsa	?	D, R
		flavocoronatus	Omnivore	D
		fremblii	Omnivore	R, C
		guentheri	Omnivore	R, C
		guttatissimus	Omnivore	C, S
		guyotensis	Omnivore	
		interruptus	Omnivore	C
		jayakari	?	D, R
		kleinii	Omnivore	R, C
		interruptus	Corallivore	C
		litus	Omnivore	R, C, A
		madagaskariensis	Omnivore	C
		mertensii	Omnivore	R, C
		miliaris	Omnivore	C
		mitratus	Omnivore	C
		modestus	?	R
		multicinctus	Corallivore	C
		nippon	Omnivore	R
		paucifasciatus	Omnivore	RU, C
		pelewensis	Omnivore	C
		punctatofasciatus	Omnivore	C
		quadrimaculatus	Corallivore	R, C
		sanctaehelenae	Omnivore	R
		sedentarius	Omnivore	C
		smithi	Omnivore	R
		tinkeri	Omnivore	C, D
		trichrous	Omnivore	C
		unimaculatus	Corallivore (soft)	C
		xanthurus	Omnivore	C
	Chaetodon (Clade 3)	*andamanensis*	Corallivore	C
		aureofasciatus	Corallivore	C, A
		austriacus	Corallivore	C
		baronessa	Corallivore	C
		bennetti	Corallivore	C
		larvatus	Corallivore	C
		lunulatus	Corallivore	C
		melapterus	Corallivore	C, S

Table 1.1 contd....

Table 1.1 contd.

Major clades	Genus/ minor clade	Species	Diet	Habitat
		meyeri	Corallivore	C
		octofasciatus	Corallivore	C, S
		ornatissimus	Corallivore	C
		plebius	Corallivore	C
		rainfordi	Corallivore	C
		reticulatus	Corallivore	C
		speculum	Corallivore	C
		triangulum	Corallivore	C
		tricinctus	Omnivore	C
		trifascialis	Corallivore	C
		trifasciatus	Corallivore	C
		zanzibariensis	Corallivore	C
	Chaetodon (Clade 4)	*adiergastos*	Omnivore	C
		auriga	Omnivore	C
		auripes	Omnivore	R, C, CW
		capistratus	Omnivore	C
		collare	Omnivore	C
		decussatus	Omnivore	R, C
		ephippium	Omnivore	C
		falcula	Omnivore	C
		fasciatus	Omnivore	C
		flavirostris	Omnivore	C, R
		gardineri	Omnivore	CO
		humeralis	Omnivore	R
		leucopleura	Omnivore	C, S, RU
		lineolatus	Omnivore	D, C
		lunula	Omnivore	R, C
		melannotus	Corallivore (soft)	C
		mesoleucos	Omnivore	C
		nigropunctus	Omnivore	R, C
		ocellatus	Omnivore	C
		ocellicaudus	Corallivore (soft)	C
		oxycephalus	Omnivore	C
		pictus	Omnivore	
		rafflesi	Omnivore	C
		selene	Omnivore	S, C
		semeion	Omnivore	C

Table 1.1 contd....

Table 1.1 contd.

Major clades	Genus/ minor clade	Species	Diet	Habitat
		semilarvatus	Omnivore	C
		striatus	Omnivore	R, C
		ulietensis	Omnivore	C
		vagabundus	Omnivore	C
		wiebeli	Omnivore	R, C
		xanthocephalus	Omnivore	C, I
	Prognathodes	*aculeatus*	Omnivore	C, D
		aya	Omnivore	D
		brasiliensis	Omnivore	R
		dichrous	Omnivore	R
		falcifer	Omnivore	R, D
		guezei	Omnivore	D
		guyanensis	Omnivore	R, D
		guyotensis	?	O
		marcellae	Omnivore	D, R, C
		obliquus	Omnivore	R

This strong association between bannerfish and the Australian region is mirrored by a close association with temperate and rocky reef habitats. Among the bannerfishes, 3 out of 8 lineages are found on temperate subtropical rocky reefs and many species in the other lineages are found in rocky, deep, or coastal waters (Table 1.1). Moreover, it is noteworthy that there are no bannerfishes in the Atlantic, despite the presence of *Johnrandallia* in the East Pacific, which split from *Heniochus* at about 9.0 Ma (Bellwood et al., 2010). This early division suggests that the bannerfish clade may have been present in the Americas (East Pacific and tropical Atlantic) prior to the closure of the Isthmus of Panama. In this scenario, the absence of bannerfishes from the Caribbean today would be the result of regional extinction, as in the surgeonfish genus *Naso* (Bellwood and Wainwright, 2002). However, colonisation of the East Pacific by relatively old clades that arose in the Australian region of the West Pacific remains a distinct possibility (cf. Burridge and Smolenski, 2004).

Diverging from the *Chaetodon* lineage at about 23 Ma, *Prognathodes* closely resembles the bannerfishes, with a long snout and deep rocky reef associations. It also shares their limited diversity. *Prognathodes* is most speciose in the Atlantic with 6 of the 10 species. This genus is relatively large compared to the bannerfishes where *Heniochus* is the largest genus with just 8 species. However, it is in *Chaetodon* that we see the most spectacular diversification with 93 species. It is in this genus that we first see a distinct move onto coral reefs and where there is the most corallivory. Analyses of relative rates of diversification strongly suggest that it was the habitat shift onto coral reefs, and not a dietary shift to corallivory, that underpinned the significant diversification within the genus (Bellwood et al., 2010). This pattern appears to be a relatively widespread phenomenon, with a number of reef fish groups diversifying after moving onto reefs, including the Tetradontiformes, Labridae, Pomacentridae and Apogonidae (Alfaro et al., 2007; Bellwood and Wainwright, 2002; Cowman and Bellwood, 2011).

Phenotypic Diversity

In terms of morphological innovation, the most remarkable changes are seen within the bannerfishes. The bannerfishes exhibit several extreme morphologies. Some, such as the tubemouthed *Coradion,* are of unknown functional significance (although the similarity with tubemouthed labrids may suggest mucous feeding or corallivory), while others have been carefully examined. Foremost among these is the unique suspensorial rotation seen in *Forcipiger* and *Chelmon* (Ferry-Graham et al., 2001a, b). In these taxa the usually stationary suspensorium (the bones that support the mouth closing muscles in the 'cheek') is articulated enabling the fishes to project their jaws forward at high speed. In *Forcipiger longirostris* the suspensorial rotation enables to jaws to move forward more than twice as far as other chaetodontid species. This extreme protrusion of the jaws enables them to snatch small caridean shrimps from benthic substrata (Ferry-Graham et al., 2001a, b).

In contrast to the bannerfishes, the butterflyfishes are morphologically conservative; The major axis of variation being the extent of intermandibular rotation (Konow et al., 2008; Konow and Ferry-Graham, Chapter 2) and increasing jaw strength or gut length seen in certain coral feeding specialists (Elliott and Bellwood, 2003; Berumen et al., 2011). However, what they lack in phenotypic diversity they make up for in species diversity. Each of the four distinct clades with the genus *Chaetodon* has its own characteristic biogeographic and ecological features. *Chaetodon* Clade 1, the African butterflyfishes, contains just 3 species; all are restricted to coastal and marginal reefal systems on the west coast of Africa. *Chaetodon* Clade 2, containing omnivores, is pantropical with a wide diversity of species including several interesting and distinctive lineages. This includes: a) a deep dwelling lineage (*tinkeri* lineage), b) three separate corallivore lineages (*quadrimaculatus, multicinctus* and *unimaculatus*) that eat both soft and hard corals, and c) a relatively young Atlantic lineage (*sedentarius, sanctahelenae*). The age of this latter lineage at about 2 Ma and the location of the two component species suggest that this may be an example of colonisation of the Atlantic from the Indian Ocean via the Cape of Good Hope (cf. Bellwood et al., 2004; Cowman and Bellwood, 2012; reviewed by Floeter et al., 2008).

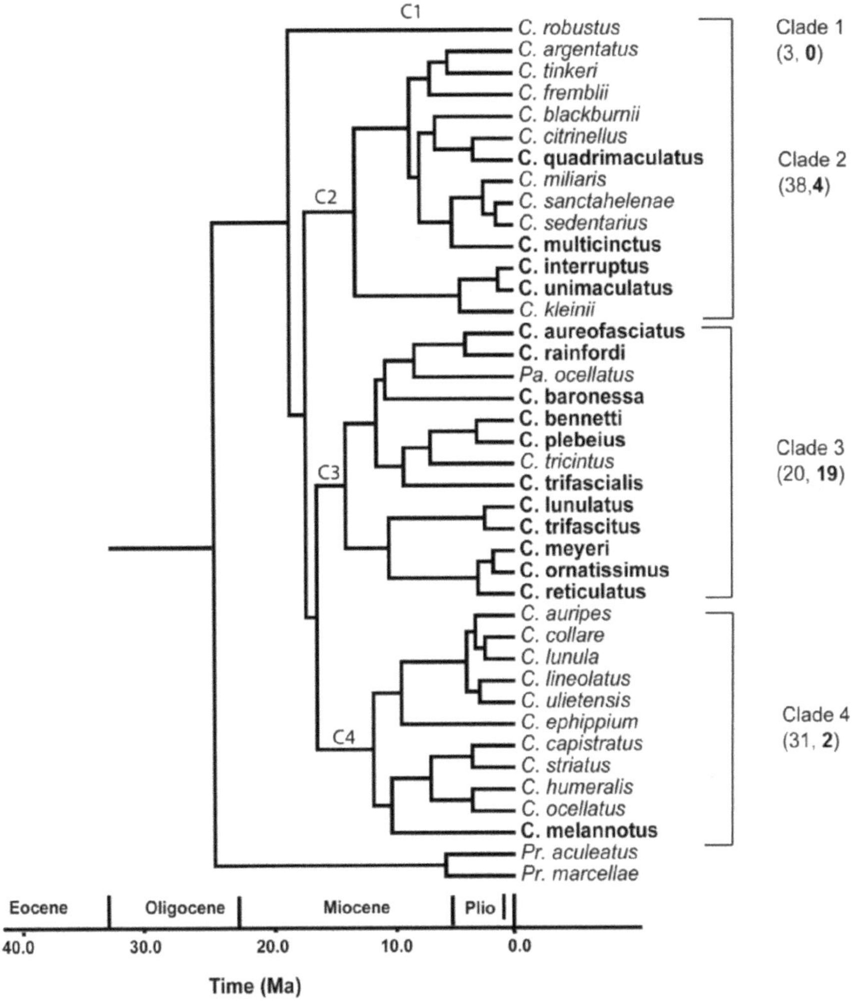

Fig. 1.4 Major clades within the genus *Chaetodon*. Species were selected to represent all traditional genera and subgenera (following Bellwood et al., 2010). The number of species in each clade and the number of obligate corallivores (in bold) is given in parentheses.

Chaetodon Clade 3, the coral feeders, is restricted to the Indo-Pacific. This is an unusual clade comprised almost entirely of obligate coral feeding species. This clade contains the most abundant and colourful butterflyfishes found on coral reefs. Within the group are almost all of the hard coral feeding specialists, including those which feed on just one or two coral species (Berumen and Pratchett, 2008), or that ingest specific parts of corals or just coral mucous (Cole et al., 2008). It is noteworthy that there is little evidence of any changes to the relatively generalised *Chaetodon* morphology in association with this switch to corallivory. Documented morphological changes are restricted to an increase in the length, and concomitant decrease in the diameter, of the intestine in some corallivorous species (Elliot and Bellwood, 2003) and slight modifications to the jaws. The largest degree of modification in the oral jaws, in terms of intramandibular flexion, is recorded in clade 3, in the subgenus *Citharoedus* (Konow et al., 2008, shown in the cladogram by the *C. meyeri-ornatissimus-reticulatus* clade. This appears to be an example of a specialised powerful biting coral feeding mode. Species within *Chaetodon* Clade 3 have the strongest links with coral reefs and it is these species that exhibit the greatest changes in response to the global decline in coral reef condition (Wilson et al., 2006; Pratchett et al., 2008).

Chaetodon Clade 4 contains mainly omnivores. It shares many similarities with clade 2. It has a circum-tropical distribution and contains numerous omnivorous species. There is only one corallivorous lineage (*C. melannotus-ocellicaudus*), but interestingly this is restricted to feeding on soft corals. As in the rest of the butterflyfishes all lines of evidence suggest that the main constraint on dietary changes is at the physiological level. This predation of soft corals appears to require specific biochemical adaptations to cope with secondary metabolites (Wylie and Paul, 1989; Coll et al., 1989). The sister lineage to the *melannotus* lineage contains almost all of the Atlantic chaetodontid species (there are no Atlantic bannerfishes).The remaining species in clade 4 are all restricted to the Indo-Pacific and are characterised by generalised dietary and habitat requirements.

Clade 4 contains the majority of Atlantic chaetodontid species. Like all other species in this clade, they are found on coral reefs, but many extend into non-reef habitats, including rocky reefs and coastal systems. The tropical Atlantic supports fewer chaetodontid species than the Indo-Pacific and contains no corallivores; it is the coral specialising clade 3 that is missing in the Atlantic. This absence of corallivores is consistent with the suggestion that reefs in the region support a more temperate fauna and that modern coral reef fish faunas in the Atlantic have a significant non-reefal composition (Bellwood and Wainwright, 2002). This may reflect repeated colonisation by taxa that refuge in or colonised from non-reef habitats (Bellwood, 1997; Floeter et al., 2008). In this regard, it is noteworthy that every Atlantic chaetodontid has been reported from non-reef habitats, and that the most diverse Atlantic chaetodontid genus (*Prognathodes*) is most frequently reported from deep rocky slopes and is rarely found on shallow coral reefs.

CONCLUSIONS

The evolutionary history of the family Chaetodontidae is a long one with close ties to coral reefs. Separating from their most recent common ancestor with the angelfishes about 50 Ma, the chaetodontids saw an early division into two groups (bannerfishes and butterflyfishes) at about 33 Ma (Bellwood et al., 2010). At this time the biodiversity hot spot for tropical marine taxa was in the European-Arabian region of the globe and it is from here that we find the earliest chaetodontid fossils. Shortly after this time we saw a move onto coral reefs (at about 18 Ma) which underpinned a significant increase in the diversity of the group. In the Atlantic however, these reefal links were limited or lost, and today they remain relatively weak. It was in the Indo-Pacific that the close relationship between coral reefs and chaetodontids (butterflyfishes in particular), proliferated. This close association is reflected by at least five separate occurrences of corallivory, starting at about 15.7 Ma in clade 3, with the youngest origination in the *C. quadricmaculatius* lineage in clade 2 at about 3.2 Ma (Bellwood et al., 2010). This increasing interdependency between the reef and its associated fishes coincides with the Miocene relocation of the world's largest marine biodiversity hotspot to the Indo Australian Archipelago (Bellwood et al., 2005; Renema et al., 2008). This location appears to have been the focus for the concatenation and development of modern coral reef faunas and, more importantly, the establishment of modern coral reef ecosystem processes (Cowman et al., 2009; Bellwood et al., 2010; Cowman and Bellwood, 2011, 2013).

The Miocene expansion of coral reef specialists fits in with escalating patterns of marine predation, but may mark a distinct third phase. Initially, the marine Mesozoic revolution was characterised by increased predation with resultant infaunalisation and armouring of prey (Vermeij, 1977). Following the K/T boundary, coral reefs experienced a Cenozoic grazing revolution, with the appearance of numerous groups of piscine herbivores (Bellwood, 2003). Finally, in the Miocene we see the rise of 'precision predation' with specialised taxa feeding on small flesh fragments, mucous or particulate material (Cowman et al., 2009; Bellwood et al., 2010). In this latter phase, butterflyfishes stand as one of the best examples of this novel expression of precision predation. Coral reefs and their associated fishes have shared a long history (Bellwood and Wainwright, 2002). However, few families have forged such a strong interdependency as the chaetodontids. The chaetodontids and most especially the butterflyfishes of clade 3 stand as the epitome of coral reef fishes.

ACKNOWLEDGEMENTS

We thank P. Cowman, P. Forey, D. Gouget, N. Konow, C. Patterson, B. Rosen, L. Sorbini, J. Tanner, P. Wainwright, R. Zorzin and colleagues in the ARC Centre of Excellence at James Cook University for their kindness and generous gifts of time and knowledge in helping with the preparation of this chapter.

REFERENCES

Alfaro, M.E., F. Santini and C.D. Brock. 2007. Do reefs drive diversification in marine teleosts? Evidence from the pufferfish and their allies (Order Tetraodontiformes). *Evolution* 61: 2104–2126.

Allen, G.R., R. Steene and M. Allen. 1998. *A guide to angelfishes and butterflyfishes*, Odyssey Publishing, Perth, Australia.

Arambourg, C. 1927. Les poissons fossilesd' Oran. *Matériaux Pour la Carte Géologique de l'Algérie, 1er Sèrie-Paléontologie* 6: 1–218.

Bannikov, A. 2004. Fishes from the Eocene of Bolca, northern Italy, previously classified with the Chaetodontidae (Perciformes). *Studie Ricerchesui Giacimenti Terziari di Bolca, Museo Civico di Storia Naturale di Verona* 10: 13–92.

Barber, P.H. and D.R. Bellwood. 2005. Biodiversity hotspots: evolutionary origins of biodiversity in wrasses (*Halichoeres*: Labridae) in the Indo-Pacific and new world tropics. *Molecular Phylogenetics and Evolution* 35: 235–253.

Bellwood, D.R. 1994. A phylogenetic study of the parrofishes family Scaridae (Pisces: Labroidei), with a revision of genera. *Records of the Australian Museum Supplement* 20: 1–86.

Bellwood, D.R. 1996. The Eocene fishes of Monte Bolca: the earliest coral reef fish assemblage. *Coral Reefs* 15: 11–19.

Bellwood, D.R. 1997. Reef fish biogeography: habitat associations, fossils and phylogenies. *Proceedings of the 8th International Coral Reef Symposium, Panama,* pp. 379–384.

Bellwood, D.R. 2003. Origins and escalation of herbivory in fishes: a functional perspective. *Paleobiology* 29: 71–83.

Bellwood, D.R. and O. Schultz. 1991. A review of the fossil record of the parrotfishes (Labroidei: Scaridae) with a description of a new *Calotomus* species from the Middle Miocene (Badenian) of Austria. *Annalen Naturhistorisches Museum Wien* 92: 55–71.

Bellwood, D.R. and L. Sorbini. 1996. A review of the fossil record of the Pomacentridae (Teleostei: Labroidei) with a description of a new genus and species from the Eocene of Monte Bolca, Italy. *Zoological Journal of the Linnean Society* 117: 159–174.

Bellwood, D.R. and P.C. Wainwright. 2002. The history and biogeography of fishes on coral reefs. In *Coral reef fishes: dynamics and diversity in a complex ecosystem,* P.F. Sale (ed.) Academic Press, San Diego, USA pp. 5–32.

Bellwood, D.R., L. van Herwerden and N. Konow. 2004. Evolution and biogeography of marine angelfishes (Pisces: Pomacanthidae). *Molecular Phylogenetics and Evolution* 33: 140–155.

Bellwood, D.R., T.P. Hughes, S.R. Connolly and J. Tanner. 2005. Environmental and geometric constraints on Indo-Pacific coral reef biodiversity. *Ecology Letters* 8: 643–651.

Bellwood, D.R., S. Klanten, P.F. Cowman, M.S. Pratchett, N. Konow and L. van Herwerden. 2010. Evolutionary history of the butterflyfishes (f: Chaetodontidae) and the rise of coral feeding fishes. *Journal of Evolutionary Biology* 23: 335–349.

Bernardi, G., G. Bucciarelli, D. Costagliola, D.R. Robertson and J. Heiser. 2004. Evolution of coral reef fish *Thalassoma* spp. (Labridae). 1. Molecular phylogeny and biogeography. *Marine Biology* 144: 369–375.

Berumen, M.L. and M.S. Pratchett. 2008. Trade-offs associated with dietary specialisation in corallivorous butterflyfishes (Chaetodontidae: *Chaetodon*). *Behavioral Ecology and Sociobiology* 62: 989–994.

Berumen, M.L., M.S. Pratchett and B. Goodman. 2011. Relative gut lengths of coral reef butterflyfishes (Pisces: Chaetodontidae). *Coral Reefs* 30: 1005–1010.

Blot, J. 1980. La faune ichthyologique des gisements du Monte Bolca (Province de Vérone, Italie): Catalogue systématique présentant l'état actuel des recherches concernant cette faune. *Bulletin Muséum national d'Histoire naturelle Paris, 4e série C* 2: 339–396.

Blum, S.D. 1989. Biogeography of the Chaetodontidae: an analysis of allopatry among closely related species. *Environmental Biology of Fishes* 25: 9–31.

Bogachev, V.V. 1964. Paleoichthyological notes *Chaetodon pennigern.* sp. *Geologic-geophysic researches of the oil-and gas-bearing regions of Azerbaijan. Trudy Azerbaydzhanskogo nauchno-issledovatel'skogo institua podobyche nefti.* 13: 126–133 (in Russian).
Burridge, C.P. and A.J. Smolenski. 2004. Molecular phylogeny of the Cheilodactylidae and Latridae (Perciformes: Cirrhitoidea) with notes on taxonomy and biogeography. *Molecular Phylogenetics and Evolution* 30: 118–127.
Carnevale, G. 2006. Morphology and biology of the Miocene butterflyfish *Chaetodon ficheuri* (Teleostei: Chaetodontidae). *Zoological Journal of the Linnean Society* 146: 251–267.
Cole, A.J., M.S. Pratchett and G.P. Jones. 2008. Diversity and functional importance of coral-feeding fishes on tropical coral reefs. *Fish and Fisheries* 9: 286–307.
Coll, J., B. Bowden, M. Porifirio, A. Heaton, G. Konig, R. de Nys, R. Willis, P. Sammarco and M. Clayton. 1989. Chemically mediated interactions between marine organisms. *Chemica Scripta* 29: 383–388.
Cowman, P.F. and D.R. Bellwood. 2011. Coral reefs as drivers of cladogenesis: expanding coral reefs, cryptic extinction events, and the development of biodiversity hotspots. *Journal of Evolutionary Biology* 24: 2543–2562.
Cowman, P.F. and D.R. Bellwood. 2013. The historical biogeography of coral reef fishes: global patterns of origination and dispersal. *Journal of Biogeography* 40: 209–224.
Cowman, P.F., D.R. Bellwood and L. van Herwerden. 2009. Dating the evolutionary origins of wrasse lineages (Labridae) and the rise of trophic novelty on coral reefs. *Molecular Phylogenetics and Evolution* 52: 621–631.
Elliott, J.P. and D.R. Bellwood. 2003. Alimentary tract morphology and diet in three coral reef fish families. *Journal of Fish Biology* 63: 1598–1609.
Fauvelot, C., G. Bernardi and S. Planes. 2003. Reductions in the Mitochondrial DNA Diversity of Coral Reef Fish Provide Evidence of Population Bottlenecks Resulting from Holocene Sea-Level Change. *Evolution* 57: 1571–1583.
Ferry-Graham, L.A., P.C. Wainwright and D.R. Bellwood. 2001a. Prey capture in long-jawed butterflyfishes (Chaetodontidae): the functional basis of novel feeding habits. *Journal of Experimental Marine Biology and Ecology* 256: 167–184.
Ferry-Graham, L.A., P.C. Wainwright, C.D. Hulsey and D.R. Bellwood. 2001b. Evolution and mechanics of long jaws in butterflyfishes (Family Chaetodontidae). *Journal of Morphology* 248: 120–143.
Fessler, J.L. and M.W. Westneat. 2007. Molecular phylogenetics of the butterflyfishes (Chaetodontidae): Taxonomy and biogeography of a global coral reef fish family. *Molecular Phylogenetics and Evolution* 45: 50–68.
Floeter, S., L.A. Rocha, D. Robertson, J. Joyeux, W.F. Smith-Vaniz, P. Wirtz, A. Edwards, J. Barreiros, C. Ferreira and J. Gasparini. 2008. Atlantic reef fish biogeography and evolution. *Journal of Biogeography* 35: 22–47.
Frickhinger, K.A. 1991. *Fossilien Atlas: Fische,* Mergus (in German), Melle, Germany.
Gosline, W.A. 1985. Relationships among some relatively deep-bodied percoid fish groups. *Japanese Journal of Ichthyology* 31: 351–357.
Konow, N., D.R. Bellwood, P.C. Wainwright and A.M. Kerr. 2008. Evolution of novel jaw joints promote trophic diversity in coral reef fishes. *Biological Journal of the Linnean Society* 93: 545–555.
Kuiter, R.H. 2002. *Butterflyfishes, Bannerfishes and their Relatives. A comprehensive Guide to Chaetodontidae and Microcanthidae.*The Marine Fish Families Series, TMC Publishing, Chorleywood, UK.
Near, T.J., R.I. Eytan, A. Dornburg, K.L. Kuhn, J.A. Moore, M.P. Davis, P.C. Wainwright, M. Friedman and W.L. Smith. 2012. Resolution of ray-finned fish phylogeny and timing of diversification. *Proceedings of the National Academy of Sciences* 109: 13698–13703.
Patterson, C. 1993. Osteichthyes: Teleostei. In *The Fossil Record 2,* M.J. Benton (ed.) Chapman and Hall, London, UK pp. 621–656.

Pratchett, M.S., P.L. Munday, S.K. Wilson, N.A.J. Graham, J.E. Cinner, D.R. Bellwood, G.P. Jones, N.V.C. Polunin and T.R. McClanahan. 2008. Effects of climate-induced coral bleaching on coral-reef fishes—Ecological and economic consequences. *Oceanography and Marine Biology: An Annual Review* 46: 251–296.

Reese, E.S. 1977. Coevolution of corals and coral feeding fishes of the family Chaetodontidae. *Proceedings of the 3rd International Coral Reef Symposium, Miami* pp. 267–274.

Reese, E.S. 1981. Predation on corals by fishes of the family Chaetodontidae; implications for conservation and management of coral reef ecosystems. *Bulletin of Marine Science* 31: 594–604.

Renema, W., D.R. Bellwood, J.C. Braga, K. Bromfield, R. Hall, K.G. Johnson, P. Lunt, C.P. Meyer, L.B. McMonagle, R.J. Morley, A. O'dea, J.A. Todd, F.P. Wesselingh, M.E.J. Wilson and J.M. Pandolfi. 2008. Hopping hotspots: global shifts in marine biodiversity. *Science* 321: 654–657.

Smith, W.L. and W.C. Wheeler. 2006. Venom evolution widespread in fishes: a phylogenetic road map for the bioprospecting of piscine venoms. *Journal of Heredity* 97: 206–217.

Smith, W.L., J.F. Webb and S.D. Blum. 2003. The evolution of the laterophysic connection with a revised phylogeny and taxonomy of butterflyfishes (Teleostei: Chaetodontidae). *Cladistics* 19: 287–306.

Tyler, J.C., G.D. Johnson, I. Nakamura and B.B. Collette. 1989. Morphology of *Luvarus imperialis* (Luvaridae), with a phylogenetic analysis of the Acanthuroidei (Pisces). *Smithsonian Contributions to Zoology* 485: 1–78.

Vermeij, G.J. 1977. The Mesozoic marine revolution: evidence from snails, predators and grazers. *Paleobiology* 3: 245–258.

Wallace, C.C. and B.R. Rosen. 2006. Diverse staghorn corals (*Acropora*) in high-latitude Eocene assemblages: implications for the evolution of modern diversity patterns of reef corals. *Proceedings of the Royal Society B: Biological Sciences* 273: 975–982.

Wilson, S.K., N.A.J. Graham, M.S. Pratchett, G.P. Jones and N.V.C. Polunin. 2006. Multiple disturbances and the global degradation of coral reefs: are reef fishes at risk or resilient? *Global Change Biology* 12: 2220–2234.

Wood, R. 1998. The ecological evolution of reefs. *Annual Review of Ecology Evolution and Systematics* 29: 179–206.

Wylie, C.R. and V.J. Paul. 1989. Chemical defenses in three species of *Sinularia* (Coelenterata, Alcyonacea): effects against generalist predators and the butterflyfish *Chaetodon unimaculatus* Bloch. *Journal of Experimental Marine Biology and Ecology* 129: 141–160.

Yagishita, N., M. Miya, Y. Yamanoue, S.M. Shirai, K. Nakayama, N. Suzuki, T.P. Satoh, K. Mabuchi, M. Nishida and T. Nakabo. 2009. Mitogenomic evaluation of the unique facial nerve pattern as a phylogenetic marker within the percifom fishes (Teleostei: Percomorpha). *Molecular Phylogenetics and Evolution* 53: 258–266.

CHAPTER 2

Functional Morphology of Butterflyfishes

Nicolai Konow[1,*] and *Lara A. Ferry*[2]

INTRODUCTION

Butterflyfishes (family Chaetodontidae) have historically been grouped with several deep-bodied reef fish families into the squamipinnes, or *'scaly-finned'* fishes (Mok and Shen, 1982; Gosline, 1985; Blum, 1988; Tyler et al., 1989). However, it is presently uncertain whether this grouping is monophyletic (Konow et al., 2008). Apart from butterflyfishes (128 species worldwide; Fig. 2.1), and their purported sister-family (Burgess, 1974), the angelfishes, family Pomacanthidae (86 species), the squamipinnes are comprised of acanthuroid surgeonfishes (Acanthuridae), rabbitfishes (Siganidae) and the Moorish Idol (Zanclidae), the Kyphosidae (incl. microcanthids and girellids) and the fairly species-depauperate Ephippidae, Drepanidae and Scatophagidae (Tyler et al., 1989; Froese and Pauly, 2012).

Butterflyfish morphology has, in the past decades, primarily been investigated for the purpose of systematic classification (Mok and Shen, 1982; Smith et al., 2003; Blum, 1988) (Fig. 2.1). Until very recently (Littlewood et al., 2004; Hsu et al., 2007; Fessler and Westneat, 2007), phylogenetic

[1]Department of Ecology and Evolutionary Biology, Brown University, Providence 02906, Rhode Island, USA.
E-Mail: nkonow@brown.edu
[2]School of Mathematical and Natural Sciences, Arizona State University, Phoenix, Arizona 85069, USA.
E-mail: lara.ferry@asu.edu
*Corresponding author

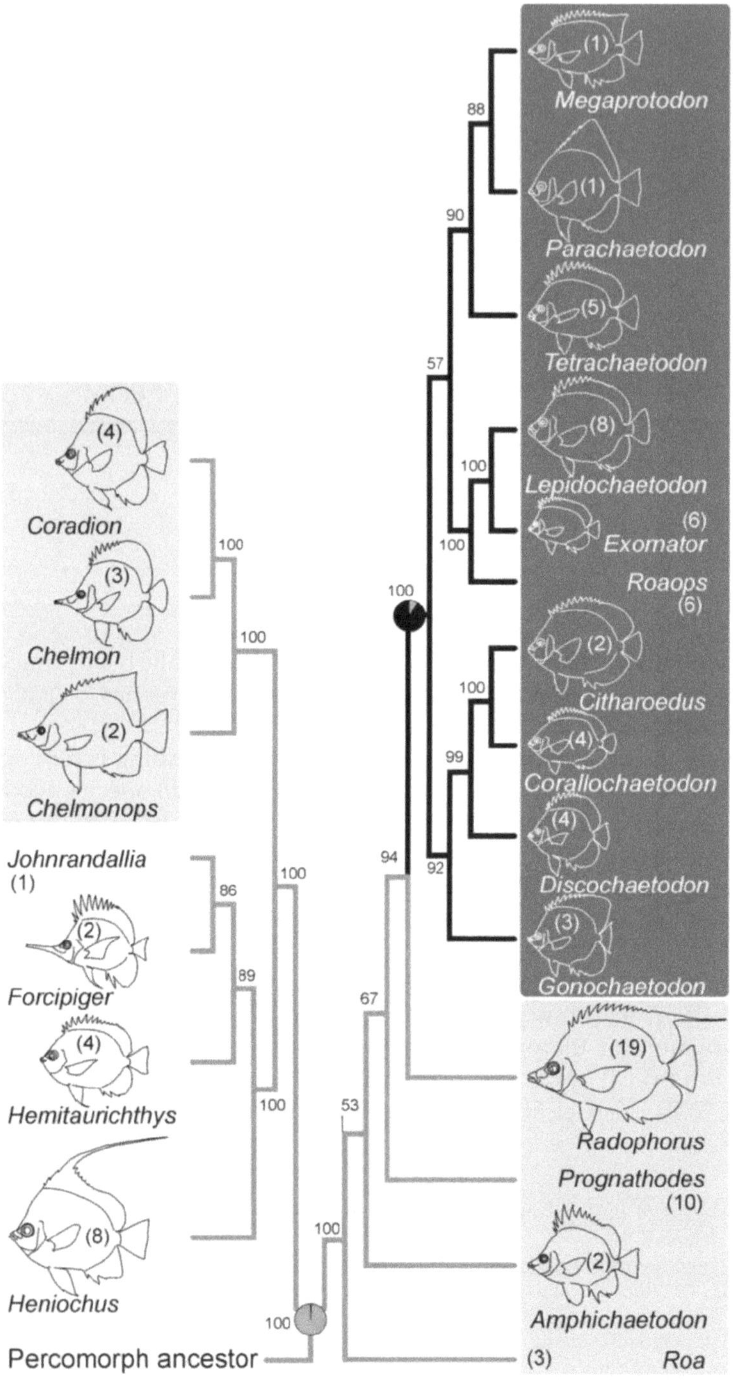
Megaprotodon
(1)
88
Parachaetodon
(1)
90
Tetrachaetodon
(5)
57
Lepidochaetodon
(8)
100
(6)
Exornator
100
Roaops
(6)
100
Citharoedus
(2)
100
Corallochaetodon
(4)
99
Discochaetodon
(4)
92
Gonochaetodon
(3)
94
Radophorus
(19)
67
Prognathodes
(10)
53
Amphichaetodon
(2)
100
(3)
Roa
Coradion
(4)
100
Chelmon
(3)
100
Chelmonops
(2)
Johnrandallia
(1)
86
Forcipiger
(2)
89
Hemitaurichthys
(4)
100
100
Heniochus
(8)
100
Percomorph ancestor

research on the family based on molecular data also lagged behind most of the other squamipinnes families (c.f. Tang et al., 1999; Clements et al., 2003; Bellwood et al., 2004). The lack of information on evolutionary morphology is puzzling, considering that this family has attained iconic status as coral reef fishes and marine ornamentals. This status has been assigned primarily based upon the key role of butterflyfishes as predators on stony (scleractinian) corals. While ecological studies of this derived feeding guild exist (for example Irons, 1989; Tricas, 1989; Alwany et al., 2003; Pratchett, 2005; Berumen and Pratchett, 2007), these tended to focus on regional patterns, and only very recently attempted to understand how butterflyfish accomplish their evolutionarily novel foraging tasks (Motta, 1985, 1989; Ferry-Graham et al., 2001a, b; Konow et al., 2008). Currently, the information relating morphology to ecology concerns a highly specialised pelagic *tholichthyes* larval morphology, the derived laterophysic canal structures involved in balance-maintenance and sound production (i.e., a joint locomotor and behavioural specialisation), the functional morphology of the locomotor apparatus itself, and a range of feeding specialisations based on novel origins of joints within the feeding apparatus.

Fig. 2.1 Interrelationships of the Chaetodontidae, reconstructed using data from Blum (1988), Ferry-Graham et al. (2001b), and Smith et al. (2003), which was modified using the super-tree technique Matrix-recombination with Parsimony (MRP). Fish icons are scaled to the mean of reported maximum body-sizes for those particular subgenera, the species number of which is given in brackets on the fish body. Branch-lengths are chosen for clarity of presentation only and numbers at branch nodes are bootstrap-values from the MRP analysis. Note how butterflyfishes naturally divide into two groups of banner and forceps fishes (left), and butterflyfishes (right). The following ecomechanic traits were mapped and optimised onto the tree in the Mesquite phylogenetics package module under maximum parsimony: Character-states for intramandibular joint (IMJ) possession are optimised to branches using black for presence and grey for absence of the joint. For the IMJ, the likelihood of the ancestral state (presumably the lack of an IMJ) was reconstructed as posterior probabilities and is reported using pie-charts at relevant nodes leading to the cladogenesis of the genus *Chaetodon*. Pie-chart shading corresponds with the branch optimising colour-scheme, and the second most probable state is indicated when its probability is greater than 0.05. Feeding mode is mapped onto the major clades using shaded boxes to delineate obligate and facultative coral-biting taxa (dark grey), those utilising both invertebrate-picking and ram-suction feeding guilds (light grey) and pure ram-suction feeders (white). Note that the combination of obligate biting strategies and possession of an IMJ coincides in the genus *Chaetodon* only. Moreover, butterflyfish jaw-lengths are typically intermediary to long in taxa that prioritise the ram-feeding end of the feeding mode continuum (see text). An exception to this rule is among members of the bannerfish clade (i.e., *Heniochus* and *Hemitaurichthys*), which commonly engage in suction-feeding planktivory. The biting *Chaetodon* butterflyfishes are, apart from their IMJ, characterised by having relatively short jaws that are mechanically efficient for biting. However, standing out are the forcepsfishes (*Chelmon* and *Chelmonops*) in their possession of ram-suction feeder traits (long jaws, suspensorial flexion; Fig. 2.4), while almost exclusively feeding using biting. For details on tree-construction, see Konow et al. (2008), from which this figure was modified with permission from the publisher.

Given the authors' expertise and the prevalent emphasis in the present volume on trophic ecology and dietary specialisations of these fishes, this chapter will treat the areas of hearing and sound production, locomotion, and feeding specialisations. In this chapter, we will take a functional morphological approach, specifically to summarise our present understanding of butterflyfish ecomechanics. We define ecomechanics as the link between organismal functional morphology (i.e., a biomechanical apparatus such as the fins or the jaws), and ecological performance, being the relative capacity and capability of the organism to use said apparatus to complete vital everyday tasks (in these cases, swimming and feeding).

Most of the recent studies of swimming and feeding have involved strong experimental components and have taken a comparative approach, and these studies can therefore be used to evaluate the relative advantage of certain structures over others in performing ecological tasks. Earlier swimming functional morphology studies (Webb, 1982; Gerstner, 1999a, b; Blake, 2004) proposed that species group into guilds depending on their swimming mode. More recently, Fulton (2007) measured the swimming performance of reef fishes empirically, in a flow tank, and conducted habitat-based validations of swimming performance. This approach served to identify characteristics of the fin apparatus that potentially explained prominent interspecific differences in swimming capability, which was then verified on the coral reef (Fulton, 2007). Similarly, in early analyses of feeding functional morphology, Motta (1982–1989) identified morphological characters within the feeding apparatus that differed across taxa and appeared to characterise different feeding guilds. Kinematics of the jaw apparatus were only measured more recently, using motion analyses of high speed video, which served to validate some functional hypotheses and refute others (Ferry-Graham et al., 2001a, b; Konow et al., 2008; but see pioneering attempts by Motta, 1985). Thus, while the earlier swimming and feeding work provided important morphological baseline knowledge, they were not really successful in terms of identifying links between morphology and ecology (i.e., eco-morphology; Motta, 1988; Wainwright, 1991). A major reason for this lack of early success could be that several salient ecomorphological relationships were only identified in later analyses, when a more integrative experimental approach could be incorporated.

Where comparative studies are lacking, we cannot place too much weight on inferred performance consequences. However, we can draw on extensive personal observations, including video documentation, from reefs and aquaria in order to make inference about organismal function. In the following, we are careful to place those inferences within an appropriate cautionary framework, adhering to the adage that one cannot infer function without directly measuring it (Motta, 1988; Wainwright, 1991).

THE MECHANOSENSORY SYSTEM

Butterflyfish have been well studied in the context of mating systems and the associated behavioural ecology of mating (e.g., Yabuta and Berumen, Chapter 8). Recently, we have begun to understand the unique functional morphological underpinnings of these behaviours. This work falls into two general areas: the study of chaetodontid hearing and the associated laterophysic canal system, and the concomitant study of chaetodontid sound production.

The Laterophysic System

The laterophysic system is unique to chaetodontids and involves a pair of projections, on the right and left sides of the body, extending from the swim bladder to the posterior region of the neurocranium. These projections create a connection with the lateral line canal located within the supracleithrum (Fig. 2.2A), although the specifics of the structure varies among species (Webb and Smith, 2000; Smith et al., 2003; Webb et al., 2006). This connection was named the laterophysic connection due to its morphological similarly to otophysic connections (Webb, 1998a; Webb and Smith, 2000); which are connections found in other fishes (i.e., otophysans) that have specialised hearing (reviewed in Webb et al., 2006).

It is hypothesised that the function of the laterophysic connection is to increase the sensitivity of the accoustico-lateralis system, the inner ear plus the lateral line, to sound pressures. Receiving sound is typically the role of the pars inferior, or the ventral portion of the inner ear. The inner ear is composed of three semi-circular canals projecting in the three dimensions. The ventral portions of the inner ear contain three chambers that each house a dense crystalline structure; these are collectively called otoliths. These otoliths rest on a bed of sensory hair cells that have an afferent connection directly to the nervous system. As a fish is nearly the same density as water, sound waves tend to pass through the fish. The dense otoliths, however, vibrate when sound waves impact them. These vibrations set the hair cells in motion and thus the sound information is transferred to the brain for processing. The lateral line also contains sensory hair cells, also referred to as lateral line neuromasts, which are similar in structure and function to the inner ear hair cells (although there are both afferent and efferent neuronal connections). The lateral line is typically used for receiving far field sound; low pressure waves that transmit at lower frequencies and over longer distances (Kalmijn, 1989).

The swim bladder contains air and is therefore also of a different density than the fish, and the surrounding medium. Therefore, it can also act as a receiver of sound waves being transmitted through the water. By extending

A.

B.

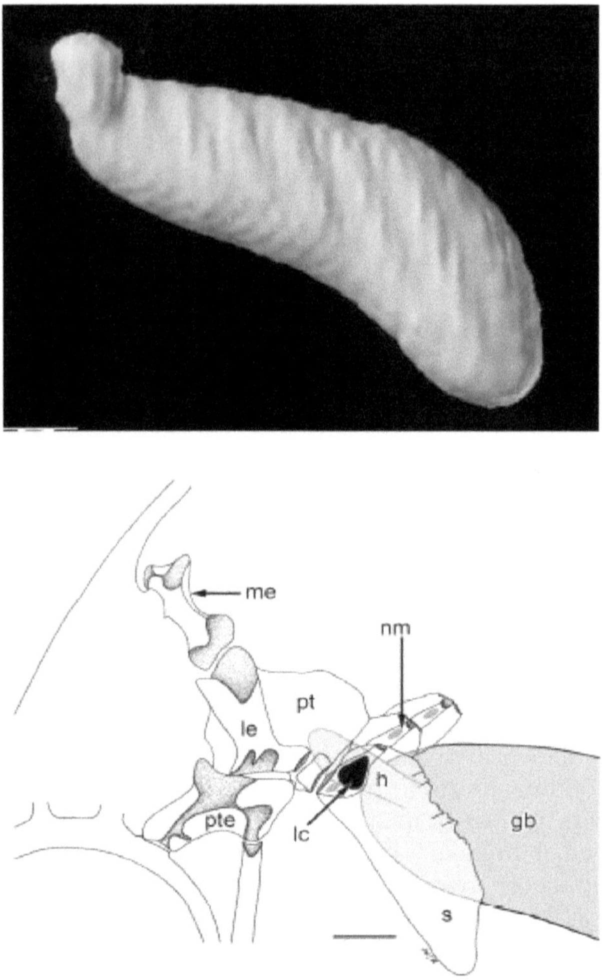

Fig. 2.2 3D CT reconstruction of the volume of the swim bladder and swim bladder horns in *Chaetodon punctofasciatus* **(A)** periodic indentations in lateral surface correspond to ribs, and **(B)** camera lucida drawing of skeletal elements posterior to the orbit at the posterior margin of the skull in *Chaetodon octofasciatus.* gb, swim bladder; h, horn; le, lateral extrascapular; me, medial extrascapular; nm, neuromast; pt, posttemporal; pte, pterotic; s, supracleithrum. Scale bars 1 mm. Modified after Webb et al. (2006) with permission from the publisher.

projections towards the sensory apparatus of the accoustico-lateralis system, the sound waves intercepted by the swim bladder are transmitted to the nervous system for processing (Schellart and Popper, 1992). Variants on the nature of the connection between the swim bladder horns and the lateral line in species of *Chaetodon* are thought to be directly related to the degree of enhancement of sound reception.

The variations range from a direct to an indirect connection between the projections of the swim bladder and the lateral line (Fig. 2.2B). A direct connection consists of layers of tissue that act as a tympanic membrane resting between the fluid filled lateral line and the air-filled swim bladder. Thus, sound waves received by the swim-bladder are putatively converted to fluid flow in the lateral line system, where they can be processed by the nervous system (Webb et al., 2006). This tympanum may be well developed, consisting of up to 4 layers of mucoid tissue, or less strongly developed and formed of only two layers of non-mucoid tissue. Generally, if the tympanic connection is less developed, the swim bladder is subdivided into two sections anteriorly and posteriorly. The projections of the swim bladder tend to be relatively long to facilitate this connection, though they vary in width. Indirect laterophysic connections contain a physical space between the swim bladder projections and the lateral line, ranging from 0.2 mm to 1 mm. Mucoid tissue may or may not be present, and the projections of the swim bladder may be long or short.

Sound Production

The presence of such elaborations for sound reception led researchers to speculate that sound production might also be present in *Chaetodon* species. Indeed, a single series of field experiments verified that sounds were produced in a variety of social contexts, including territorial displays and alert calls (Tricas et al., 2006), and this study is summarised here. Sounds were evoked by placing a single fish within the territory of a pair of *Chaetodon multicinctus*. This species is monogamous, has strong site fidelity, and is aggressive. It is also known that visual signals are fundamentally important in this species as a means of communication on the reef. From a functional morphological perspective, the production of these sounds is interesting because they assign a function to a particular mechanical movement.

The sounds recorded from this species can be grouped into motor-based and acoustic-based sounds (Tricas et al., 2006). Four motor-based sounds were recorded that were associated with actual movement of portions of the body; the tail-slap, the jump, the pelvic-fin flip, and the dorsal-anal fin erect (Fig. 2.3). Each of these movements, produced by resident fishes in this context, presumably sends a visual signal to conspecifics. However, the motion of the body also produced a recordable sound within the hearing range of chaetodontid fishes. These were low frequency, hydrodynamic sounds associated largely with the flow of water induced by the fin movement. The sounds were typically between 50 and 200 Hz peak frequency, and most lasted from 20–150 milliseconds in duration. The jump, in particular, was associated with a pulse train of four to eight pulses,

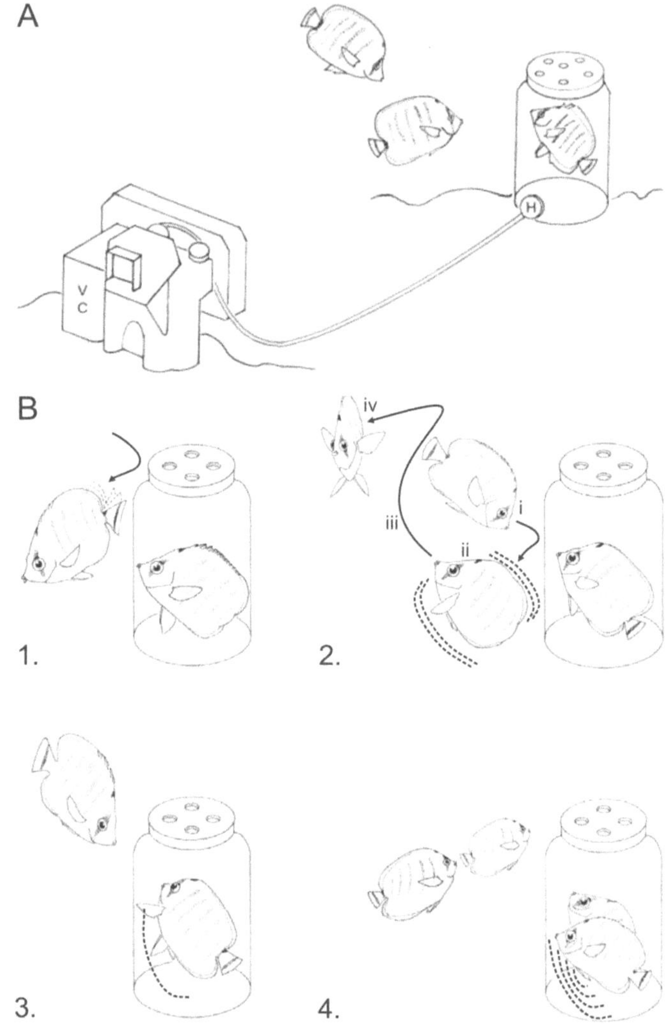

Fig. 2.3 Diagram of experimental trials to elicit sound production **(A)**, and **(B)** behaviours associated with recorded sounds in the multi-banded butterflyfish, *Chaetodon multicinctus*. Broken lines indicate sound production. (1) The tail slap behaviour, exhibited after escalated displays and aggression by territorial residents towards bottled intruders, produces both a low frequency hydrodynamic pulse and a brief broadband acoustic click. (2) The jump behaviour is displayed by resident fish and involves four parts: 1) the approach and face, 2) and rapid turn (produces a low frequency hydrodynamic pulse followed by several rapid acoustic pulses), 3) short swimming ascent, and 4) intense lateral display. (3) The pelvic fin flick behaviour, produced by both residents and bottled intruders, involves the extension of the pelvic fins and a single acoustic pulse. (4) The grunt train sound, produced only by bottled fish when approached by territory residents; no body movements were observed during the production of this sound. After Tricas et al. (2006) with permission from the publisher.

and was of higher peak frequency (400 Hz). The individual pulse lasted only about 20 ms, but a train could take up to about 350 ms.

An acoustic "click" sound was recorded from resident fish during tail slaps (Tricas et al., 2006). This was a high-frequency sound, over 3000 Hz, of short duration, 10 ms. A lower frequency grunt train was also recorded from intruder fish when fish were approached, during direct confrontations, and during displays. The peak frequency of grunt trains was about 150 Hz, and these consisted of about 20 pulses per train, lasting over 5 s. Each individual pulse was fairly long, lasting about 40 ms.

The actual mechanism of sound production remains unknown, but Tricas et al. (2006) offered some hypotheses. The grunts, for example, were produced when no visible external movement was recorded, thusly they may be produced by structures inside the body, such as the swim bladder and modifications thereof. The high-frequency of the click suggests stridulation, but cavitation of water, as is seen in many shrimp species (e.g., Patek et al., 2007), cannot be eliminated. The pulse-trains associated with jumps were probably generated by the visible vibrations of the entire body during the sequence of behaviours associated with the jump. The pelvic and dorsal-anal fin behaviours are likely the result of skeletal anatomy and contact between bony structures in some fashion. All of these hypotheses, however, have yet to be tested.

BUTTERFLYFISH LOCOMOTORY MORPHOLOGY

The emergence of benthic feeding habits among coral reef fishes led to a need for navigating in close proximity with a complex, potentially noxious and abrasive reefal substratum. Therefore, fine-scale maneuverability and rapid braking became a priority. Butterflyfishes appear to be exceptional swimming performers in this capacity in that they can hover inverted or in other abnormal postures close to the substratum and perform precise and repetitive feeding strikes within a microhabitat—area restricted to the circumference of a single scleractinian zooid or polyp (N. Konow, pers. obs.). Below, we review the surprisingly limited existing evidence that either directly or indirectly investigated this scenario in an ecomechanical context.

Girdle Rotation and Fine-scale Maneuverability

Among spiny-rayed fishes (the acanthopterygians), the pelvic fin-bearing girdle has, over the course of evolution, gradually been rotated anteriorly (Fig. 2.4A) and moved into a position immediately ventral to the pectoral fin-bearing girdle (Webb, 1982; Blake, 2004). This rotation is pronounced among both labroid and squamipinnes taxa and appears to reach its extreme within

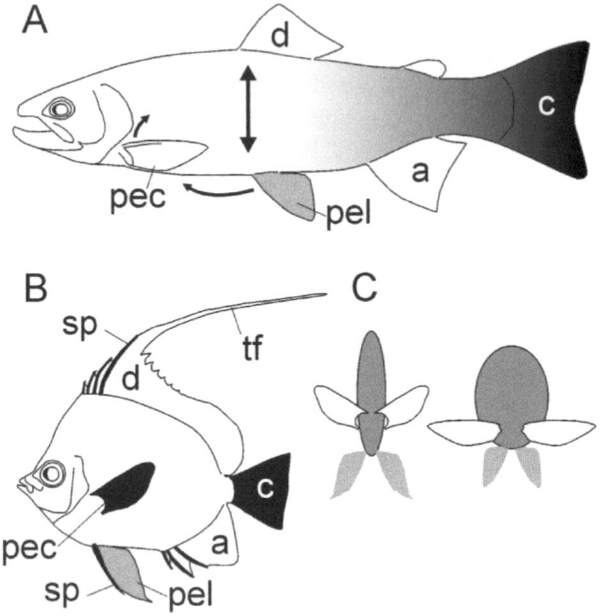

Fig. 2.4 Fin placement and fin use during swimming in **(A)** basal teleost fishes, exemplified by a protacanthopterygian trout (*Salmo*), and **(B)** in derived acanthomorph reef fishes, exemplified by a bannerfish (*Heniochus*). Abbreviations: a, anal fin; c, caudal (tail) fin; d, dorsal fin; pec, pectoral fin; pel, pelvic fin; sp, spine; tf, training filament. **(A)** The trout is a characteristic BCF (body-caudal fin) swimmer, involving undulations of the entire caudal body region in addition to the caudal fin itself, as illustrated by the shaded body-area. **(B)** In contrast, the bannerfish uses a combination of lift-based pectoral fin and undulatory caudal fin propulsion, as illustrated by the fins in black. Users of this Chaetodontiform swimming mode also have several modifications of the fin apparatus, including rotation of the pectoral and pelvic girdles (see curved arrows in **A**) to a constellation where the pelvic girdle is positioned directly ventral to the pectoral girdle **(B)**. Additional specialisations include trailing fin edge filaments and leading edge reinforcements by spiny rays. The typical deepening of the body in Chaetodontiform swimmers relative to BCF swimmers, as seen in the cross-sections in **(C)** is considered an adaptation towards a more stable body (keel-effect) and a paired-fin constellation that is more efficient for braking [Figure generated de-novo].

chaetodontoid fishes, which encompasses butterflyfishes and angelfishes (Fig. 2.4B). In these taxa, the pelvic fins are positioned directly ventral to, and sometimes even immediately anterior to the pectoral fins.

As a result, these two fin pairs, with an ancestral ventral placement, are brought into a novel constellation among bony fishes, yielding a combined fin surface area that permits efficient braking (Gerstner, 1999a). Furthermore, when this fin constellation is combined with a deepened body shape, as seen in the squamipinnes (Fig. 2.4C), fine-scale maneuverability is strongly enhanced (Gerstner, 1999b). A stiff body, as seen in chaetodontoids and most other squamipinnes, is a trait that theoretically adds to the optimisation of maneuverability during 'unsteady' swimming, *sensu* Webb (1982, 1984).

However, most predictions presented by early authors either lacked empirical evidence or were not supported by experimental data (Blake, 2004). Nevertheless, the work of Fulton (2007) supports this notion, as elaborated upon in the following section.

The Chaetodontiform Swimming Mode

Fish swimming modes are generally divided into undulatory and oscillatory mechanisms, of which typically the undulatory modes dominate among reef fishes (Webb, 1998b; Blake, 2004). Undulatory modes can be classified along a gradient of fin-use modes ranging from drag-based and caudal-fin dominated, or sub-carangiform propulsion (Fig. 2.4A), named after trevallies and other carangid fishes, and anguilliform or eel-like propulsion relying on an elongated tail section (Webb, 1984). Meanwhile, at the opposite end of the fin-use continuum is lift-based, or labriform propulsion, named after wrasses and parrotfishes (f. Labridae), which typically use their pectoral fins almost exclusively in a "flapping" manner analogous to aerial flight to provide the means of steady swimming.

The general locomotor mode among butterflyfishes is characterised by use of both the median and paired fins, and is thus often termed medial-paired fin locomotion (MPF) (Webb, 1984; Gerstner, 1999a, b; Blake, 2004). A series of characteristic body morphologies in animals using MPF propulsion were initially identified by Webb (1984). These included the pectoral fins being placed mid-lateral, the pelvic fins being placed ventrolateral, symmetrical and soft-rayed dorsal and anal fins, spine-reinforcements of fin leading edges and a short, deep (i.e., saucer-shaped), and laterally compressed body (Webb, 1984). All these traits in combination are almost exclusively observed among chaetodontoid fishes (Fig. 2.4B), thus making it appropriate to coin the associated swimming mode *chaetodontiform locomotion* (Webb, 1984; Webb and Weihs, 1986).

However, experimental evidence remained unavailable to determine whether chaetodontoid taxa equipped with these traits, indeed, used their fins differently than fishes swimming using alternative propulsive modes. It was only recently demonstrated empirically that chaetodontoids, and a very few other taxa, including some pomacenthrids (damselfishes) and nemipterids (threadfin breams), use a novel pairing of body and caudal fin (BCF) and MPF propulsion (Fulton, 2007). When examined empirically, the swimming speeds achieved by a range of butterflyfish species came close to matching those achieved by highly efficient labriform swimmers (Fulton, 2007). Also, butterflyfish field cruising speeds were maintained at a very high percentage of the maximum prolonged speeds achieved in flow-tank trials. By utilising drag-based caudal fin undulation in combination with lift-based pectoral fin rowing, chaetodontoid swimmers arrive at an effective means of negotiating high-energy, wave-swept habitats while maintaining the maneuverability and braking capability required for effective close range negotiation of complex habitat topographies. The energetically and physically efficient propulsion-mode, paired with a fin constellation ideal for fine-scale maneuvering and braking, yield predators with intimately specialised locomotory capabilities for taking advantage of the novel and complex resource opportunities on the reef (Pratchett et al., Chapter 13).

BUTTERFLYFISH FEEDING MECHANICS AND FUNCTIONAL MORPHOLOGY

It is difficult to think of a fish group with as diverse and varied feeding related morphology as the butterflyfishes (Fig. 2.5). Yet, there are only three major mechanisms of prey capture that are used by fishes (e.g., Liem, 1980), and chaetodontids are no exception to this rule: (1) "suction feeding" in which the fish expands its oral cavity, thus generating a pressure gradient that draws water and prey into the mouth, (2) "ram feeding" in which the prey remains stationary and the predator overtakes and engulfs the prey in the oral cavity, and (3) "biting" in which the fish directly applies its jaws to the prey, removing it from the substratum with a scraping or biting action. As far as is known, all teleost fish prey capture events can be described by one, or a combination of these three behaviours (Motta, 1982; Ferry-Graham et al., 2001a, b; Konow et al., 2008).

Butterflyfishes typically have short, robust jaws (Fig. 2.1) that are used for biting corals and other attached prey, and even parasites off the bodies of marine macrofauna (Fig. 2.5A-E). Indeed, biting is the most common feeding mode in the family (Harmelin-Vivien and Bouchon-Navaro, 1983; Sano, 1989). The jaw mechanics associated with biting have been described (Motta, 1985, 1989); as have the associated foraging behaviours (e.g., Harmelin-Vivien and Bouchon-Navaro, 1983; Tricas, 1989; Cox, 1994). Planktivory is also common across the family (Fig. 2.5B), and several short-jawed species have been studied in the context of how their jaws function to capture mid-water prey (Table 2.1; Motta, 1982, 1984b). While corallivorous species presumably have retained a robust jaw, and typically also strong teeth, from a biting ancestor (Motta, 1989), planktivorous species may secondarily have lost some of these features, while many species use modified behaviours to engage novel feeding guilds, such as cleaning behaviours (Table 2.1; Motta, 1988, 1989).

In all cases, the butterflyfish feeding mechanism by and large resembles the generalised perciform (perch-like fish) condition in terms of mechanical movements. And, there is a basic series of movements of the cranial region that characterises prey capture: 1) the ventral head region, or hyoid

Fig. 2.5 Prey-capture diversity in the Chaetodontidae. **(A)** Facultative cleaning: *Heniochus acuminatus* cleaning Giant Sunfish at Nusa Lembogan, Indonesia; **(B)** Facultative planktivory: *Hemitaurichthys polylepis* plankton-feeding off Osprey Reef, Australia; **(C)** Invertivory: *Chelmonops curiosus* feeding on attached invertebrates at Rapid Bay Jetty, South Australia; **(D)** Soft-coral feeders: *Chaetodon capistratus* feeding on soft coral in the Caribbean. **(E)** Obligate hard-coral feeders: *Chaetodon trifascialis* feeding on scleractinian coral polyps (*Acropora hyacinthus*) in Moorea, French Polynesia [All photos have been released from copyright by the authors and editors].

Colour image of this figure appears in the colour plate section at the end of the book.

Table 2.1 Non-reefal feeding guilds in the Chaetodontidae.

Species	Plankton feeding	Cleaning activity	Source
Chaetodon striatus	+	+	Sazima and Sazima (2001)
Chaetodon litus	+	+	Allen et al. (1989)
Chaetodon smithi	+		Allen (1985)
Chaetodonkleinii	+		Hobson (1974)
Chaetodon milliaris	+	+	Hobson (1991); Motta (1982); Ralston (1981)
Chaetodon sanctahelenae	+		Hourigan (1989)
Chaetodon sedentarius	+		Sazima and Sazima (2001)
Heniochus dipreutes	+	+	Konow et al. (2006)
Hemitaurichthys polylepis	+		Allen et al. (1989)
Hemitaurichthys zoster	+		Allen et al. (1989)
Hemitaurichthys multispinnis	+		Allen et al. (1989)
Hemitaurichthys thompsoni	+		Allen et al. (1989)
Johnrandallia nigrirostris		+	Allen et al. (1989)
Forcipiger flavissimus		+	Ludwig (1984)

(essentially the fish 'tongue-bone'), is rapidly lowered concomitantly with elevation of the skull. This causes expansion of the mouth and gill chambers, causing a negative pressure gradient between the inside and outside of the fish; 2) The lower jaw is then depressed to open the mouth, typically with the additional contribution of the upper jaw being pushed out (i.e., protruded). This facilitates a release of the pressure gradient, pulling water and prey into the mouth (these traits may be less fully developed in primarily biting taxa); 3) the opercular region is expanded to move the water through the head and out the gill openings; while 4) the mouth is closed onto or around the prey.

However, there appear to be several examples among butterflyfishes of rather extreme modifications to the basic bony fish mechanism for prey capture (Fig. 2.6). Below, we examine two major axes of variation in feeding specialisations, which appear to have evolved in separate major branches of the butterflyfish phylogeny, i.e., in two distinct butterflyfish lineages (Blum, 1988; Ferry-Graham et al., 2001b; Smith et al., 2003; Fessler and Westneat, 2007; Konow et al., 2008). Our first case-study concerns the evolution of suspensorial (cheek) flexion and associated lower jaw protrusion in banner (*Heniochus*) and forcepsfishes (*Chelmon, Forcipiger*). These taxa catch their prey using ram-suction feeding, which is the basal predacious feeding mode among jawed fishes. The second case-study concerns the iconic corallivorous *Chaetodon* butterflyfishes. These fishes have abandoned ram-suction feeding in the water column, engaged the complex reef-matrix and adopted biting strategies, an entirely novel bony fish feeding mode, in order to feed on

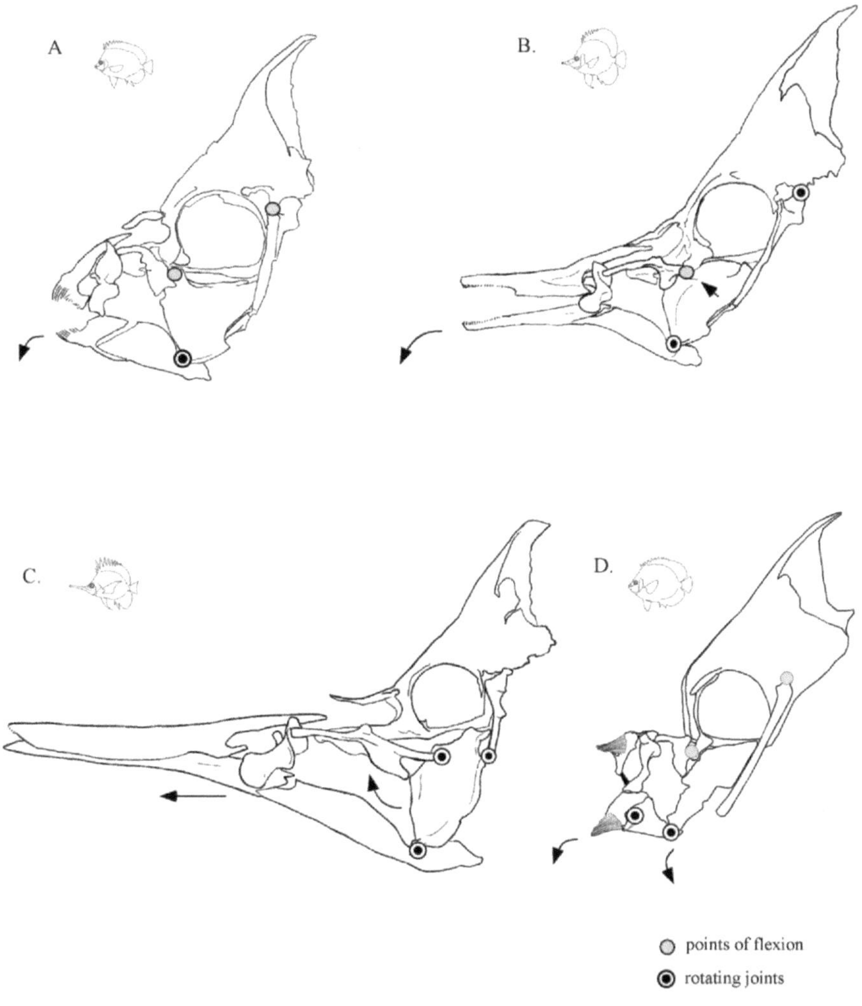

Fig. 2.6 Skull morphology in the Chaetodontidae. Specific aspects of the cranial anatomy of **(A)** *Chaetodon xanthurus* illustrating lower jaw motion with one joint at the quadrate, **(B)** *Chelmon rostratus* illustrating jaw motion when two joints are present, one within the suspensorium (note that the joint between the palatine and the quadrate complex is a sliding joint), and **(C)** *Forcipiger longirostris* illustrating jaw motion when three joints are present, two within the suspensorium. Points of flexion are indicated by grey points, rotating joints are indicated by bulls-eyes and the direction of movement is indicated by arrows. Scale bars are 1.0 cm. After Ferry-Graham et al. (2001a). **(D)** *Chaetodon ornatissimus* the position of joints and flexion resembles *C. xanthurus* **(A)**, yet, this and other members of the subgenera *Citharoedus* and *Corallochaetodon* have an extra, intramandibular joint in the position marked by the anterior-most bulls-eye, at the junction between the distal-most dentary and proximal-most articular bones forming the lower jaw [Fig. 2.6D was generated de novo].

coral tissue. In the process, the biting butterflyfishes have also evolved a series of novel joints in their feeding apparatus. We summarise how joints in various regions of the skull promote different themes in butterflyfish ecomechanics.

Elongate Jaws, Jointed Heads

Several lineages of butterflyfishes have an exceptionally elongate upper jaw (premaxilla) and lower jaw (mandible), compared with their sister species and with other bony fishes (see Figs. 2.1 and 2.6). In fact, elongate jaws are fairly widespread in the butterflyfish family occurring in all members of the genera *Forcipiger, Chelmon*, and *Chelmonops*. Slightly elongate jaws are also found in some members of *Prognathodes* and even some *Chaetodon (Radophorus)*. Common names assigned within the general literature, such as "forceps fish" (e.g., Allen et al., 1998; Kuiter, 2002), suggest a function of the elongate jaws that is similar to how biting short jaws might work, except that the jaws are longer. Motta (1988) however noted that there is rotation of the suspensorium during feeding in *Forcipiger* species. The result of this rotation is that, during feeding, both the upper and lower jaws are protruded anteriorly. Indeed, detailed studies of the anatomy, and high-speed video analyses of several species capturing live prey, confirmed that the protrusion of the upper and lower jaws is achieved through rotation of the suspensorial elements via the addition of joints to the existing range of flexion-points within this mechanical unit (Ferry-Graham et al., 2001a, b).

Up to three distinct joints may be involved in lower jaw motion; two of which are derived within the Chaetodontidae. Depending on the number of joints present, there are different consequences for the path of motion of the lower jaw. The cranial anatomy of *Chaetodon xanthurus* is drawn (Fig. 2.6A) to demonstrate the condition found in short-jawed butterflyfishes, including genera such as *Prognathodes, Heniochus*, and *Johnrandallia*. This condition is analogous to that found in generalised perciforms. The suspensorial bones are fixed such that there is no rotation during jaw depression, and no movement of the jaw joint. The lower jaw rotates on the fixed quadrate and the jaw rotates ventrally through an arc (arrow).

Chelmon rostratus is drawn illustrating the intermediate modifications found in this species (Fig. 2.6B). The hyomandible moves with the quadrate complex, thus a posterior point of limited rotation is at the articulation of the hyomandible with the skull. The quadrate complex slides under the palatine due to the loose articulation between the two. The palatine itself is largely fixed, but slight movement of the quadrate relative to the palatine provides the freedom necessary for the quadrate to rotate a small amount on the lower jaw during depression, thus the lower jaw moves both anteriorly and ventrally.

Forcipiger longirostris is shown to illustrate the condition in both *Forcipiger* species. There is a total of three joints; two novel joints in the suspensorium and one at the quadrate-articular jaw joint. The two suspensorial joints facilitate rotation relative to the fixed neurocranium (Fig. 2.6C). The rotating quadrate complex is shown pivoting on the hyomandible and the palatine. Anterior rotation of the quadrate facilitates anterior motion of the jaw joint, and therefore protrusion of the lower jaw. When rotation occurs simultaneously at the hyomandibular-metapterygoid joint and the quadrate-lower jaw joint, the lower jaw will follow an anterior course, with little dorsal or ventral motion. *F. flavissimus* exhibits a less mobile version of this model than *F. longirostris* due to the constraints outlined in the previous section.

The result of these changes in feeding apparatus functional morphology is that species with elongate jaws are afforded a feeding advantage in terms of absolute protrusion (Fig. 2.7). The addition of extra joints within the suspensorium provides for increased mobility and therefore increased

Eco-mechanic traits	Biters			Ram-suction feeders	
	Chaetodon	*Radophorus*	*Chelmon*	*Forcipiger*	*Heniochus*
Intramandibular Joint	+	(+)	-	-	-
Suspensorial joint(s)	+	-	+	++	-
Protrusible jaws	(u)	u	u,l	u,l	u(l)
Jaw length	-	++	+++	+++	+
Substratum surface	cv	cc	cc	f-cc	f

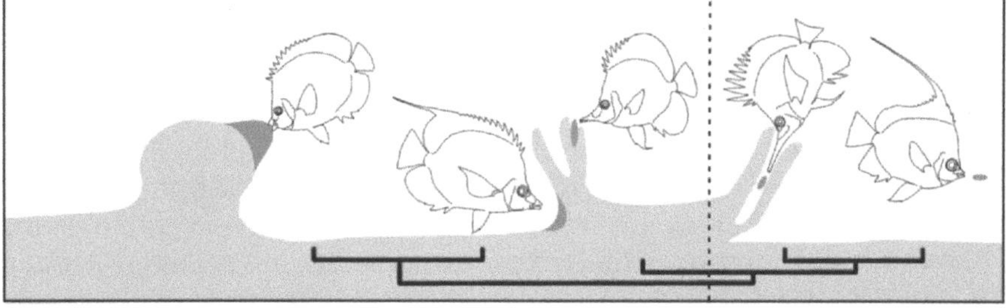

Fig. 2.7 Ecomechanic feeding guilds and functional specialisations in the Chaetodontidae. Diagram depicting previously studied feeding guilds and the functional specialisations that are thought to be underpinning these diverse guilds. Note that the diagram is segregated into exemplification of biters (L) and ram-suction feeders (R) with the phylogenetic interrelationships outlined at the bottom of the diagram for comparison (for phylogenetic interrelationships of the butterflyfishes, see also Fig. 2.1 and Bellwood and Pratchett–Chapter 1). The eco-mechanical traits listed over the fish images are important functional attributes that are treated in this chapter. These include the intramandibular joint (+, joint presence; (+), flexion presence; –, joint absence), which allows for increased gape expansion; (number of) novel joints in the suspensorium that enables protrusion of the lower jaw (l), in addition to the upper (u); variation in oral jaw length (–, minute; +, short; ++, intermediary; +++, long) and the prevalent shape of the microhabitat-types, being either concave (cc), convex (cv), utilisation of free-living prey (f) or combinations of the three. For each of the treated taxa, the unique combinations of traits characterise drastically differing feeding guilds that butterflyfishes utilise on reefs. Dark shading = prey [Figure generated de novo].

anteriorly-directed protrusion of the upper and lower jaws (Ferry-Graham et al., 2001a). The most modified of the long-jawed species, *F. longirostris*, has the advantage that it can initiate a strike significantly farther from the prey than any other species (Ferry-Graham et al., 2001b). This species covers the distance between predator and prey using protrusion of the jaws, or a *jaw-ram* based attack, as opposed to a body-ram based attack, where the predator "over-swims" the prey, or strong inertial suction, where the predator draws the prey into its mouth. The capacity for generation of suction in this species is no better, and interestingly, not less pronounced either, than in any other butterflyfish that has been studied to date (Ferry-Graham et al., 2001b). The long distance that the prey has to travel between the oral aperture and the esophagus certainly puts a premium on suction generation in order to prevent the escape of elusive prey. Considering that the natural diet of *F. longirostris* almost exclusively consists of highly elusive calaenoid copepods (Ludwig, 1984; Motta, 1988; Ferry-Graham et al., 2001a), it is however highly interesting that the strike in this species is significantly slower than in the other *Forciper, Chelmon, Heniochus*, and planktivorous *Chaetodon* species studied (Ferry-Graham et al., 2001b).

Protrusion of the lower jaw is unusual, both among butterflyfishes and among teleosts in general. Most fishes protrude only the upper jaw (premaxilla) when they feed, and not nearly to the extent seen in *Forcipiger* (Motta, 1984a). The only other quantitative descriptions of anteriorly directed protrusion of the lower jaw, thus accomplishing 'whole-mouth' protrusion, are for the sling-jaw wrasse *Epibulus insidiator*, the cichlid *Petenia splendida* and pomacanthid angelfishes. *Epibulus* also possesses a novel joint within the suspensorium that facilitates anterior translation of the jaw joint and hence extensive jaw protrusion (Westneat and Wainwright, 1989; Westneat, 1990). *Petenia,* and to some degree the closely related *Caquetaia* species, have similarly evolved two joints within the suspensorium to facilitate rotation of the unit and anteriorly-directed protrusion of the lower jaw (Walzek and Wainwright, 2003). Finally, in angelfishes (f. Pomacanthidae), the purported sister group to the butterflyfishes (Burgess, 1974), a mechanism involving suspensorial rotation that facilitates lower jaw protrusion has also evolved (Konow and Bellwood, 2005).

Suspensorial Flexion at the Palatoethmoid Junction

Among the key-characters in previous morphology-based taxonomical analyses was palatoethmoid flexion or anterior loosening of the jaw apparatus (mandibular and palatine arch) bones from the suspensorium, with accompanying separation of the ligaments holding the palatine bone in place on the vomerine/ethmoid bones of the neurocranium. This trait was among the principal diagnostics that led to separation of the butterflyfishes from the angelfishes (Burgess, 1974), and later to separate the coral-feeding *Chaetodon* butterflyfishes, which have palatoethmoid flexion, from their non-*Chaetodon* sisters (Blum, 1988; Ferry-Graham et al., 2001b; Smith et al., 2003; Littlewood et al., 2004). Separation and reduction of the otherwise tight and stout ligaments connecting the jaw apparatus with the nose-end of the skull in non-*Chaetodon* butterflyfishes enabled the jaws in *Chaetodon* butterflyfishes to move their jaws more freely during feeding and thus may be a key trait in augmenting the capabilities of coral-feeding butterflyfishes, allowing them to move their jaws over an intricately shaped substratum (Figs. 2.7 and 2.8). Loosening of the suspensorium, however, would theoretically lead to a functional compromising of the rapid and precise jaw-protrusion movements involved with ram-suction feeding among banner and forcepsfishes (Ferry-Graham et al., 2001a, b; Konow et al., 2008).

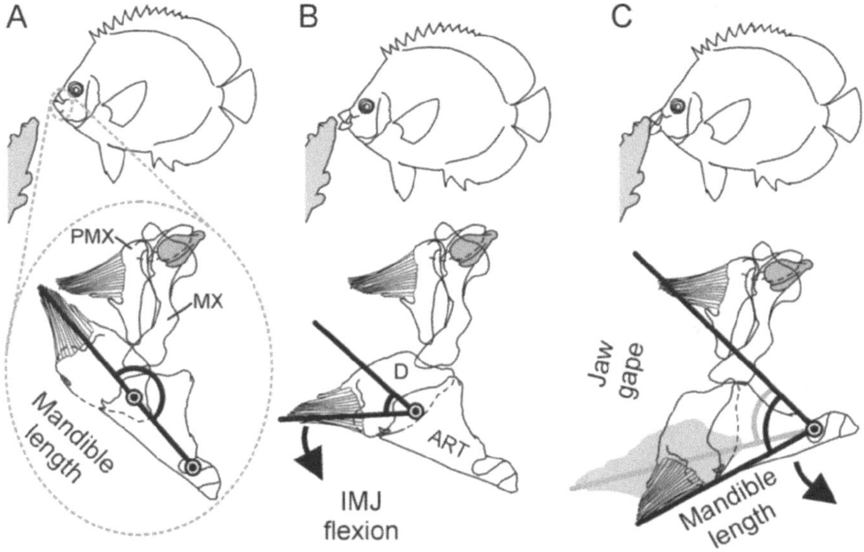

Fig. 2.8 Biomechanical function of the intramandibular joint in *Chaetodon*. Sequential illustrations of IMJ function during substratum scraping in *C. ornatissimus*. Diagrams were created by superimposing outlines of jaw structures onto video frames recorded at the time of bite onset **(A)**, maximum IMJ rotation (bulls-eye in **B)** and maximum lower jaw joint rotation (bulls-eye in **C**), coinciding with prey-contact **(C** then returns to **A** in a scraping lunge, *sensu* Motta, 1988). The shaded outline of the dentary bone **(C)** indicates the hypothetical position of this bone in a lower jaw where intramandibular flexion is absent. By comparing the rotation angle in black with the hypothetical angle corresponding with a non-jointed lower jaw in grey, the augmented jaw-gape in an IMJ-bearing mandible is made evident. By comparing the lower leg lengths of the black and grey angles, the shortening of the mandible out-lever caused by rotation around the IMJ is illustrated. Abbreviations: PMX, premaxilla; MX, maxilla; D, dentary; ART, articular [Figure generated de novo].

Intramandibular Flexion and Biting Mechanics

When a fish adopts biting strategies, entirely novel challenges are placed on the feeding apparatus. Whereas suction feeding activity clearly benefits from an expansive skull, a biting, or prey-dislodging feeding mode, requires highly articulate jaws to excavate, scrape, or nip prey off its attachment. This gradient of highly robust excavators to relatively gracile nippers is found in its entirety among *Chaetodon* butterflyfishes.

Biologists commonly use engineering principles to obtain a more sophisticated understanding of how a muscle-skeleton system operates and moves to accomplish an ecological task (e.g., Westneat, 1990). From a biomechanical perspective, the function of a fish lower jaw system is described by a third-order lever mechanism in which the rate of opening and closing trades off with the ability to generate a forceful action. It is a reasonable assumption that when the feeding target is prey attached to the substratum, then speed is typically not a priority. However, given the sturdiness of substratum attachment common to reef-dwelling invertebrates, a forceful jaw closure certainly could be important. In terms of speed-force relationships in a lower jaw system, a shorter lower jaw out-lever will, every thing else being equal, provide the fish with a greater force transmission advantage (Konow et al., 2008; Ferry-Graham and Konow, 2010). This mechanical theorem provides at least a hypothetical explanation for the apparent evolutionary selection against the long jaws of many ram-suction feeding butterflyfishes in favour of the short, stout jaws that characterise *Chaetodon* butterflyfishes (Blum, 1988; Motta, 1988, 1989; Ferry-Graham et al., 2001a, b; Konow et al., 2008).

However, an alternative mechanism exists that is capable of dynamically altering the lower jaw out-lever length at a critical point in time when the tooth-bearing surfaces of the upper and lower jaws engage the substratum, and jaw muscles contract to retract the jaws, along with the prey (Fig. 2.6D). This mechanism has only recently been comprehensively described and quantified (Konow et al., 2008), although its presence in a wide variety of coral reef fishes that bite their prey of the substratum has been anecdotally mentioned in publications spanning the past century (angelfishes, Gregory, 1933; sea-chubs, Vial and Ojeda, 1990; parrotfishes, Bellwood, 1994; surgeonfishes, Purcell and Bellwood, 1993). A novel intramandibular lower jaw joint (IMJ), placed between the dentary and articular of the mandible (lower jaw), allows the dentary tooth surfaces to rotate, or move dorso-ventrally relative to the articular bone (Konow and Bellwood, 2005, 2011). This extra joint doubles the degree-of-freedom in the lower jaw system, which means that the toothed surface of the dentary can move in radically different ways compared with a generalised single-hinged lower jaw (Fig. 2.8).

One of the few known biting reef fish groups where the IMJ had not previously been identified was among the corallivorous butterflyfishes.

Using manipulations of dissected specimens and analyses of feeding movements in the lower jaw obtained from high-speed video recordings, it was observed that flexion at the junction between the distal dentary and proximal articular bones of the mandible (lower jaw) is a basal trait for *Chaetodon*. This flexion increases gradually from subgenus to subgenus towards *Corallochaetodon* and *Citharoedus* (Fig. 2.9), the *Chaetodon* crown taxa (Fig. 2.1). In these obligate corallivorous taxa; more pronounced flexion at the IMJ is seen than in any other known IMJ-bearing coral reef fish.

Via rotation of the IMJ, in concert with rotation in the generalised lower jaw articulation, coral-feeding *Chaetodon* butterflyfishes can not only shorten the lower jaw out-lever instantaneously (Fig. 2.8B-C), but also displace the lower jaw tooth rows further away from the upper jaw tooth rows than if the IMJ had not been present (Fig. 2.8C). This function, analogous to how

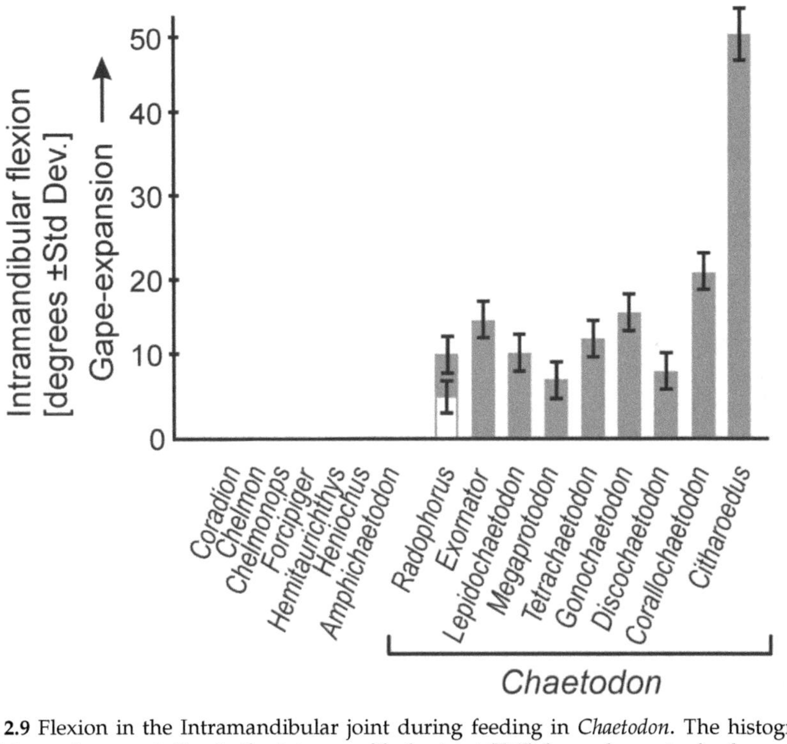

Fig. 2.9 Flexion in the Intramandibular joint during feeding in *Chaetodon*. The histogram depicts maximum rotation in the intramandibular joint (IMJ) for each nominal subgenus in the genus Chaetodon (*sensu* Blum, 1988; Ferry-Graham et al., 2001b; Smith et al., 2003; Konow et al., 2008). The measurements were obtained via motion analyses of high-speed video, measuring the angle in Fig. 2.8A, of live fish feeding in aquaria or in the wild, or of direct manipulations of the IMJ in sacrificed or anaesthetised specimens. For *Chaetodon* [*Radophorus*], the grey column represents *Chaetodon* [*Radophorus*] *melannotus*, an obligate corallivore, while the white column represents the mean IMJ rotation in other [*Radophorus*] taxa. Modified from Konow et al. (2008).

the IMJ functions in most acanthuroid taxa, allows the fish to produce a wider gape and thus contact a larger area of substratum per bite (Motta, 1989) while the lower jaw is shortened for maximised mechanical efficiency (Konow et al., 2008).

Interestingly, intramandibular flexion does not appear to be a basal trait for the butterflyfishes, as it is in both acanthuroid surgeonfishes and chaetodontoid angelfishes. Instead, the IMJ in butterflyfishes has evolved through a gradual increase in flexion (Fig. 2.9), seemingly in concert with the adoption of biting prey capture modes, including corallivory (Fig. 2.1). Documentation of such a close evolutionary correlation between IMJ flexion and biting prey capture previously only existed for the labroid parrotfishes. During parrotfish evolution, an IMJ has evolved at least on two separate occasions (Bellwood, 1994; Streelman et al., 2002).

So far, the exact sequential relationship between the origin of specialised and obligate corallivory, and acquisition of the IMJ, remains unresolved. Because flexion within the lower jaw is not present in all butterflyfishes, and a true IMJ in fact only is found in a few *Chaetodon* crown taxa (Fig. 2.9), it is a reasonable assumption that the IMJ is a fairly recent functional innovation within the Chaetodontidae (Fig. 2.1). Confident determination of the time since cladogenesis of such traits requires analyses of morphological evolution that are ancestry-corrected. Such analyses can only be carried out when a chronogram has been acquired by time-calibration of existing phylogenetic data (e.g., Fessler and Westneat, 2007) with available fossil (Carnevale, 2006) and biogeographical evidence (Bellwood and Pratchett –Chapter 1). Such an approach would be an important first step towards answering several ecomorphology questions in butterflyfish evolutionary biology.

Teeth and Guts—Food Procurement and Throughput

In a series of publications, Motta (1984b, 1987, 1989) and co-workers (Sparks et al., 1990) documented butterflyfish oral jaw dentition and associated gross morphological and ultrastructural traits. These studies focused on the evolution of dentition and tooth types among butterflyfish species utilising divergent feeding guilds. In an evolutionary context, butterflyfish dentition is generally conservative, with all species studied to date having retained a tooth shape resembling the hooked component of Velcro™. However, changes in the attachment angle of the tooth to the jaw, and a significant reduction in tooth diameter (Fig. 2.10A), was observed when comparing ram-suction feeding omnivores with biting corallivores (Motta, 1984b, 1989). Moreover, the highest levels of iron-reinforced enamel were measured from the most obligate corallivores (Motta, 1987; Sparks et al., 1990). This result

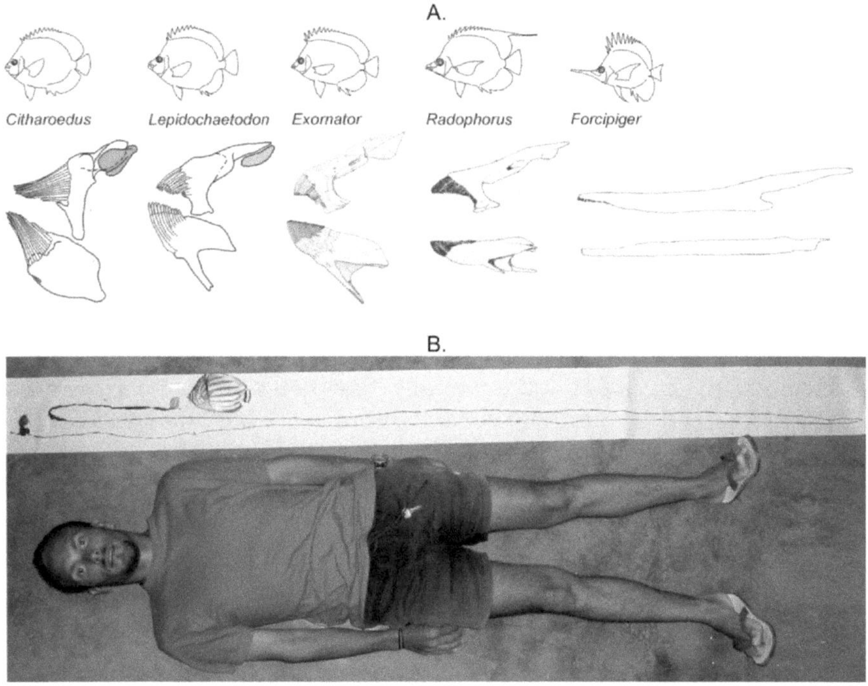

Fig. 2.10 (A) Tooth morphology and dentition diversity in the Chaetodontidae. Both tooth-bearing bony elements of the upper and lower jaws are relatively conserved within genus *Chaetodon* while dentition morphology ranges from slender villiform teeth in *Citharoedus ornatissimus*, via robust spatulate teeth in *Lepidochaetodon unimaculatus*, to diminutive brush-like teeth in *Exornator milliaris*. Basal to the genus *Chaetodon*, the length of upper and lower jaw bony elements increase, as seen in both *C. Radophorus*; in the clade comprised by *Chelmon* and *Chelmonops*, and in *Forcipiger*. Tooth morphology in the latter taxa ranges from the robust hook-like teeth in *Radophorus* to diminutive dentition in *Forcipiger*. While the teeth in *Chaetodon* taxa generally are arranged in multiple-tiered arrays in parallel with the tooth-bearing surface of the bony jaw elements, the teeth in intermediary and long-jawed taxa generally insert at a steeper angle. **(B)** Gut length in *Chaetodon* [*Citharoedus*] *ornatissimus*. Example of dramatic regional intraspecific differences in feeding-related morphology. This obligate corallivorous *C. ornatissimus* from Kaneohe Bay, Hawaii has a 4500 mm long alimentary tract, and the gut-body index averages 28 in Hawaii. On the outer Great Barrier Reef, Australia (Northern section) the same species has a much more modest 11 body lengths of alimentary tract. One author (N. K., TL= 176 cm) is depicted for scale [Figure generated *de novo*].

matched similar findings from perciform lineages that encompass obligate substratum-biting taxa, namely acanthurids, balistoids and cichlids (Suga et al., 1992), and the obligately corallivorous tetraodontid boxfishes (Suga et al., 1989). Planktivorous taxa (Table 2.1) appear to trade-off prominent dentition in terms of distribution and tooth size with increased suction-feeding efficiency. In other words, oral jaws adorned with too many, or too prominent, teeth could potentially hinder the passage of prey into the oral cavity during inertial suction feeding (Motta, 1982). However, even among obligate planktivores the Velcro™ hook-like dentition, which appears to be highly useful for efficient prey-handling, is still retained (Motta, 1984b, 1989).

Tight links between tooth morphology and diet of butterflyfishes, however, still remain largely obscure and this may to some extent be caused by intraspecific regional variation in variables associated with feeding guild specialisation. One of several characteristic examples is that of the tear drop butterflyfish *C. unimaculatus*. This butterflyfish species possesses the singularly most robust set of jaws found within all butterflyfishes. According to field census data collected by Motta (1988), this species typically uses its sturdy jaws for scraping scleractinian hard corals (*Montipora*) in Hawaii (Motta, 1988; Cox, 1994), and in Moorea (Pratchett, Chapter 6). However, at Lizard Island on the Northern Great Barrier Reef (GBR), this species utilises an entirely different prey type, browsing on alcyonean soft corals (Wylie and Paul, 1989; Pratchett, 2005).

Similar intraspecific discrepancies are also encountered at the level of alimentary tract morphology (Fig. 2.10B). It has been suggested that *C. ornatissimus* (along with the closely related species, *C. meyeri*) feed on coral mucous, rather than live coral tissues (Hobson, 1974; Reese, 1977). In the Hawaiian Islands, the obligate hard-coral feeder *C. ornatissimus* has an alimentary tract that averages 28 times the standard body length (Fig. 2.9; see also Motta, 1988). The gut contents of *C. ornatissimus* from Hawaii have also been shown to include a high proportion of calcium carbonate (coral skeleton). Meanwhile, guts from specimens off the Central GBR, Australia are almost 3-fold shorter (e.g., Berumen et al., 2011) and do not contain calcium carbonate. The extent to which butterflyfishes actually feed on coral mucous versus coral tissue may vary geographically, and this in turn may be reflected in their gut morphology.

Alimentary tract morphology in butterflyfishes has primarily been investigated for systematic purposes (see Mok and Chen, 1982). However, recent studies (Elliot and Bellwood, 2003; Berumen et al., 2011) have established a strong relationship between gut length and trophic guild (i.e., corallivores, herbivores, carnivores) in butterflyfishes, and demonstrated that gut lengths of corallivorous butterflyfish exceed those seen in both pomacentrid and labrid corallivores (Elliott and Bellwood, 2003). The fact that corallivores often have even longer guts than herbivores has puzzled researchers, including Motta (1988), who found *C. trifascialis*, an obligate stony coral polyp picker with restricted flexion between the lower jaw bones, to have a short gut, analogous with carnivores. In contrast, *C. ornatissimus* (above) uses its scraping lunges of the IMJ-bearing lower jaw to browse on corals and has an extremely long gut. It has been postulated that this trend is related in some way to a distinct distribution of zooxanthellae and symbiotic algae in separate regions of the coral tissue (i.e., mantle vs. zooids; see Motta, 1988). However, such trends in butterflyfish ecomechanics and trophic physiology have not been investigated.

FUTURE ECOMORPHOLOGY WORK

Intraspecific regional discrepancies like those described in the previous section can have the potential to strongly bias the results of future biogeographical region-spanning quantitative analyses. On the other hand, such tantalising, albeit sporadic, patterns of variation illustrate that a considerable wealth of evidence stands to be obtained via exhaustive functional morphological studies of butterflyfishes, analogous to the work of Motta (1982–1989), but expanded to a larger regional scale. Such undertakings will be required in order to confidently link morphology, function, behaviour and resource-use via ecomorphological analyses (e.g., Motta, 1988; Wainwright, 1991).

Similarly, jaw morphology data should be explored to test an ecomorphological hypothesis remaining after Motta's (1988) work; namely that key feeding morphological traits are linked to the feeding guild of butterflyfishes. Considering recently discovered functional novelties in the feeding apparatus of corallivorous butterflyfishes, post-dating his work, Motta's (1988) hypothesis should be revisited, by testing whether acquisition of novel joints in their feeding apparatus has resulted in eco-mechanically predictable changes to the associated jaw bones in butterflyfishes.

Given the relative ease with which a broad species-range of butterflyfishes of relatively large body-size can be obtained from the aquarium trade, it would be a worthy avenue of research to use recordings of muscle activity (electromyography) to examine if a shift in muscle activity patterning has accompanied the transition from ram-suction feeding to biting in butterflyfishes, especially in conjunction with IMJ acquisition. Given the prevalence of hypotheses and results from other fish groups suggesting that motor patterns are largely conserved (e.g., Wainwright, 2002), it would be interesting to know if a muscle activity pattern shift accompanies the trophic shifts within the butterflyfish family, and if said changes to motor pattern are at all consistent with changes recorded from other fish families.

ACKNOWLEDGEMENTS

Our thanks to P. C. Wainwright and D.R. Bellwood, for collecting footage of feeding kinematics in some of the long-jawed butterflyfish species. Aspects of some figures were generously made available from Ian Hart, Jackie Webb, and Tim Tricas. Original work herein was supported by the Danish Research Agency (NK).

REFERENCES

Allen, G.R. 1985. *Butterfly and angelfishes of the world.* Vol. 2, Mergus Publishers, Melle, West Germany.

Allen, G.R., R. Steene and M. Allen. 1998. *A guide to angelfishes and butterflyfishes*, Odyssey Publishing, Perth, Australia.

Alwany, M., E. Thaler and M. Stachowitsch. 2003. Food selection in two corallivorous butterflyfishes, *Chaetodonaustriacus* and *Chaetodon trifascialis,* in the northern Red Sea. *Marine Ecology* 24: 165–177.

Bellwood, D.R. 1994. A phylogenetic study of the parrofishes family Scaridae (Pisces: Labroidei), with a revision of genera. *Records of the Australian Museum Supplement* 20: 1–86.

Bellwood, D.R., L. van Herwerden and N. Konow. 2004. Evolution and biogeography of marine angelfishes (Pisces: Pomacanthidae). *Molecular Phylogenetics and Evolution* 33: 140–155.

Berumen, M.L. and M.S. Pratchett. 2008. Trade-offs associated with dietary specialisation in corallivorous butterflyfishes (Chaetodontidae: *Chaetodon*). *Behavioral Ecology and Sociobiology* 62: 989–994.

Berumen, M.L., M.S. Pratchett and B.A. Goodman. 2011. Relative gut lengths of coral reef butterflyfishes (Pisces: Chaetodontidae). *Coral Reefs* 30: 1005–1010.

Blake, R.W. 2004. Fish functional design and swimming performance. *Journal of Fish Biology* 65: 1193–1222.

Blum, Jr., S.D. 1988. The osteology and phylogeny of the Chaetodontidae (Teleostei: Perciformes). PhD dissertation, University of Hawaii, Honolulu, USA.

Burgess, W.E. 1974. Evidence for the elevation to family status of the angelfishes (Pomacanthidae), previously considered to be a subfamily of the butterflyfish family, Chaetodontidae. *Pacific Science* 28: 57–71.

Carnevale, G. 2006. Morphology and biology of the Miocene butterflyfish *Chaetodon ficheuri* (Teleostei: Chaetodontidae). *Zoological Journal of the Linnean Society* 146: 251–267.

Clements, K.D., R.D. Gray and J. Howard Choat. 2003. Rapid evolutionary divergences in reef fishes of the family Acanthuridae (Perciformes: Teleostei). *Molecular Phylogenetics and Evolution* 26: 190–201.

Cox, E.F. 1994. Resource use by corallivorous butterflyfishes (Family Chaetodontidae) in Hawaii. *Bulletin of Marine Science* 54: 535–545.

Elliott, J.P. and D.R. Bellwood. 2003. Alimentary tract morphology and diet in three coral reef fish families. *Journal of Fish Biology* 63: 1598–1609.

Ferry-Graham, L.A. and N. Konow. 2010. The intramandibular joint in *Girella*: a mechanism for increased force production? *Journal of Morphology* 271: 271–279.

Ferry-Graham, L.A., P.C. Wainwright and D.R. Bellwood. 2001a. Prey capture in long-jawed butterflyfishes (Chaetodontidae): the functional basis of novel feeding habits. *Journal of Experimental Marine Biology and Ecology* 256: 167–184.

Ferry-Graham, L.A., P.C. Wainwright, C.D. Hulsey and D.R. Bellwood. 2001b. Evolution and mechanics of long jaws in butterflyfishes (Family Chaetodontidae). *Journal of Morphology* 248: 120–143.

Fessler, J.L. and M.W. Westneat. 2007. Molecular phylogenetics of the butterflyfishes (Chaetodontidae): Taxonomy and biogeography of a global coral reef fish family. *Molecular Phylogenetics and Evolution* 45: 50–68.

Froese, R. and D. Pauly, eds. 2012. *FishBase*. World Wide Web electronic publication: www.fishbase.org, version (08/2012).

Fulton, C. 2007. Swimming speed performance in coral reef fishes: field validations reveal distinct functional groups. *Coral Reefs* 26: 217–228.

Gerstner, C.L. 1999a. Maneuverability of four species of coral-reef fish that differ in body and pectoral-fin morphology. *Canadian Journal of Zoology* 77: 1102–1110.

Gerstner, C.L. 1999b. The function of a deep body morphology in coral reef fishes: enhanced maneuverability or protection from gape-limited predators? *Dissertation Abstracts International Part B: Science and Engineering* 59: 3192.

Gosline, W.A. 1985. Relationships among some relatively deep-bodied percoid fish groups. *Japanese Journal of Ichthyology* 31: 351–357.

Gregory, W.K. 1933. Fish skulls: a study of the evolution of natural mechanisms. *The American Philosophical Society* 23: 1–468.

Harmelin-Vivien, M. and Y. Bouchon-Navaro. 1983. Feeding diets and significance of coral feeding among chaetodontid fishes in Moorea (French Polynesia). *Coral Reefs* 2: 119–127.

Hobson, E.S. 1974. Feeding relationships of teleostean fishes on coral reefs in Kona, Hawaii. *Fishery Bulletin* 72: 915–1031.

Hobson, E.S. 1991. Trophic relationships of fishes specialized to feed on zooplankters above coral reefs. In *The Ecology of Fishes on Coral Reefs*, P.F. Sale (ed.) Academic Press, Inc., San Diego, USA pp. 69–95.

Hourigan, T. 1989. Environmental determinants of butterflyfish social systems. *Environmental Biology of Fishes* 25: 61–78.

Hsu, K.C., J.P. Chen and K.T. Shao. 2007. Molecular phylogeny of *Chaetodon* (Teleostei: Chaetodontidae) in the Indo-West Pacific: evolution in geminate species pairs and species groups. *The Raffles Bulletin of Zoology* 14: 77–86.

Irons, D.K. 1989. Temporal and areal feeding behavior of the butterflyfish, *Chaetodon trifascialis*, at Johnston Atoll. *Environmental Biology of Fishes* 25: 187–193.

Kalmijn, A.J. 1989. Functional evolution of lateral line and inner ear sensory systems. In *The mechanosensory lateral line: neurobiology and evolution*, S. Coombs, P. Gorner and H. Munz (eds.). Springer-Verlag, New York, USA, pp. 187–215.

Konow, N. and D.R. Bellwood. 2005. Prey-capture in *Pomacanthus semicirculatus* (Teleostei, Pomacanthidae): functional implications of intramandibular joints in marine angelfishes. *Journal of Experimental Biology* 208: 1421–1433.

Konow, N. and D.R. Bellwood. 2011. Evolution of high trophic diversity based on limited functional disparity in the feeding apparatus of marine angelfishes (f. Pomacanthidae). PLoS ONE 6(9): e24113. doi:10.1371/journal.pone.0024113.

Konow, N., R. Fitzpatrick and A. Barnett. 2006. Adult emperor angelfish (*Pomacanthus imperator*) clean giant sunfishes (*Molamola*) at Nusa Lembongan, Indonesia. *Coral Reefs* 25: 208–208.

Konow, N., D.R. Bellwood, P.C. Wainwright and A.M. Kerr. 2008. Evolution of novel jaw joints promote trophic diversity in coral reef fishes. *Biological Journal of the Linnean Society* 93: 545–555.

Kuiter, R.H. 2002. *Butterflyfishes, Bannerfishes and their Relatives. A comprehensive Guide to Chaetodontidae and Microcanthidae.* The Marine Fish Families Series, TMC Publishing, Chorleywood, UK.

Liem, K.F. 1980. Acquisition of energy by teleosts: adaptive mechanisms and evolutionary patterns. *Environmental physiology of fishes* 35: 299–334.

Littlewood, D.T.J., S.M. McDonald, A.C. Gill and T. Cribb. 2004. Molecular phylogenetics of*Chaetodon* and the Chaetodontidae (Teleostei: Perciformes) with reference to morphology. *Zootaxa* 779: 1–20.

Ludwig, G.M. 1984. Constrasts in morphology and life history among Hawaiian populations of two longnose butterflyfishes, *Forcipiger longirostris* and *F. flavissimus*: a possible case of character displacement. PhD dissertation, University of Hawaii, Honolulu, USA.

Mok, H.K. and S.C. Shen. 1982. Phylogeny of the chaetodontids on the basis of kidney and intestinal differentiations. *Japanese Journal of Ichthyology* 29: 43–61.

Motta, P.J. 1982. Functional morphology of the head of the inertial suction feeding butterflyfish, *Chaetodonmiliaris* (Perciformes, Chaetodontidae). *Journal of Morphology* 174: 283–312.

Motta, P.J. 1984a. Mechanics and functions of jaw protrusion in teleost fishes: a review. *Copeia*: 1–18.

Motta, P.J. 1984b. Tooth attachment, replacement, and growth in the butterfly fish, *Chaetodonmiliaris* (Chaetodontidae, Perciformes). *Canadian Journal of Zoology* 62: 183–189.
Motta, P.J. 1985. Functional morphology of the head of Hawaiian and Mid-Pacific butterflyfishes (Perciformes, Chaetodontidae). *Environmental Biology of Fishes* 13: 253–276.
Motta, P.J. 1987. A quantitative analysis of ferric iron in butterflyfish teeth (Chaetodontidae, Perciformes) and the relationship to feeding ecology. *Canadian Journal of Zoology* 65: 106–112.
Motta, P.J. 1988. Functional morphology of the feeding apparatus of ten species of Pacific butterflyfishes (Perciformes, Chaetodontidae): an ecomorphological approach. *Environmental Biology of Fishes* 22: 39–67.
Motta, P.J. 1989. Dentition patterns among Pacific and Western Atlantic butterflyfishes (Perciformes, Chaetodontidae): relationship to feeding ecology and evolutionary history. *Environmental Biology of Fishes* 25: 159–170.
Patek, S., B. Nowroozi, J. Baio, R. Caldwell and A. Summers. 2007. Linkage mechanics and power amplification of the mantis shrimp's strike. *Journal of Experimental Biology* 210: 3677–3688.
Pratchett, M.S. 2005. Dietry overlap among coral-feeding butterflyfishes (Chaetodontidae) at Lizard Island, northern Great Barrier Reef. *Marine Biology* 148: 373–382.
Purcell, S.W. and D.R. Bellwood. 1993. A functional analysis of food procurement in two surgeonfish species, *Acanthurusnigrofuscus* and *Ctenochaetus striatus* (Acanthuridae). *Environmental Biology of Fishes* 37: 139–159.
Ralston, S. 1981. Aspects of the reproductive biology and feeding ecology of *Chaetodon miliaris*, a Hawaiian endemic butterflyfish. *Environmental Biology of Fishes* 6: 167–176.
Reese, E.S. 1977. Coevolution of corals and coral feeding fishes of the family Chaetodontidae. *Proceedings of the 3rd International Coral Reef Symposium* 1: 267–274.
Sano, M. 1989. Feeding habits of Japanese butterfyfishes (Chaetodontidae). *Environmental Biology of Fishes* 25: 195–203.
Sazima, C. and I. Sazima. 2001. Plankton-feeding aggregation and occasional cleaning by adult butterflyfish, *Chaetodon striatus* (Chaetodontidae), in southwestern Atlantic. *Cybium* 25: 145–151.
Schellart, N.A.M. and A.N. Popper. 1992. Functional aspects of the evolution of the auditory system of actinopterygian fish. In *The Evolutionary Biology of Hearing*, D.B. Webster, R.R. Fay and A.N. Popper (eds.). Springer-Verlag, New York, USA pp. 295–322.
Smith, W.L., J.F. Webb and S.D. Blum. 2003. The evolution of the laterophysic connection with a revised phylogeny and taxonomy of butterflyfishes (Teleostei: Chaetodontidae). *Cladistics* 19: 287–306.
Sparks, N.H., P.J. Motta, R.P. Shellis, V.J. Wade and S. Mann. 1990. An analytical electron microscopy study of iron-rich teeth from the butterflyfish (*Chaetodon ornatissimus*). *Journal of Experimental Biology* 151: 371–385.
Streelman, J., M. Alfaro, M. Westneat, D. Bellwood and S. Karl. 2002. Evolutionary history of the parrotfishes: biogeography, ecomorphology, and comparative diversity. *Evolution* 56: 961–971.
Suga, S., K. Wada, Y. Taki and M. Ogawa. 1989. Iron concentration in teeth of tetra-odontiform fishes and its phylogenetic significance. *Journal of Dental Research* 68: 1115–1123.
Suga, S., Y. Taki and M. Ogawa. 1992. Iron in the enameloid of perciform fish. *Journal of Dental Research* 71: 1316–1325.
Tang, K.L., P.B. Berendzen, E. Wiley, J.F. Morrissey, R. Winterbottom and G.D. Johnson. 1999. The phylogenetic relationships of the suborder Acanthuroidei (Teleostei: Perciformes) based on molecular and morphological evidence. *Molecular Phylogenetics and Evolution* 11: 415–425.
Tricas, T.C. 1989. Prey selection by coral-feeding butterflyfishes: strategies to maximize the profit. *Environmental Biology of Fishes* 25: 171–185.

Tricas, T.C., S.M. Kajiura and R.K. Kosaki. 2006. Acoustic communication in territorial butterflyfish: test of the sound production hypothesis. *Journal of Experimental Biology* 209: 4994–5004.

Tyler, J.C., G.D. Johnson, I. Nakamura and B.B. Collette. 1989. Morphology of *Luvarus imperialis* (Luvaridae), with a phylogenetic analysis of the Acanthuroidei (Pisces). *Smithsonian Contributions to Zoology* 485: 1–78.

Vial, C.I. and F.P. Ojeda. 1990. Cephalic anatomy of the hebivorous fish *Girella laevifrons* (Osteichthyes: Kyphosidae): mechanical considerations of its trophic function. *Revista Chilena de Historia Natural* 63: 247–260.

Wainwright, P.C. 1991. Ecomorphology: experimental functional anatomy for ecological problems. *American Zoologist* 31: 680–693.

Wainwright, P.C. 2002. The evolution of feeding motor patterns in vertebrates. *Current Opinion in Neurobiology* 12: 691–695.

Waltzek, T.B. and P.C. Wainwright. 2003. Functional morphology of extreme jaw protrusion in neotropical cichlids. *Journal of Morphology* 257: 96–106.

Webb, J.F. 1998a. Laterophysic connection: a unique link between the swimbladder and the lateral line system in *Chaetodon* (Perciformes: Chaetodontidae). *Copeia* 1998: 1032–1036.

Webb, J.F. and W.L. Smith. 2000. The laterophysic connection in chaetodontid butterflyfish: morphological variation and speculations on sensory function. *Philosophical Transactions of the Royal Society of London. Series B: Biological Sciences* 355: 1125–1129.

Webb, J.F., W.L. Smith and D.R. Ketten. 2006. The laterophysic connection and swim bladder of butterflyfishes in the genus *Chaetodon* (Perciformes: Chaetodontidae). *Journal of Morphology* 267: 1338–1355.

Webb, P.W. 1982. Locomotor patterns in the evolution of actinopterygian fishes. *American Zoologist* 22: 329–342.

Webb, P.W. 1984. Body form, locomotion and foraging in aquatic vertebrates. *American Zoologist* 24: 107–120.

Webb, P.W. 1998b. Swimming. In *The physiology of fishes*, D.H. Evans (ed.) CRC Press, Boca Raton, USA, pp. 3–24.

Webb, P.W. and D. Weihs. 1986. Functional locomotor morphology of early life history stages of fishes. *Transactions of the American Fisheries Society* 115: 115–127.

Westneat, M.W. 1990. Feeding mechanics of teleost fishes (Labridae; Perciformes): A test of four-bar linkage models. *Journal of Morphology* 205: 269–295.

Westneat, M.W. and P.C. Wainwright. 1989. Feeding mechanism of *Epibulusinsidiator* (Labridae; Teleostei): evolution of a novel functional system. *Journal of Morphology* 202: 129–150.

Wylie, C.R. and V.J. Paul. 1989. Chemical defenses in three species of *Sinularia* (Coelenterata, Alcyonacea): effects against generalist predators and the butterflyfish *Chaetodon unimaculatus* Bloch. *Journal of Experimental Marine Biology and Ecology* 129: 141–160.

CHAPTER 3

Hybridisation among Butterflyfishes

Jean-Paul A. Hobbs,[1,*] *Lynne van Herwerden,*[2] *Morgan S. Pratchett*[3] and *Gerald R. Allen*[4]

INTRODUCTION

Hybridisation is defined as the interbreeding of individuals from two distinct populations (*sensu stricto* species), which are distinguishable on the basis of one or more heritable characters, following Harrison (1993). Hybrids have been recognised, based on intergrading of characteristic features from parent species, since at least the 1700's and have been documented among plants, corals, gastropods, crustaceans, insects, amphibians, reptiles, birds, mammals and fish (Barton and Hewitt, 1985; Schwenk, 1993; Bierne et al., 2003; van Oppen and Gates, 2006). At least 10% of animals and 25% of plant species are known to hybridise, although the true proportion is probably higher because hybridisation often goes unnoticed (Mallet, 2005, 2007). The fact that hybridisation is not a rare phenomenon but occurs in a considerable

[1]The Oceans Institute and School of Plant Biology, The University of Western Australia, Crawley 6009, Australia.
E-mail: jean-paul.hobbs@uwa.edu.au
[2]School of Marine and Tropical Biology, James Cook University, Townsville QLD 4811, Australia.
E-mail: lynne.vanherwerden@jcu.edu.au
[3]ARC Centre of Excellence for Coral Reef Studies, James Cook University, Townsville QLD 4811, Australia.
E-mail: morgan.pratchett@jcu.edu.au
[4]Western Australian Museum Locked Bag 49, Welshpool DC, Perth, Western Australia 6986, Australia.
E-mail: gerald.allen@wa.gov.au
*Corresponding author

proportion of species, presents a significant challenge to the fundamental biological definition of a species (Barton and Hewitt, 1985; Mallet, 2005).

In addition to the taxonomic implications, the study of hybridisation is crucial to evolutionary biology for three reasons. Firstly, it provides unique insights into reproductive isolation and the speciation process. For example, a fundamental tenet of sympatric speciation is that there are behavioural or semi-permeable physical barriers to reproduction among distinct components of a population, leading through time to the eventual divergence of these populations into recognisably distinct species (e.g., Bush, 1969; Munday et al., 2004). The question then becomes what is it that limits interbreeding, and what level of interbreeding can occur while species continue to diverge? Secondly, hybridisation can increase genetic diversity and produce novel genotypes that enables hybrids to exploit new, unoccupied or altered habitats and these hybrids can evolve into new species (Seehausen, 2004; van Oppen and Gates, 2006; Mallet, 2007). This process has been linked to adaptive radiations of several groups of plants and animals (Seehausen, 2004). Thirdly, hybridisation can also decrease species richness through extinction (Rhymer and Simberloff, 1996) or reverse-speciation (Seehausen, 2006). Therefore, hybridisation can play a significant role in the generation and loss of species.

Hybridisation usually occurs in narrow and specific geographic areas where genetically distinct populations or species interbreed and produce offspring (Hewitt, 1988; Harrison, 1993; Arnold, 1997). These areas are known as hybrid zones and are the "natural laboratories" for studying hybridisation (Barton and Hewitt, 1989). Hybrid zones can arise within the overlapping geographic distributions of sympatric species (or populations) (e.g., van Herwerden et al., 2006; Yaakub et al., 2006), or through secondary contact of allopatric species (or populations) at or beyond the edge of a species, geographic range (Mayr, 1963; Hewitt, 1988, 2000). The latter (i.e., secondary hybrid zones) are believed to be the most common (Barton and Hewitt, 1985; Hewitt, 1988). Hybrid zones are not evenly distributed around the world and frequently occur in clusters (Hewitt, 2000). Clusters of hybrid zones often form where regional biotas, which have been separated by vicariant events, re-expand and come into secondary contact at biogeographic borders (termed a "suture zone": Remington, 1968; Hewitt, 2000; but see Swenson and Howard, 2004). In the marine environment hybrid zones often group on biogeographic borders (Gardner, 1997), for example, the Florida peninsula (America; Avise, 2000), the entrance of the Baltic Sea (Johannesson and Andre, 2006), and the Indo-Pacific border in the tropical eastern Indian Ocean (Hobbs et al., 2009).

In addition to secondary contact of allopatric species, hybridisation is also promoted by abundance disparities (Arnold, 1997), environmental gradients (Hewitt, 1988; Gardner, 1997), disruption to habitat (Rhymer and

Simberloff, 1996; Arnold, 1997; Seehausen et al.,1997; Coyne and Orr, 2004) and introduced species, including exotic predators (Rhymer and Simberloff, 1996; Roush, 1997; Taylor et al., 2006). Although these factors have been examined in detail in the terrestrial and freshwater environments we are only beginning to learn how these factors affect hybridisation in the marine environment (McMillan et al., 1999; Bierne et al., 2003; van Oppen and Gates, 2006; van Herwerden et al., 2006; Frisch and van Herwerden, 2006; Yaakub et al., 2006). The applicability of terrestrial based models to the marine environment may be limited given the distinct differences in the life history of terrestrial and marine organisms. For instance, dispersal ability plays a central role in the formation of hybrid zones and the dispersive larval phase coupled with high fecundities of most marine organisms means that their dispersal ability is many times greater than that of terrestrial organisms, thereby creating much wider hybrid zones (e.g., McMillan et al., 1999). Therefore, we cannot assume that the conditions conducive to hybridisation in the terrestrial environment are the same for the marine environment.

Hybridisation is most common among recently diverged taxa and there is a taxonomic bias with fishes being one of the most common groups of animals to hybridise (Mayr, 1979; Avise, 2004; Mallet, 2005). Traditionally, it was thought that freshwater fishes hybridise more than marine fishes, and temperate species more than tropical species, thus hybridisation was considered to be lacking in tropical marine fishes (Hubbs, 1955). However, this generalisation appears to be premature and reflected the lack of research on tropical marine fishes at the time. By 1956, there appeared to be only six instances of hybridisation in marine fishes, compared to 129 cases in freshwater fishes (Randall, 1956). Following the advent of SCUBA diving and an increase in marine research there has been a surge of new reports of hybrid marine fishes. By 1997 at least 172 marine species had been reported to hybridise, and bony fishes were the most common group accounting for 56% of the reported hybrids (Gardner, 1997; Fig. 3.1). Of the bony fishes, coral reef fishes have the most reported hybrids with at least 75 species implicated in hybridisation (Yaakub et al., 2006).

Among coral reef fishes, butterflyfishes (family Chaetodontidae) have the highest number of reported hybrids of any family, with approximately 30 species reported to form hybrids with congenerics (Yaakub et al., 2006). Butterflyfishes are also among the best-studied group of coral reef fishes, with considerable data available on species distributions (McMillan et al., 1999), phylogenetic relationships (Fessler and Westneat, 2007; Bellwood et al., 2009), ecology (e.g., Pratchett, 2005; Cole et al., 2008; Cole and Pratchett, Chapter 5), and behavioural interactions (Robert and Ormond, 1992; Wrathall et al., 1992; Berumen and Pratchett, 2006), thereby providing a unique opportunity to test terrestrially-derived theories of hybridisation in the world's most diverse vertebrate community. The primary aim of this

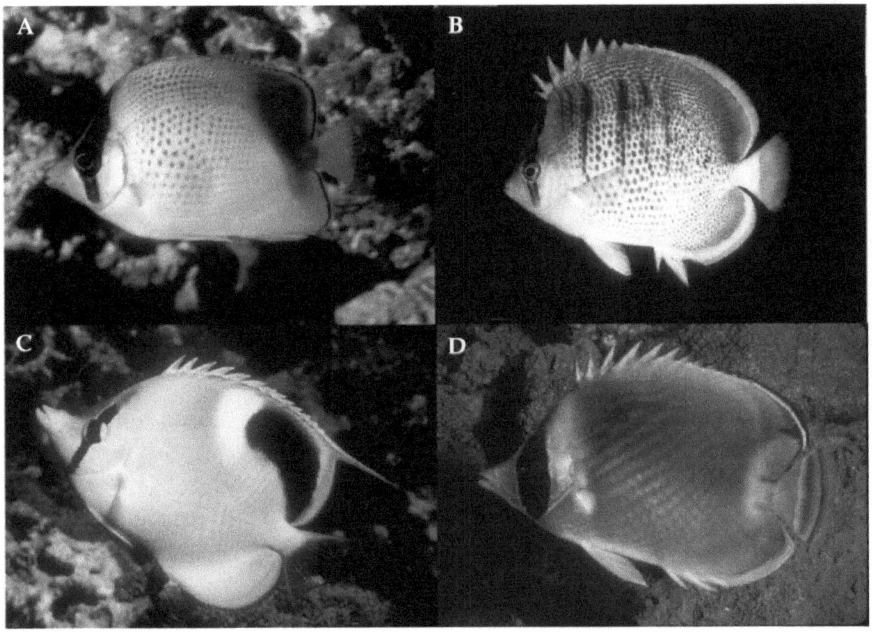

Fig. 3.1 Hybrid butterflyfishes. **(A)** *C. milliaris* x *C. tinkeri*, **(B)** *C. multicinctus* x *C. quadrimaculatus*, **(C)** *C. ephippium* x *C. semion*, **(D)** *C. auriga* x *C. lunula* (All photos: G. Allen).

Colour image of this figure appears in the colour plate section at the end of the book.

chapter is to document the range of hybrid butterflyfishes that have been recorded and consider whether there is a biogeographical, phylogenetic, ecological or behavioural basis to explain the occurrence of hybrids within this family.

In this review we use the term "hybrid" to include individuals that have been identified as hybrids based on intermediate colour (Fig. 3.1). Although genetic analyses are usually required to confirm hybridisation, these techniques have only been applied relatively recently for reef fish hybrids (McMillan et al., 1999; van Herwerden and Doherty, 2006; van Herwerden et al., 2006; Yaakub et al., 2006, 2007; Kuriiwa et al., 2007; Marie et al., 2007). Of the few molecular studies conducted so far on butterflyfishes, all confirm the status of hybrids identified based on intermediate colouration (McMillan et al., 1999; Montanari et al., 2012; Montanari et al., unpublished data). Therefore, intermediate colouration appears to be a good proxy for hybridisation in butterflyfish, although further genetic studies are required to confirm this assumption. This review also regards formation of heterospecific pairs (Fig. 3.2) as tentative evidence of hybridisation, because butterflyfishes mostly breed monogamously (Yabuta, 1997) and

Fig. 3.2 Heterospecific pairs of butterflyfishes; **(A)** *C. trifasciatus* and *C. lunulatus* at Christmas Island (Indian Ocean), **(B)** *C. decussatus* and *C. vagabundus* at Ashmore Reef, Western Australia. More than 10 heterospecific pairs of these species were observed at each of the respective locations. Hybrid *C. trifasciatus* x *C. lunulatus* have also been observed at Christmas Island (Photos: JP Hobbs).

Colour image of this figure appears in the colour plate section at the end of the book.

pair-formation tends to correspond with the onset of sexual maturation and is likely to be for the purpose of reproduction (Pratchett et al., 2006a). As for pairings within species, heterospecific pairs of butterflyfishes examined so far are heterosexual with mature gonads (Hobbs, unpublished data) and are therefore probably interbreeding. Moreover, heterospecific pairs are often seen in locations where hybrids of these putative parental species are recorded (Hobbs et al., 2009). The formation of heterospecific pairs is a critical step in hybridisation because it shows a breakdown in assortative mating and identifies the conditions that enable this reproductive barrier to be overcome. For some heterospecific pairs, resulting hybrids have not been seen. This could be because the hybrids are rare, or they have the colour pattern of one of the parent species (McMillan et al., 1999), or interbreeding does not produce viable hybrid offspring. Failure to produce hybrids from interbreeding is important because it provides unique insights into the mechanism that establishes reproductive isolation (and therefore speciation) in butterflyfishes.

INCIDENCE OF HYBRIDISATION

Pyle and Randall (1994) reported that butterflyfishes are the family of marine fishes with the greatest number of hybrids, based on 15 reported crosses and at least 12 other known crosses at that time. As a proportion of species, Yakuub et al. (2006) suggested that the angelfishes (family Pomacanthidae) were the predominant family of marine fishes to hybridise, but they conceded that in absolute terms, there are more species of butterflyfishes than angelfishes that form hybrids. We know of 34 naturally occurring crosses, mostly based on observations of hybrid individuals with intermediate colouration (Table 3.1). There are also 6 further combinations of species that have been seen to form heterospecific pairs, which are likely to lead to interspecific reproduction even though hybrid offspring have not yet been reported. Excluding heterospecific pairs, there are 44 species of butterflyfishes that represent putative parental species of reported hybrids. Since there are nominally 114 species of butterflyfishes (Fessler and Westneat, 2007), the proportion of species that form hybrids within this family is 38.6%, which is much higher than has been reported within other families of marine fishes (Yakuub et al., 2006), and higher than the average proportion of hybridising plant (25%) and animal (10%) species (Mallet, 2005). The true proportion of hybridising butterflyfish species is likely to be even higher because hybrids often go undetected (Mallet, 2005, 2007) or observed crosses are yet to be reported.

For most coral reef fishes, hybridisation is restricted to a few genera within each family (Yakuub et al., 2006). Within the Labridae for example, most of the hybridising species belong to the genus *Thalassoma* (Walsh

Table 3.1 Parental species (and sub-genera) for reported hybrids. The feeding mode for each species is also given: O, obligate corallivores; F, facultative corallivores; and N, non-corallivore. Support for each hybrid in increasing order of significance is given: P, heterospecific pairing; C, intermediate colouration; M, morphological evidence (e.g., mersitics), and G, genetic confirmation. Citations given were selected to provide greatest support for each reported hybrid.

Species 1	**Species 2**	**Diet** [1,2]	**Support**	**Key reference**
C. adiergastos (Rhombochaetodon)	*C. collare* (Rhombochaetodon)	F, F	P	Hobbs (Pers. Obs.)
C. argentatus (Rhombochaetodon)	*C. mertensii* (Rhombochaetodon)	F, F	C	Allen et al. (1998)
C. argentatus (Rhombochaetodon)	*C. xanthurus* (Rhombochaetodon)	F, F	C	Michael (2004)
C. aureofasciatus (Discochaetodon)	*C. rainfordi* (Discochaetodon)	O, O	M	Randall et al. (1977)
C. auriga (Radophorus)	*C. ephippium* (Radophorus)	F, F	M	Randall et al. (1977)
C. auriga (Radophorus)	*C. vagabundus* (Radophorus)	F, F	P	Pratchett (Pers. Obs.)
C. auriga (Radophorus)	*C. ulientensis* (Radophorus)	F, F	P	Pratchett (Pers. Obs.)
C. auriga (Radophorus)	*C. fasciatus* (Chaetodontops)	F, F	C	Randall and Firdman (1981)
C. auriga (Radophorus)	*C. lunula* (Chaetodontops)	F, F	C	Kuiter (2002)
C. auriga (Radophorus)	*C. trifasciatus* (Corallochaetodon)	F, O	C	Schroeder (2006)
C. auriga (Radophorus)	*C. rafflesi* (Radophorus)	F, F	C	Hobbs (Pers. Obs.)
C. auripes (Chaetodontops)	*C. lunula* (Chaetodontops)	F, F	C	Michael (2004)
C. austriacus (Corallochaetodon)	*C. melapterus* (Corallochaetodon)	O, O	C	Kuiter (2002)
C. baronessa (Gonochaetodon)	*C. triangulum* (Gonochaetodon)	O, O	P	Kuiter (2002)
C. burgessi (Roaops)	*C. tinkeri* (Roaops)	N, N	C	Allen et al. (1998)
C. burgessi (Roaops)	*C. flavocoronatus* (Roaops)	N, N	C	Allen et al. (1998)
C. daedalma (Lepidochaetodon)	*C. nippon* (Lepidochaetodon)	F, F	C	Allen et al. (1998)
C. decussatus	*C. vagabundus* (Radophorus)	F, F	P	Hobbs (Pers. Obs.)
C. ephippium (Radophorus)	*C. semeion* (Radophorus)	F, N	M	Randall et al. (1977)
C. ephippium (Radophorus)	*C. xanthocephalus* (Radophorus)	F, F	C	Allen et al. (1998)

Table 3.1 contd....

C. falcula	*C. ulientensis* (Radophorus)	N, F	P	Hobbs (Pers. Obs.)
C. guentheri (Heterochaetodon)	*C. oxycephalus* (Rhaphorus)	N, F	C	Kuiter (2002)
C. guentheri (Heterochaetodon)	*C. daedallma* (Lepidochaetodon)	N, F	M	Sano et al. (1984)
C. guttatissimus (Exornator)	*C. punctatofasciatus* (Exornator)	F, F	G	S. Montanari (unpub. data)
C. kleinii (Lepidochaetodon)	*C. unimaculatus* (Lepidochaetodon)	F, F	M	Randall et al. (1977)
C. larvatus (Gonochaetodon)	*C. triangulum* (Gonochaetodon)	O, O	C	Michael (2004)
C. lineolatus (Radophorus)	*C. oxycephalus* (Radophorus)	N, F	C	Anon.
C. lunulatus (Corallochaetodon)	*C. trifasciatus* (Corallochaetodon)	O, O	G	Montanari et al., 2012
C. mertensii (Rhombochaetodon)	*C. xanthurus* (Rhombochaetodon)	F, F	C	Froese and Pauly (2009)
C. meyeri (Citharoedus)	*C. ornatissimus* (Citharoedus)	O, O	C	Allen et al. (1998)
C. miliaris (Heterochaetodon)	*C. multicinctus* (Exornator)	N, O	C	Michael (2004)
C. miliaris (Heterochaetodon)	*C. tinkeri* (Rhaphorus)	N, N	M	Randall et al. (1977)
C. multicinctus (Exornator)	*C. quadrimaculatus* (Exornator)	O, O	P	Fig 3.1B
C. multicinctus (Exornator)	*C. pelewensis* (Exornator)	O, F	P*	McMillan et al. (1999)
C. multicinctus (Exornator)	*C. punctatofasciatus* (Exornator)	O, F	P*	McMillan et al. (1999)
C. ocellatus (Chaetodon)	*C. striatus* (Chaetodon)	F, N	M	Clavijo (1985)
C. pelewensis (Exornator)	*C. punctatofasciatus* (Exornator)	F, F	G	McMillan et al. (1999)
C. rafflesi (Radophorus)	*C. vagabundus* (Radophorus)	F, F	C	Kuiter (2002)
C. reticulatus (Chaetodontops)	*C. meyeri* (Cithaoedus)	O, O	C	Senou et al. (2006)
C. reticulatus (Chaetodontops)	*C. ornatissimus* (Cithaoedus)	O, O	C	Senou et al. (2007)

P*—Allopatric species observed to form pairs in captivity.

and Randall, 2004). Similarly, all reported hybrids within the family Chaetodontidae occur among species from the genus *Chaetodon*. *Chaetodon* butterflyfishes account for the majority (87/114) of species within the family Chatodontidae, and more than half (44/87) of *Chaetodon* species appear to form hybrids. Some species, most notably *Chaetodon auriga,* appear to hybridise with a range of different species, whereas other species (e.g., *C. baronessa*) only hybridise with closely related sister species. There is little data on the occurrence of distinct hybrids, though most natural crosses have been described based on a single individual bearing distinctive colouration or markings. However, in areas where *C. pelewensis* and *C. punctatofasciatus* co-occur, up to 70% of individuals bear markings that appear to be intermediary between the two species, indicating very high levels of hybridisation (McMillan et al., 1999).

By any measure, *Chaetodon* butterflyfishes appear particularly predisposed to hybridisation. This may however, be a sampling artefact, as suggested by Randall and Fridman (1981) and Yakuub et al. (2006). The high number of reported hybrids within the family Chaetodontidae, and also Pomacanthidae, may be because they are very conspicuous reef fishes, occurring in relatively low densities, and are also very popular with underwater photographers and aquarists (Randall and Firdman, 1981; Pyle and Randall, 1994). Nominal species within these families are also fairly distinct and mostly have very consistent colouration (McMillan et al., 1999), meaning that hybrids may be more easily recognised than in most other families of coral reef fishes. If butterflyfishes really are predisposed to hybridisation, there are a number of evolutionary, ecological, and behavioural processes that may explain the high incidence of hybrids.

GEOGRAPHIC OCCURRENCE OF BUTTERFLYFISH HYBRIDS

The global distribution of marine hybrids is not uniform (Gardner, 1997), rather there are concentrations occurring in certain locations (Avise, 2000; Johannesson and Andre, 2006; Hobbs et al., 2009). Based on available reports, nearly 90% (34 of 38) of butterflyfish hybrids and heterospecific pairs are clustered in four geographic areas: southern Japan, Hawaii, Papua New Guinea-Micronesia and the eastern Indian Ocean (including Cocos (Keeling) Islands, Christmas Island, southern Indonesia and the offshore reefs of north-west Australia; Fig. 3.3). It is important to acknowledge that sampling intensity and species richness can affect geographic patterns of reported hybrids. For example, the lack of reported hybrid locations in the West Indian Ocean may be partly due to lower sampling intensity, while the dearth of hybrids in the Eastern Pacific-Atlantic area may reflect the low diversity of butterflyfishes in this region. However, there are many

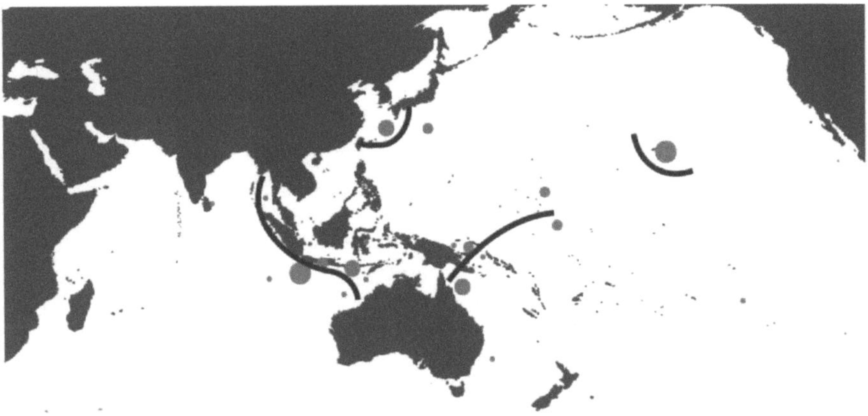

Fig. 3.3 The geographic distribution of reported locations of butterflyfish hybrids and heterospecific pairs. Some hybrids or heterospecific pairs have been observed at multiple locations and for these cases each reported location is represented by a dot (with the caveat that it is 300 km from another reported location). The size of the circle is representative of the number of species pairs that have formed heterospecific pairs or produced hybrids at a reported location, with the largest circles symbolising four cases. Not all instances from Table 3.1 are presented because locations were not always reported. The positions of four recognised biogeographic borders are illustrated with black lines (Blum, 1989; Bellwood and Wainwright, 2002; Hsu et al., 2007).

Colour image of this figure appears in the colour plate section at the end of the book.

well-studied and high diversity locations that lack reports of hybrids. Despite differences in sampling intensity and species richness, the four areas identified above as butterflyfish hybrid "hotspots" are also locations where hybrids of many other reef fish taxa have been found (Gardner, 1997; Pyle and Randall, 1994; Kuriiwa et al., 2007; Hobbs et al., 2009).

Importantly, these four areas of butterflyfish hybridisation align with recognised biogeographic borders (Blum, 1989; Bellwood and Wainwright, 2002; Hsu et al., 2007). Even the minority of cases that fall outside these four areas lie on other proposed biogeographic borders (Blum, 1989; Bellwood and Wainwright, 2002; Hsu et al., 2007). Why are butterflyfish forming heterospecific pairs and producing hybrids at biogeographic borders? This phenomenon occurs regularly in the terrestrial and freshwater environments because biogeographic borders represent areas of confluence between two regional biotas resulting in secondary contact between recently diverged, allopatric species (suture zones: Remington 1968; Avise, 2000; Hewitt, 2000). Examination of the geographic distribution of the species pairs in Table 3.1 (excluding cases involving endemics) reveals that only 13 of 35 cases involve pair formation or hybridisation between typically allopatric species. Thus, the possibility that hybridisation in butterflyfishes is due to secondary contact of allopatric species is only supported in about a third of cases. The remaining 22 cases represent hybridisation between sympatric species.

Given sympatric species co-occur throughout their geographic range, why do they only hybridise at biogeographic borders? Biogeographic borders by their very nature represent the edge of a range for many species and hybridisation is common at range edges (Mayr, 1963; Hewitt, 1988). Examination of locations where sympatric butterflyfish hybrids or heterospecific pairs have been reported revealed that in the majority of cases (14 of 17) at least one of the species is at its range edge (endemic species excluded from analysis). This indicates that butterflyfishes capable of hybridising or forming heterospecific pairs only do so under certain conditions that are present at peripheral locations. Species tend to have low abundances at the edge of their range (Brown, 1984), and therefore a lack of conspecific partners would increase the likelihood of hybridisation at the range edge. Many of these peripheral locations are also isolated islands and therefore likely to a have a relatively high proportion of both vagrants and resident species with small populations, and the lack of conspecifics would increase the chances of hybridisation (Rhymer and Simberloff, 1996; Rose and Polis, 2000; Hobbs et al., 2007, 2009). It is also possible that because species richness and abundance are generally low at peripheral and isolated locations, hybrids may be more prevalent as a result of utilising unoccupied niches (Lexer et al., 2003; Rieseberg et al., 2003; Willis et al., 2006). Either way, it is clear that hybridisation in butterflyfishes (both allopatric and sympatric species) is largely constrained to locations at the edge a species range.

PHYLOGENY OF HYBRID BUTTERFLYFISHES

Hybridisation involves the recombination of genetic material from two distinct species, requiring that parental species have compatible genotypes. As genotypes diverge, the probability of successful recombination of genetic material declines (Jiggins and Mallet, 2000), thereby maintaining stable species boundaries between species with incompatible genomes. It is expected therefore, that occurrence of hybrids is not random with respect to phylogenetic relationships within any given taxon. Rather, there would be a tendency to have the greatest occurrence of hybrids among most closely and recently diverged species pairs (Mallet, 2005). Butterflyfish are an excellent group to test this idea given the prevalence of hybridisation.

The most comprehensive phylogenetic study of butterflyfish published to date (Fessler and Westneat, 2007), includes at least 25 species pairs that are known to hybridise and another 10 species for which the hybrid partner was not sampled in the study. Four distinct phylogenetic clades were identified and hybridisation is reported within three of the four clades (Fig. 3.4). Only Clade 1 species, represented by two species, do not appear to undergo hybridisation. Clade 2 contains six closely related species pairs that are

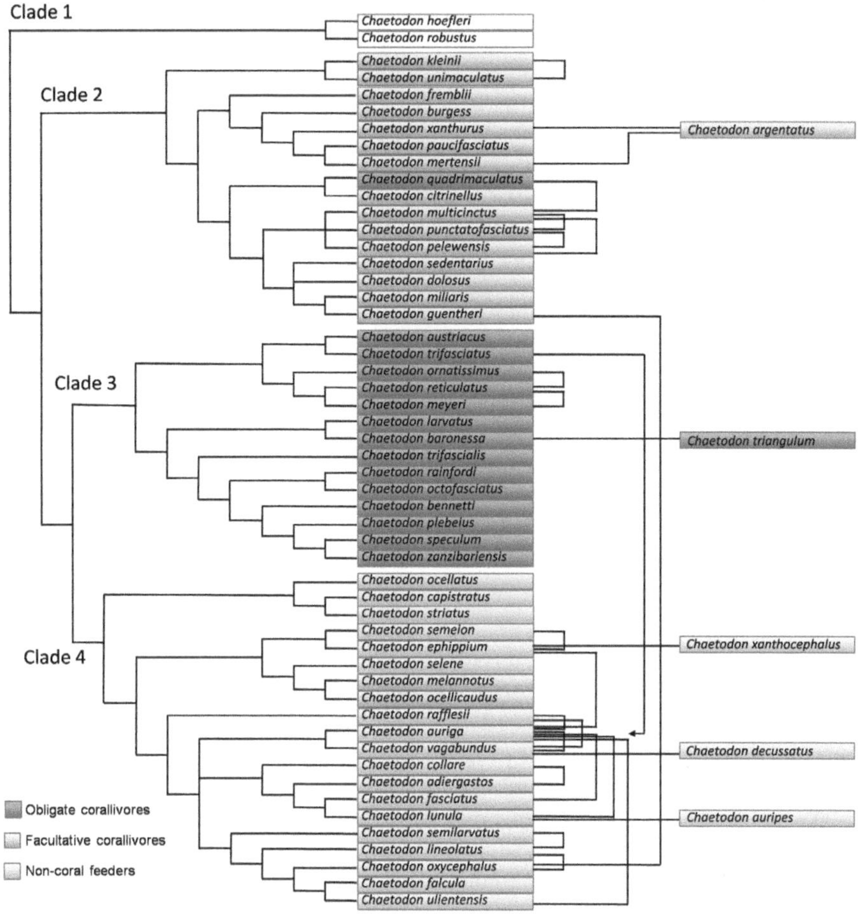

Fig. 3.4 Phylogenetic relationships among pairs of butterflyfishes known to hybridise. The phylogeny presented is based on Fessler and Westneat (2007), indicating 4 major clades that broadly correspond with major trophic guilds (indicated by colours). Hybridisations involving species within the current phylogeny are shown by linkages between species on the right side of the figure.

Colour image of this figure appears in the colour plate section at the end of the book.

hybridising and another 4 species for which their hybridising partner was not represented in the Fessler and Westneat (2007) study. Clade 3 species are less prone to hybridisation, with two species pairs crossing and another 5 species for which the other hybridising parent species was not sampled by Fessler and Westneat (2007). Hybridisation is most common in Clade 4, with 15 species pairs hybridising, not including the one species for which its hybrid partner was not sampled by Fessler and Westneat (Table 3.1;

Fig. 3.4). Only two species pairs hybridised between different clades and both of these were Clade 3 x Clade 4 crosses. One species, *C. auriga* hybridises with at least 9 other species. Overall Clade 4 contains nearly 50% of the hybridising butterflyfish species pairs (including the recently reported hybrid offspring of *C. semilarvatus* and *C. lineolatus* in the Red Sea). These observations suggest that hybridisation is most often observed between closely related butterflyfishes (as identified by molecular genetic data) and rarely between divergent species not sharing recent common ancestors, which is consistent with findings from the terrestrial environment (Mallet, 2005). Even so, it appears that certain monophyletic groups are more prone to hybridsation (e.g., Clade 4) compared to others (e.g., Clade 2).

ABUNDANCE DISPARITIES AND HABITAT OVERLAP

Hybridisation is also more likely where there are abundance disparities (Arnold, 1997) or if both parent species are rare. Based on the authors' unpublished data and observations, at least one of the parent species is rare (average density less than 2 per 1 hr dive or approximately 3000 m^2) in 11 of the 19 (58%) surveyed locations where butterflyfish hybrids or heterospecific pairs have been recorded. Thus, a lack of conspecific mates would increase the chance of hybridisation. However, the data also reveal that hybrids sometimes occur at locations where both parent species are moderately common or abundant (assuming the hybrids were produced at that location). Although a species may be common at a location, an individual of a common species may be restricted to a particular habitat (e.g., an isolated patch reef) where an absence of conspecific partners forces it to pair or mate with a congeneric. Alternatively, hybridisation between common species may occur due to sneak spawning, opportunistic interbreeding or accidental hybridisation, as has been suggested for some other common reef fishes (Frisch and van Herwerden, 2006; Yaakub et al., 2006, 2007; Walker and Ryen, 2007).

The formation of heterospecific pairs and subsequent hybridisation is more likely where two species overlap in their habitat use (Arnold, 1997). To test this, we used published estimates of the depth distribution of butterflyfishes (Allen et al., 1998) as a broad measure of habitat overlap. For species that hybridise or form heterospecific pairs, we found that in 36 of the 40 cases (90%) the species pairs overlap by 50% or more in their depth distribution. Although, some species pairs may differ in habitat use at the reef zone or microhabitat level, the overlap in depth distribution indicates that individuals are likely to have plenty of heterospecific encounters and this increases the possibility of hybridisation. Heterospecific encounters, and therefore the potential for hybridisation, would be much less likely in reef fishes where congeners occupy discrete and separate habitats (e.g., coral gobies; Munday et al., 1997).

BEHAVIOURAL TRAITS OF HYBRID BUTTERFLYFISHES

While phylogenetic affinities, disparities in abundance and overlap in the depth distribution can all increase the likelihood of hybridisation, inter-specific variation in reproductive modes and behaviour can further affect the possibility of hybridisation. Differences in the timing or mode of reproduction among sympatric species represent impermeable barriers to heterospecific breeding (e.g., Palumbi, 1994), but different modes of reproduction are also more or less likely to result in hybridisation. *Thalassoma* wrasses (family Labridae) exhibit both mass-spawning, where males group-spawn with multiple females simultaneously (e.g., Warner, 1995), as well as pair-spawning, depending on the density of individuals and the sex and phase of the individuals (Sara et al., 2005). However, hybridisation between *Thalassoma* species (*T. quinquevittatum* and *T. jansenii*) at Holmes Reef in the Coral Seas is solely attributed to group-spawning, which enables initial phase males of *T. jansenii* to "sneak spawn" with harems of *T. quinquevittatum* (Yakuub et al., 2006). Sneak spawning has also been implicated in hybridisations among mass-spawning salmonid species (Garcia-Vazquez et al., 2002). However, Allen (1979) noted that it is the pair-spawning species of butterflyfishes and angelfishes that typically form hybrids. While some butterflyfishes (e.g., *C. melannotus*) are known to aggregate in certain locations and at specific times in the year, possibly for the purpose of mass-spawning (Pratchett et al., 2006a), the typical model of reproduction for butterflyfishes involves strong long-term pair formation and monogamous mating (Yabuta, 1997; Pratchett et al., 2006a; Yabuta and Berumen, Chapter 8). In this case, successful reproduction results from mutual consent by both partners. Monogamous mating combined with strong and consistent partner recognition (*sensu* Yabuta, 2000) should therefore provide significant barriers to hybridisation.

Assortative mating within and among species, whereby individuals choose mates of similar size and appearance, is considered to be an important process resulting in speciation and divergence of sympatric species that lack sexual dimorphism (e.g., Lande, 1981). Strong assortative mating and interspecific aggression will tend to reduce opportunities for hybridisation. However, many butterflyfishes form heterospecific pairs (Table 3.1), suggesting that assortative mating does not occur, or is relaxed, in some situations. McMillan et al. (1999) found that *C. pelewensis* and *C. punctatofasciatus*, which hybridise in a large contact zone, do not display assortative mating. When given a choice of mates, *C. pelewensis* and *C. punctatofasciatus* did not distinguish between conspecifics, congenerics or hybrids. For these species, it appears that a different selective force is responsible for maintaining colour differences between the species, despite extensive levels of introgression beyond the hybrid zone (McMillan and Palumbi, 1995). Assortative mating has not been specifically tested for any other species of butterflyfishes, but the low incidence of hybrids in most locations suggest that strong assortative mating is important in maintaining species differences. It is probably only when conspecific mates are particularly scarce, that the desire to form pairs and reproduce overcomes strict adherence to conspecific mating. Even in these situations, innate aggression among some species may prevent pairing and interbreeding.

While partly confounded by phylogeny, butterflyfishes that feed exclusively on live corals (obligate corallivores) appear less likely to hybridise compared to facultative corallivores and non-coral feeders (Fig. 3.4). Obligate corallivores are highly territorial and very aggressive towards congenerics, whereas facultative corallivores and non-coral feeders tend to have large home-ranges, and rarely or only weakly defend these ranges against congenerics (Berumen and Pratchett, 2006). Strong interspecific aggression among obligate corallivores also begins almost immediately after settlement, which would largely prohibit formation of heterospecific pairs, which is fundamental to successful reproduction by monogamous pair-spawning species. Consequently, the number of hybrid butterflyfishes with one or more of their putative parents being an obligate corallivores is much less than expected base on the relative abundance of obligate corallivorous species versus facultative corallivores and non-coral feeders (Fig. 3.5). When obligate corallivores do hybridise, it is almost

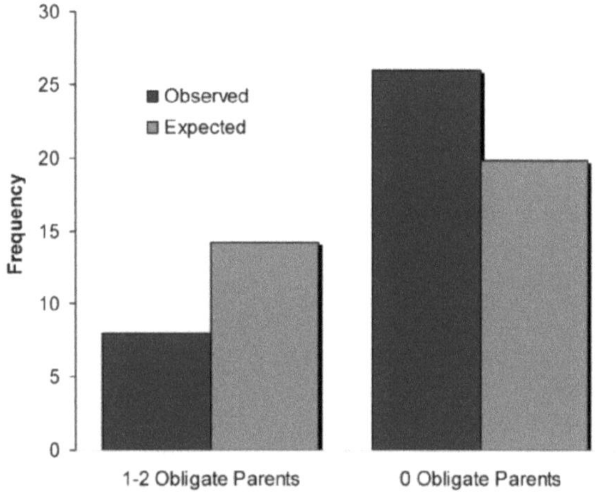

Fig. 3.5 Number of hybrid butterflyfishes with 1 or 2 versus 0 parental species that are obligate corallivores. There are significantly fewer than expected hybrids involving obligate corallivores ($\chi^2 = 28.7$, df = 1, $p<0.05$), possibly because strong aggression among obligate corallivores prevents inter-specific breeding.

always with their most recently diverged sibling species, which have similar appearance and habits, such as *C. trifasciatus* and *C. lunulatus*, *C. baronessa* and *C. trangulum*, or *C. austriacus* and *C. melapterus*. This species almost certainly evolved in allopatry and hybridisation resulted from secondary contact in small areas of overlap. Limited hybridisation among obligate corallivores may be due to genetic incompatibilities among more distantly related species, but this seems unlikely given that obligate coral-feeding species are no older than more generalist counterparts (Bellwood et al., 2009; Bellwood and Pratchett, Chapter 1). Rather, strong interspecific aggression may account for limited hybridisation within obligate corallivores and may even contribute to sympatric speciation within this group.

CONCLUSIONS

By combining data on biogeography, phylogeny, ecology and behaviour we can examine the conditions that facilitate the high level of hybridisation in butterflyfishes. Hybridisation occurs most often between recently diverged species at locations on the edge of their geographic range, where one or both parent species are rare and both species occupy the same habitat and do not aggressively defend territories from congenerics (e.g., Montanari et al., 2012). Any of these factors can promote hybridisation, however if these factors are associated then their combination can greatly increase the chance of hybridisation. For example, hybridisation amongst closely related species also means they probably have similar ecology, feeding and behavioural traits. The first three factors are also interrelated in that the edge of a species range often represents a biogeographic border and this may put a species in contact with a closely related sister species (Hobbs and Salmond, 2008). Species at the edge of their range also tend to have low abundances (Brown, 1984) and isolated locations also have a high proportion of vagrant species with low abundances (Rhymer and Simberloff, 1996; Rose and Polis, 2000; Hobbs et al., 2007, 2009). Thus it is not the locations *per se* that promote hybridisation, but rather the conditions that are associated with the biogeographic position of these locations.

Given the above conditions, the most likely scenario for hybridisation to occur is that at the edge of its geographic range a species has low abundance and therefore an individual may have difficulty finding a conspecific partner. However, because range edges usually coincide with biogeographic borders a species may come into contact with an allopatric sister species. For sympatric species at the edge of their range, searching for conspecifics may put them in contact with another closely related species with similar ecological and behavioural traits. The chances of heterospecific contact are increased if two species overlap in habitat use and the formation of a heterospecific pair is more likely if the two species are not territorial or aggressive to congenerics (i.e., facultative corallivores and non-coral feeders). Once the assortative mating barrier has been overcome, heterospecific pairs are likely to interbreed and the production of hybrids occurs if there are no post-zygotic barriers and if the hybrids have sufficient fitness to survive and reproduce in a particular niche. This scenario would fit with most cases of butterflyfish hybridisation, however accidental hybridisation, opportunistic interbreeding or sneak spawning may be occurring at some locations where both parent species are common.

Hybridisation in butterflyfishes (and reef fishes in general) may become more common in the future as climate change and other anthropogenic impacts modify the marine the environment. Increasing habitat destruction due to multiple disturbances (particularly coral bleaching) can reduce the abundance of butterflyfishes in affected locations (Wilson et al., 2006; Pratchett et al., 2006b, 2008), and a lack of conspecifics partners would promote interbreeding and hybridisation. Anthropogenic modification of habitat structure may also increases hybridisation (Rhymer and Simberloff, 1996; Arnold, 1997; Seehausen et al., 1997; Coyne and Orr, 2004). In coastal marine habitats, the construction of break walls has created new contact zones leading to hybridisation between fish species that normally occupy different habitats within distinct depth ranges (Kimura and Munehara, 2009). Increased turbidity has also impaired mate recognition and led to hybridisation in freshwater fishes (Seehausen, 2006), and a similar scenario may occur on coral reefs as a consequence of sustained and ongoing sedimentation and eutrophication adjacent to major urban centres. The forecasted acidification of the ocean can also affect the sensory system of some reef fish and inhibit the ability to discriminate between suitable mates (Munday et al., 2009). Therefore, the conditions known to promote hybridisation (e.g., abundance disparities, overlap in habitat use, breakdown of assortative mating) are predicted to become more common in the future because of the increasing alteration of the physical and chemical properties of the marine environment brought about by a multitude of human impacts.

FUTURE DIRECTIONS

The combination of distinct colour patterns of adults and hybrids, a pair-forming social system, well known geographic distribution, established phylogeny and evidence of unprecedented levels of hybridisation make butterflyfish a fitting group in which to explore hybridisation in coral reef fishes. Because we know so little about hybridisation in reef fishes, areas of future interest are numerous. As a starting point we need to test if terrestrial theories apply to the marine environment. One area that sparks immediate interest is the overcoming of pre-zygotic reproductive isolation mechanisms through the formation of heterospecific pairs. Assuming that these heterospecifc pairs interbreed and produce viable offspring (which needs to be confirmed) the breakdown of assortative mating is crucial in the formation of hybrids. A combination of mate choice experiments, the collection and sexing of heterospecific pairs and hybrids from natural populations, in combination with molecular evidence of unidirectional or bidirectional matings will elucidate how these pre-zygotic barriers are overcome.

Another area of importance relates to post-zygotic isolating mechanisms and determining if the hybrid offspring are viable, fertile and what their fitness is relative to parent species within and outside of the hybrid zone. This will determine whether hybrids are an evolutionary "dead-end" or whether they play a potentially significant role in the evolution of species. To ascertain fitness will require a combination of in-depth genetic analyses and some estimates of fitness (e.g., growth and fecundity) for hybrids and parent species. Tank experiments and *in vitro* fertilisation can be used to test if hybrids produce viable and compatible gametes (Frisch and van Herwerden, 2006). Genetic analysis will be able to determine whether the hybrids are successfully backcrossing with the parent species and the level and direction of introgression.

It has been argued that rapid radiations of species have occurred in the terrestrial and freshwater environments (and possibly in marine corals) because hybrids have occupied unexploited niches, often at the edge of their ranges where the abundances of the parent species are low (Seehausen, 2004; van Oppen and Gates, 2006; Willis et al., 2006). Given that butterflyfishes fit this pattern (i.e., a speciose group with hybridisation often occurring at range edges where abundances of parent species is often low) it is worth examining if hybrids exploit unoccupied niches. This could be achieved by comparing diet and habitat use of the hybrids and parent species, following Montanari et al. (2012).

Ultimately, genetic studies using both nuclear and mitochondrial (maternal) markers will allow us to confirm the hybrid status of putative hybrids and the parental contributors, the presence of uni- and/or bi-directional hybridisation and the extent of introgression. Previous studies have shown that producing viable first generation (F1) hybrids is most difficult, but once produced, the door is opened and may allow for reinforcement of reproductive barriers (Marshall et al., 2002), hybrid speciation (Hegarty and Hiscock, 2004), extinction (Rhymer and Simberloff, 1996) or reverse-speciation (Seehausen, 2006; Taylor et al., 2006). Molecular studies of butterflyfish hybrids provide the ideal opportunity to determine how hybridisation impacts on the generation, maintenance and loss of coral reef fish biodiversity.

REFERENCES

Allen, G.R. 1979. *Butterfly and angelfishes of the world.* Vol. 2, John Wiley and Sons, New York, USA.

Allen, G.R., R. Steene and M. Allen. 1998. *A guide to angelfishes and butterflyfishes*, Odyssey Publishing, Perth, Australia.

Arnold, M.L. 1997. *Natural hybridisation and evolution*, Oxford University Press, New York, USA.

Avise, J.C. 2000. *Phylogeography*, Harvard University Press, Cambridge, USA.

Avise, J.C. 2004. *Molecular markers, natural history, and evolution*, Sinauer Associates, Sunderland, USA.

Barton, N.H. and G.M. Hewitt. 1985. Analysis of hybrid zones. *Annual Review of Ecology Evolution and Systematics* 16: 113–148.

Barton, N.H. and G.M. Hewitt. 1989. Adaptation, speciation and hybrid zones. *Nature* 341: 497–503.

Bellwood, D.R. and P.C. Wainwright. 2002. The history and biogeography of fishes on coral reefs. In *Coral reef fishes: dynamics and diversity in a complex ecosystem*, P.F. Sale (ed.). Academic Press, San Diego, USA pp. 5–32.

Bellwood, D.R., S. Klanten, P.F. Cowman, M.S. Pratchett, N. Konow and L. van Herwerden. 2009. Evolutionary history of the butterflyfishes (f: Chaetodontidae) and the rise of coral feeding fishes. *Journal of Evolutionary Biology* 23: 335–349.

Berumen, M.L. and M.S. Pratchett. 2006. Effects of resource availability on the competitive behaviour of butterflyfishes (Chaetodontidae). *Proceedings of the 10th International Coral Reef Symposium, Okinawa* pp. 644–650.

Bierne, N., P. Borsa, C. Daguin, D. Jollivet, F. Viard, F. Bonhomme and P. David. 2003. Introgression patterns in the mosaic hybrid zone between *Mytilusedulis* and *M. galloprovincialis*. *Molecular Ecology* 12: 447–461.

Blum, S.D. 1989. Biogeography of the Chaetodontidae: an analysis of allopatry among closely related species. *Environmental Biology of Fishes* 25: 9–31.

Brown, J.H. 1984. On the relationship between abundance and distribution of species. *American Naturalist* 124: 255–279.

Bush, G.L. 1969. Sympatric host race formation and speciation in frugivorous flies of the genus *Rhagoletis* (Diptera, Tephritidae). *Evolution* 23: 237–251.

Cole, A.J., M.S. Pratchett and G.P. Jones. 2008. Diversity and functional importance of coral-feeding fishes on tropical coral reefs. *Fish and Fisheries* 9: 286–307.

Coyne, J.A. and H.A. Orr. 2004. *Speciation*, Sinauer Associates, Sunderland, USA.

Fessler, J.L. and M.W. Westneat. 2007. Molecular phylogenetics of the butterflyfishes (Chaetodontidae): Taxonomy and biogeography of a global coral reef fish family. *Molecular Phylogenetics and Evolution* 45: 50–68.

Frisch, A.J. and L. van Herwerden. 2006. Field and experimental studies of hybridisation between coral trouts, *Plectropomus leopardus* and *Plectropomus maculatus* (Serranidae), on the Great Barrier Reef, Australia. *Journal of Fish Biology* 68: 1013–1025.

Froese, R. and D. Pauly, eds. 2009. *FishBase*. World Wide Web electronic publication: www.fishbase.org, version (07/2009).

Garcia-Vazquez, E., P. Moran, J. Perez, J. Martinez, J. Izquierdo, B. De Gaudemar and E. Beall. 2002. Interspecific barriers between salmonids when hybridisation is due to sneak mating. *Heredity* 89: 288–292.

Gardner, J. 1997. Hybridisation in the sea. *Advances in Marine Biology* 31: 1–78.

Harrison, R.G. 1993. Hybrids and hybrid zones: historical perspective. In *Hybrid zones and the evolutionary process*, R.G. Harrison (ed.). Oxford University Press, Oxford, UK pp. 3–12.

Hegarty, M.J. and S.J. Hiscock. 2004. Hybrid speciation in plants: new insights from molecular studies. *New Phytologist* 165: 411–423.

Hewitt, G.M. 1988. Hybrid zones-natural laboratories for evolutionary studies. *Trends in Ecology & Evolution* 3: 158–167.
Hewitt, G.M. 2000. The genetic legacy of the Quaternary ice ages. *Nature* 405: 907–913.
Hobbs, J.P.A. and J. Salmond. 2008. Cohabitation of Indian and Pacific ocean species at Christmas and Cocos (Keeling) islands. *Coral Reefs* 27: 933–933.
Hobbs, J.P.A., A.J. Frisch, J. Hender and J.J. Gilligan. 2007. New records of Angelfishes (Pomacanthidae) and Butterflyfishes (Chaetodontidae) from Christmas and Cocos (Keeling) Islands, Indian Ocean. *Journal of the Royal Society of Western Australia* 90: 107–109.
Hobbs, J.P.A., A.J. Frisch, G.R. Allen and L. van Herwerden. 2009. Marine hybrid hotspot at Indo-Pacific biogeographic border. *Biology Letters* 5: 258–261.
Hsu, K.C., J.P. Chen and K.T. Shao. 2007. Molecular phylogeny of *Chaetodon* (Teleostei: Chaetodontidae) in the Indo-West Pacific: evolution in geminate species pairs and species groups. *The Raffles Bulletin of Zoology* s14: 77–86.
Hubbs, C.L. 1955. Hybridisation between fish species in nature. *Systematic Zoology* 4: 1–20.
Jiggins, C.D. and J. Mallet. 2000. Bimodal hybrid zones and speciation. *Trends in Ecology & Evolution* 15: 250–255.
Johannesson, K. and C. André. 2006. Life on the margin: genetic isolation and diversity loss in a peripheral marine ecosystem, the Baltic Sea. *Molecular Ecology* 15: 2013–2029.
Kimura, M.R. and H. Munehara. 2010. The disruption of habitat isolation among three Hexagrammos species by artificial habitat alterations that create mosaic-habitat. *Ecological Research* 25: 41–50.
Kuiter, R.H. 2002. *Butterflyfishes, Bannerfishes and their Relatives. A comprehensive Guide to Chaetodontidae and Microcanthidae*. The Marine Fish Families Series, TMC Publishing, Chorleywood, UK.
Kuriiwa, K., N. Hanzawa, T. Yoshino, S. Kimura and M. Nishida. 2007. Phylogenetic relationships and natural hybridisation in rabbitfishes (Teleostei: Siganidae) inferred from mitochondrial and nuclear DNA analyses. *Molecular Phylogenetics and Evolution* 45: 69–80.
Lande, R. 1981. Models of speciation by sexual selection on polygenic traits. *Proceedings of the National Academy of Sciences* 78: 3721–3725.
Lexer, C., M.E. Welch, O. Raymond and L.H. Rieseberg. 2003. The origin of ecological divergence in *Helianthus paradoxus* (Asteraceae): selection on transgressive characters in a novel hybrid habitat. *Evolution* 57: 1989–2000.
Mallet, J. 2005. Hybridisation as an invasion of the genome. *Trends in Ecology & Evolution* 20: 229–237.
Mallet, J. 2007. Hybrid speciation. *Nature* 446: 279–283.
Marie, A.D., L. van Herwerden, J.H. Choat and J.P.A. Hobbs. 2007. Hybridization of reef fishes at the Indo-Pacific biogeographic barrier: a case study. *Coral Reefs* 26: 841–850.
Marshall, J.L., M.L. Arnold and D.J. Howard. 2002. Reinforcement: the road not taken. *Trends in Ecology & Evolution* 17: 558–563.
Mayr, E. 1963. *Animal Species and Evolution*, Belknap Press of Harvard University Press, Cambridge, USA.
Mayr, E. 1979. *Animal Species and Evolution.* 6 ed, Belknap Press, Cambridge, USA.
McMillan, W.O. and S.R. Palumbi. 1995. Concordant evolutionary patterns among Indo-West Pacific butterflyfishes. *Proceedings of the Royal Society of London. Series B: Biological Sciences* 260: 229–236.
McMillan, W.O., L.A. Weigt and S.R. Palumbi. 1999. Color pattern evolution, assortative mating, and genetic differentiation in brightly colored butterflyfishes (Chaetodontidae). *Evolution* 247—260.
Michael, S.W. 2004. *Angelfishes & Butterflyfishes*, TFH Publications, Neptune City, USA.
Montanari, S.R., L. van Herwerden, M.S. Pratchett, J.A. Hobbs and A. Fugedi. 2012. Reef fish hybridisation: lessons learnt from butterflyfishes (genus *Chaetodon*). *Ecology and Evolution* 2: 310–328

Munday, P.L., G.P. Jones and M.J. Caley. 1997. Habitat specialisation and the distribution and abundance of coral-dwelling gobies. *Marine Ecology Progress Series* 152: 227–239.

Munday, P.L., L. van Herwerden and C.L. Dudgeon. 2004. Evidence for sympatric speciation by host shift in the sea. *Current Biology* 14: 1498–1504.

Munday, P.L., D.L. Dixson, J.M. Donelson, G.P. Jones, M.S. Pratchett, G.V. Devitsina and K.B. Doving. 2009. Ocean acidification impairs olfactory discrimination and homing ability of a marine fish. *Proceedings of the National Academy of Sciences* 106: 1848–1852.

Palumbi, S.R. 1994. Genetic divergence, reproductive isolation, and marine speciation. *Annual Review of Ecology Evolution and Systematics* 25: 547–572.

Pratchett, M.S. 2005. Dietry overlap among coral-feeding butterflyfishes (Chaetodontidae) at Lizard Island, northern Great Barrier Reef. *Marine Biology* 148: 373–382.

Pratchett, M.S., O.A. Pradjakusuma and G.P. Jones. 2006a. Is there a reproductive basis to solitary living versus pair-formation in coral reef fishes? *Coral Reefs* 25: 85–92.

Pratchett, M.S., S.K. Wilson and A.H. Baird. 2006b. Declines in the abundance of *Chaetodon* butterflyfishes following extensive coral depletion. *Journal of Fish Biology* 69: 1269–1280.

Pratchett, M.S., P.L. Munday, S.K. Wilson, N.A.J. Graham, J.E. Cinner, D.R. Bellwood, G.P. Jones, N.V.C. Polunin and T.R. Mcclanahan. 2008. Effects of climate-induced coral bleaching on coral-reef fishes—Ecological and economic consequences. *Oceanography and Marine Biology: An Annual Review* 46: 251–296.

Pyle, R.L. and J.E. Randall. 1994. A review of hybridization in marine angelfishes (Perciformes: Pomacanthidae). *Environmental Biology of Fishes* 41: 127–145.

Randall, J.E. 1956. *Acanthurus rackliffei*, a possible hybrid surgeon fish (*A. achilles* x *A. glaucopareius*) from the Phoenix Islands. *Copeia* 1: 21–25.

Randall, J.E. and D. Fridman. 1981. *Chaetodon auriga* x *Chaetodon fasciatus*, a hybrid butterflyfish from the Red Sea. *Revue. fr. Aquariol* 7: 113–116.

Randall, J.E., G.R. Allen and R.C. Steene. 1977. Five probable hybrid butterfly fishes of the genus *Chaetodon* from the central and western Pacific. *Records of the Australian Museum* 6: 3–26.

Remington, C.L. 1968. Suture-zones of hybrid interaction between recently joined biotas. *Evolutionary Biology* 2: 321–428.

Rhymer, J.M. and D. Simberloff. 1996. Extinction by hybridization and introgression. *Annual Review of Ecology Evolution and Systematics* 27: 83–109.

Rieseberg, L.H., O. Raymond, D.M. Rosenthal, Z. Lai, K. Livingstone, T. Nakazato, J.L. Durphy, A.E. Schwarzbach, L.A. Donovan and C. Lexer. 2003. Major ecological transitions in wild sunflowers facilitated by hybridization. *Science* 301: 1211–1216.

Roberts, C.M. and R.F.G. Ormond. 1992. Butterflyfish social behaviour, with special reference to the incidence of territoriality: a review. *Environmental Biology of Fishes* 34: 79–93.

Rose, M.D. and G.A. Polis. 2000. On the insularity of islands. *Ecography* 23: 693–701.

Roush, W. 1997. Hybrids consummate species invasion. *Science* 277: 316–317.

Sano, M., M. Shimizu and Y. Nose. 1984. Food Habits of teleostean reef fishes in Okinawa Island, Southern Japan. *University Museum University of Tokyo Bulletin* 25.

Sara, G., C.N. Bianchi and C. Morri. 2005. Mating behaviour of the newly-established ornate wrasse *Thalassomapavo* (Osteichthyes: Labridae) in the Ligurian Sea (north-western Mediterranean). *Journal of the Marine Biological Association of the UK* 85: 191–196.

Schroeder, T.A. 2006. Annual Report for Fiscal Year 2006. Joint Institute for Marine and Atmospheric Research, University of Hawaii, Honolulu, USA, pp. 110.

Schwenk, K. 1993. Interspecific hybridization in *Daphnia*: distinction and origin of hybrid matrilines. *Molecular Biology and Evolution* 10: 1289–1302.

Seehausen, O. 2004. Hybridisation and adaptive radiation. *Trends in Ecology & Evolution* 19: 198–207.

Seehausen, O. 2006. Conservation: losing biodiversity by reverse speciation. *Current Biology* 16: R334–R337.

Seehausen, O., J.J.M. van Alphen and F. Witte. 1997. Cichlid fish diversity threatened by eutrophication that curbs sexual selection. *Science* 277: 1808–1811.

Senou, H., H. Kodato, T. Nomura and K. Yunokawa. 2006. Coastal fishes of Ie-jima Island, the Ryukyu Islands, Okinawa, Japan. *Bulletin of the Kanagawa Prefectural Museum* 35: 67–92.

Senou, H., Y. Kobayashi and N. Kobayashi. 2007. Coastal fishes of the Miyako group, the Ryukyu islands, Japan. *Bulletin of the Kanagawa Prefectural Museum* 36: 47–74.

Swenson, N.G. and D.J. Howard. 2004. Do suture zones exist? *Evolution* 58: 2391–2397.

Taylor, E., J. Boughman, M. Groenenboom, M. Sniatynski, D. Schluter and J. Gow. 2006. Speciation in reverse: morphological and genetic evidence of the collapse of a three-spined stickleback (*Gasterosteusaculeatus*) species pair. *Molecular Ecology* 15: 343–355.

van Herwerden, L. and P.J. Doherty. 2006. Contrasting genetic structures across two hybrid zones of a tropical reef fish, *Acanthochromis polyacanthus* (Bleeker 1855). *Journal of Evolutionary Biology* 19: 239–252.

van Herwerden, L., J.H. Choat, C.L. Dudgeon, G. Carlos, S.J. Newman, A.J. Frisch and M.J.H. van Oppen. 2006. Contrasting patterns of genetic structure in two species of the coral trout *Plectropomus* (Serranidae) from east and west Australia: Introgressivehybridisation or ancestral polymorphisms. *Molecular Phylogenetics and Evolution* 41: 420–435.

van Oppen, M.J.H. and R.D. Gates. 2006. Conservation genetics and the resilience of reef-building corals. *Molecular Ecology* 15: 3863–3883.

Walker, S.P.W. and C. Ryen. 2007. Opportunistic hybridization between two congeneric tropical reef fish. *Coral Reefs* 26: 539–539.

Walsh, F. and J.E. Randall. 2004. *Thalassoma jansenii* x *T. quinquevittatum* and *T. nigrofasciatum* x *T. quinquevittatum*, hybrid labrid fishes from Indonesia and the Coral Sea. *Journal of Ichthyology and Aquatic Biology* 9: 69–74.

Warner, R.R. 1995. Large mating aggregations and daily long-distance spawning migrations in the bluehead wrasse, *Thalassoma bifasciatum*. *Environmental Biology of Fishes* 44: 337–345.

Willis, B.L., M.J.H. van Oppen, D.J. Miller, S.V. Vollmer and D.J. Ayre. 2006. The role of hybridization in the evolution of reef corals. *Annual Review of Ecology Evolution and Systematics* 37: 489–517.

Wilson, S.K., N.A.J. Graham, M.S. Pratchett, G.P. Jones and N.V.C. Polunin. 2006. Multiple disturbances and the global degradation of coral reefs: are reef fishes at risk or resilient? *Global Change Biology* 12: 2220–2234.

Wrathall, T., C. Roberts and R. Ormond. 1992. Territoriality in the butterflyfish *Chaetodon austriacus*. *Environmental Biology of Fishes* 34: 305–308.

Yaakub, S.M., D.R. Bellwood, L. van Herwerden and F. Walsh. 2006. Hybridization in coral reef fishes: introgression and bi-directional gene exchange in *Thalassoma* (family Labridae). *Molecular Phylogenetics and Evolution* 40: 84–100.

Yaakub, S.M., D.R. Bellwood and L. van Herwerden. 2007. A rare hybridization event in two common Caribbean wrasses (genus *Halichoeres*; family Labridae). *Coral Reefs* 26: 597–602.

Yabuta, S. 1997. Spawning migrations in the monogamous butterflyfish, *Chaetodon trifasciatus*. *Ichthyological Research* 44: 177–182.

Yabuta, S. 2000. Behaviors in agonistic interaction of the butterflyfish (*Chaetodon lunulatus*). *Journal of Ethology* 18: 11–15.

CHAPTER 4

Biogeography of Butterflyfishes: A Global Model for Reef Fishes?

Michel Kulbicki,[1,*] *Laurent Vigliola,*[2] *Laurent Wantiez,*[3] *Nicolas Hubert,*[4] *Sergio R. Floeter*[5] and *Robert F. Myers*[6]

INTRODUCTION

The biogeography of coral reef fishes is a complex issue still poorly understood largely because i) the knowledge of geographic ranges of many species is still far from complete; and ii) identification remains problematic for several taxa, especially within some families such as the Muraenidae, Ophichtidae, Apogonidae, Gobiidae and Trypterygiidae. There have been a number of attempts to understand the regional distribution of coral reef

[1]Centre IRD (Laboratoire Arago), BP 44-66651, Banyuls, France.
E-mail: michel.kulbicki@ird.fr
[2]Centre IRD (Nouméa), 101 Promenade Roger Laroque, Anse Vata, BP A5-98848 Noumea, New Caledonia.
E-mail: laurent.vigliola@ird.fr
[3]Université de la Nouvelle-Calédonie, BP R4-98851 Nouméa cedex, New Caledonia.
E-mail: laurent.wantiez@univ-nc.nc
[4]Université Montpellier II, Case courrier 63, Institut des Sciences de l'Evolution, Dpt Conservation et Domestication, 34095 Montpellier.
E-mail: nicolas.hubert@ird.fr
[5]Departamento de Ecologia e Zoologia, Universidade Federal de Santa Catarina, Florianópolis, Santa Catarina 88010-970, Brazil.
E-mail: sergio.floeter@ufsc.br
[6]Seaclicks/CoralGraphics, 9273 Silent Oak Circle, Wellington FL 33411-6634 United States.
E-mail: robmyers1423@gmail.com
*Corresponding author

fishes, in the Atlantic (e.g., Floeter et al., 2008), in the Indo-west Pacific (e.g., Springer, 1982; Bellwood and Hughes, 2001; Santini and Winterbottom, 2002; Allen, 2000, 2007) and in the Tropical Eastern Pacific (e.g., Robertson and Cramer, 2009), however there have been very few attempts to understand such patterns on a worldwide scale. A notable exception is Roberts et al. (2002), which investigated the distribution of 1700 species of coral reef fishes worldwide to compare the overlap between diversity hotspots and species geographic range. At present there is little general understanding of the distribution of coral reef fishes, except for the major patterns of diversity and endemism, and on the factors that may lead to this distribution at a global scale.

Coral reef fishes (abbreviated hereafter as "reef fish") are of particular interest for a worldwide biogeography analysis because they are distributed according to a number of large-scale gradients that can be related to several major hypotheses in ecology (see Whittaker et al., 2001 for a review of hypotheses and Ricklefs, 2004 for a comprehensive framework). Among the numerous gradients along which reef fish diversity varies, five are of particular interest: latitude, distance to biodiversity centres, island size, connectivity (specifically, the degree of exchange between populations) and large-scale coral diversity. In addition reef fishes are diverse and display a wide range of life-history traits. Such properties make these fishes ideal candidates for biogeographical and macro-ecological studies (see reviews by Kent, 2005, 2007). Coral reef fishes are by definition closely associated with coral reefs, which are in turn restricted to warm (the lower temperature limit for scleractinian corals is 17°C) oligotrophic waters. Water temperature varies substantially/markedly with latitude and depth or next to specific features such as cold currents or upwellings. Latitudinal gradients in species diversity are well established for both terrestrial and aquatic systems (Willig et al., 2003; Vázquez and Stevens, 2004), with these patterns being related to variation in temperature, all other factors being equal (Gaston, 2000). Therefore fish diversity should be highest where water temperature is high, which is usually associated with low latitudes and shallow waters. The second major gradient, at least in the Indo-Pacific, is a decrease in diversity with distance from the centre of biodiversity, represented by the Indo-Australian Archipelago (Bellwood and Hughes, 2001; Carpenter and Springer, 2005; Allen, 2007) and commonly referred to as the "Coral Triangle". This diversity centre represents an area of species accumulation more than speciation (Santini and Winterbottom, 2002; but see Mora et al., 2003 for another view).

Island size and connectivity also have important implications for reef fish biogeography. Island biogeography theory (MacArthur and Wilson, 1967) predicts that the number of habitats is a function of island size and as a consequence the number of species will increase with island size. Similarly, isolated islands will shelter less species than relatively connected ones (e.g., Moilanen and Nieminen, 2002) due to the lower rates of exchange amongst populations. Corals are the foundation species (or ecosystem engineers) of coral reefs and many reef fishes are closely associated with corals for food and/or shelter. Therefore the diversity of reef fishes should be highest where the diversity and/or cover of corals is high. In support of this Bellwood and Hughes (2001) demonstrated a correlation between reef fish diversity and coral diversity across the Indo-Pacific, and Roberts et al. (2002) reported similar relationships across a global scale. This raises the question of whether this congruence is the result of a co-evolution of the two groups or is just the result of a stochastic process (Gaston, 2000).

Butterflyfishes (family Chaetodontidae) represent an almost ideal model for the analysis of the major biogeographical trends of reef fishes: there are few identification problems with this family, their diversity is sufficient in most regions to enable statistical comparisons, the geographic range of most species is well known and there is information on the biology and phylogeny for a large proportion of species. At present there is no other coral reef fish family possessing all these advantages. Additionally, there is a good correlation between *Chaetodon* diversity and reef fish diversity in the Pacific (Kulbicki et al., 2005). Nonetheless, the Chaetodontidae are imperfect as a model group because the family is not represented by all possible reef fish life-history traits. In particular they have a limited range of maximum sizes, their diets are not very diversified compared to the other reef fishes (e.g., Labridae: Bellwood et al., 2006), and their range of social behaviours and levels of endemism are usually low. In the present chapter we will describe/quantify the worldwide distribution of the Chaetodontidae, and evaluate some of the major hypotheses underlying the distribution of these fishes and, by extension, that of coral reef fishes in general. We will first define a restricted number of biogeographical regions based on the species composition of Chaetodontidae in a large number of localities across the world. Once these regions are defined we will test the effects of large-scale factors (i.e., latitude, island size, isolation, distance to biodiversity centres, and large scale coral diversity) and life-history traits (i.e., diet, body size and behaviour) on the diversity (alpha and beta) of the Chaetodontidae. Here, we define alpha diversity as the number of species found within a single 'site'. Therefore the area covered by this diversity may vary from a small island to large continental expanses. Beta diversity is a measure of the diversity gradient between alpha diversities from various sites, in other words by how much does alpha diversity vary from one location to the next within a region. In particular, alpha diversity will be analysed as both "absolute" alpha diversity (total number of Chaetodontidae species per site), and "relative" alpha diversity (proportion of Chaetodontidae amongst reef fish species per site). Two types of beta diversities will also be considered

(Magguran, 2004), one independent of species composition (Whittaker, 1972) and another one taking into account the variations in species composition amongst regions (Legendre et al., 2005). Finally, the available information on the phylogeny of butterflyfishes will be used to try to understand some of the patterns found in the previous analyses.

BIOGEOGRAPHICAL REGIONS AND DIVERSITY PATTERNS

Worldwide diversity patterns of butterflyfishes were studied from species presence-absence datasets gathered from both the literature and our own field surveys at 175 sites spanning all of the world's oceans (Appendix 1, Fig. 4.1). Sites for which we could gather data had relatively even spatial distribution with a butterflyfish species checklist available at every ~1000 km or less, subsequently more sites were censused in the larger Pacific Ocean (108 species lists) than in the smaller Indian and Atlantic Oceans (35 lists each). Although their distribution was relatively even in space, our sampling units (i.e., sites) were of very different size, ranging from large regions (e.g., North-West Australia) to individual reefs (e.g., Tulear reef in Madagascar).

We defined a biogeographic region as a geographically continuous area inhabited by a relatively homogeneous fauna. From this definition, it results that the search for the boundaries of a biogeographic region simply consists of clustering ecologically similar sites that are part of the same continuous geographic zone. Cluster analysis requires two algorithms, a first one to measure the ecological similarity (or distance) between sites and a second one to decide when sites are similar enough to become members of the same cluster. The Raup and Crick (1979) probabilistic similarity coefficient (S_{27} in Legendre and Legendre, 1998) was selected to estimate distances (specifically, Raup and Crick's probabilistic coefficient of similarity in species composition) and Ward's (1963) minimum variance method was chosen to group sites. To ensure that ecologically similar sites were part of the same geographically continuous region and to minimise the effects of different sampling unit size and spatial overlap in our dataset, Ward's algorithm was only used as a basis for a step-by-step manual clustering procedure. First, the Raup and Crick's distance matrix among all 175 sites was calculated and a consensus tree built based on 999 bootstrap trees (+ the original tree) and the Ward's algorithm. Second, sites that were most ecologically similar (first rank clusters), geographically close (continuous region), and with high consensus indices (strong clusters) were merged and a new presence-absence dataset with fewer but larger sites was built. From this new presence-absence dataset a new consensus tree was generated, sites were merged into another new presence-absence dataset, and the procedure was repeated until all sites formed a single cluster. Once the

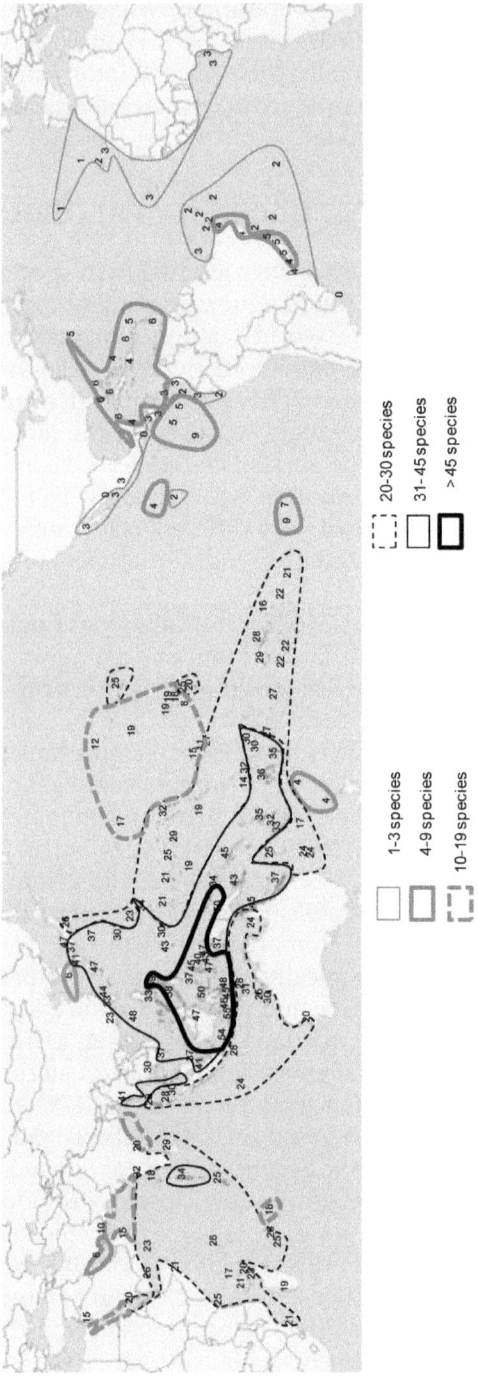

Fig. 4.1 World wide map of the alpha diversity distribution of butterflyfishes.

Colour image of this figure appears in the colour plate section at the end of the book.

biogeographic regions were defined, these were represented in a Cartesian coordinate system by multidimensional scaling (MDS), to represent distance relationships (Raup and Crick's distance) among the biogeographic regions (Legendre and Legendre, 1998).

A total of 134 species belonging to 11 genera were recorded in our dataset (Appendix II). Three undescribed species were not included in the analysis as both their status and geographical distribution are uncertain. Taxonomic richness varied across localities from a low of one genus and one species at Madeira and the Azores in the eastern Atlantic to a maximum of eight genera and 55 species at Bali, Indonesia (Fig. 4.1). Both species and genus richness were highest in the Indo-Australian Archipelago, with 6–8 genera and over 40 species at each of the Banda, Celebes, Java, Sulu, Philippines, China, Arafura, Timor, Bismark, Solomon, and Coral Sea sites. The existence of a biodiversity centre for coral reef organisms (including fishes) centred around the Indo-Australian archipelago has been well described (see Carpenter and Springer, 2005; Bellwood et al., 2005; Allen, 2007), however there is some debate over the precise boundaries of this centre. Taxonomic richness decreased both eastward and westward of this Indo-Pacific biodiversity centre, as has been described previously for many reef organisms, in particular corals and fishes (e.g., Bellwood and Hughes, 2001; Bellwood et al., 2005). Diversity was much lower in the eastern Pacific and the western Atlantic, with only 2–3 genera and less than 5–6 species at each site, and lowest in the eastern Atlantic with only 2–3 species belonging to 1–2 genera.

The cluster analysis revealed 19 biogeographic regions (Fig. 4.2) and 39 sub-regions (Appendix I). In general there is little overlap between the boundaries of these regions and the diversity isopleths in Fig. 4.1. There were nine regions in the Pacific, four in the Indian, five in the Atlantic, and one across the Indian and Pacific Oceans. Regions in the Indo-Pacific were generally inhabited by more homogeneous butterflyfish faunas than those in the Atlantic, as indicated by differences in the within-region Ward's minimum variance distance. As will be discussed later, this can be in part related to the very low levels of endemism across most of the Indo-Pacific, whereas in the Atlantic endemism is much higher. Nevertheless, heterogeneous regions also existed in the Indo-Pacific, notably in the eastern Pacific, remote Hawaii and sub-tropical New Zealand regions, where endemism levels may also be high.

In the Pacific, the three most speciose regions were those around Indonesia (Indonesia, Philippines, Indochina, Peninsula), China-Japan (China, Korea, Japan), and Melanesia (Papua New Guinea, Great Barrier Reef, Melanesia). The Indonesian and Melanesian regions were the most homogeneous as shown by a very small ($\sim10^{-13}$) within region Ward's minimum variance distance (Fig. 4.2). In comparison, the China-Japan

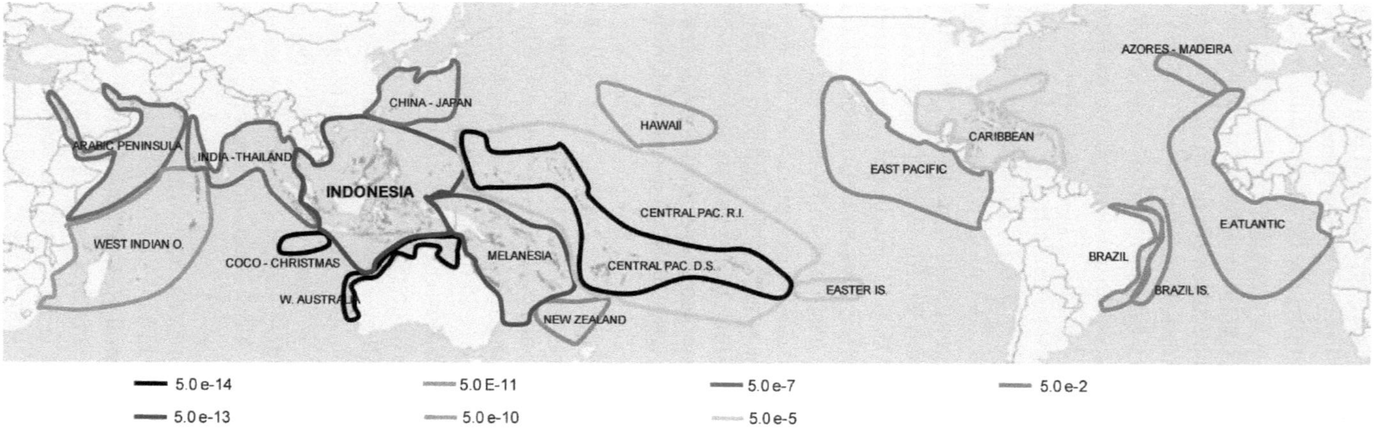

Fig. 4.2 Map of the 19 biogeographic regions determined by stepwise clustering of 175 butterflyfish species lists around the tropics. The lines indicate Ward's minimum variance distance levels.
Colour image of this figure appears in the colour plate section at the end of the book.

region was relatively heterogeneous with a clustering distance of ~10^{-7}. This region also presents the highest beta diversity of all regions, which indicates a strong variation in species richness and composition from one site to another. East of these three speciose regions lies a very large and relatively heterogeneous (~10^{-5}) region that encompasses most of the central Pacific. However, stepwise cluster analysis revealed that this large central Pacific region was formed by a very homogeneous central stripe (~10^{-14}) surrounded by a rather heterogeneous belt (~10^{-5}). The stripe was named "Central Pacific Dorsal Spine" and follows a corridor formed by the NW-SE alignment of the Mariana, central Caroline, Marshall, Gilbert, Ellice (Tuvalu), Wallis, Samoa, Fiji, Tonga, Society, Tuamotu, and Gambier Islands. Most of these islands are atolls (except the Society Islands and some of the Cook Islands), are of similar size, and have a similar coral diversity and seawater temperature regime. The belt was named "Central Pacific Remote Islands" as it was formed by a number of very scattered and geographically isolated islands on both sides of the stripe. The heterogeneity found in this "belt" may be related to the very low connectivity of these islands and also to their physical characteristics, e.g., the presence of high islands, such as the Marquesas or Rapa Islands, and atolls, such as the Line Islands. Three distinct and relatively isolated regions lie at the border of the central Pacific region, the Hawaiian region at the extreme north, the New Zealand region at the extreme south, and the Easter Island region at the extreme east. These regions are characterised by several endemic species and are at the limit of the habitat range of butterflyfishes. Separated from other Pacific regions by over 3000 km of deep waters and complex current systems, the East Pacific biogeographic region includes all sites of the western American coasts and islands. In particular, there is very little species overlap with the central and west Pacific; more than 50% of the species are endemic to the East Pacific region.

The Indian Ocean included four regions. A small and very homogeneous (~10^{-14}) "Coco-Christmas" region was formed by the Cocos-Keeling and Christmas Islands on the western border of the Indonesian Pacific region. The fauna of this region is intermediate between typical Indian Ocean fauna and Pacific fauna. Similarly, the fauna of the western Australian region contained a higher proportion of species centred on the Pacific than on the Indian Ocean. In this respect, islands off the western Australian coast (Rowley Shoals, Ashmore Reef, and Scott Reef) had a fauna that included them in the Indonesian region rather than into the western Australian region despite a closer distance to the latter. Westwards, India, the Laccadive Sea, the Bay of Bengal, the Andaman Sea, and the west coasts of Malaysia and Sumatra formed a relatively homogeneous (~10^{-7}) "India-Thailand" region. The Lakshadweep and western Indian coast share many similarities with the Maldives, but their butterflyfish fauna is less diverse. Despite the natural barrier formed by the freshwater inputs in the Bay of Bengal the similarity on both sides of this barrier remains high. This is in opposition with the other freshwater barrier formed by the Indus Delta, which creates an eastern boundary for the region grouping the Red Sea, the Persian Gulf and the western Arabian Sea. This "Arabian Peninsula" region ($\sim 10^{-7}$) is rather heterogeneous due to numerous endemic species and to the presence of an upwelling site in Oman that is believed to separate the Persian Gulf reef fish fauna from the Red Sea and Yemen (Randall, 1995; Kemp, 1998). In contrast, the southeast African coast, Madagascar, and the central Indian islands (up to the Maldives) formed a very homogeneous ($\sim 10^{-11}$) "Western Indian Ocean" region. This region could be a "sink", or accumulation centre, as species may either originate from the Arabian Peninsula or the India-Thailand regions. This is in agreement with Connolly et al. (2003) who indicated that reef fish diversity along the eastern African coast was above predicted levels. It should be noted that *Chaetodon marleyi*, which is endemic to the "Western Indian Ocean", is related to the eastern Atlantic species *Chaetodon robustus*, thus establishing a link between these oceans.

The most speciose and homogeneous ($\sim 10^{-5}$) region of the Atlantic, the "Caribbean" region, included the Caribbean Sea and the Gulf of Mexico. There were two other regions along the Atlantic coast of the Americas. The "Brazil" region included all sites along the coast of Brazil whereas the "Brazilian Islands" included remote islands offshore of South America. On the east side of the Atlantic, a large "East Atlantic" region included the western African coast and its offshore islands whereas a smaller "Azores-Madeira" region included the Azores and Madeira Islands. However for the latter two, only vagrant fish have been observed and no permanent population is known.

The consensus tree of the 19 biogeographic regions indicated highly dissimilar butterflyfish fauna in the Indo-Pacific and the Atlantic Oceans (Fig. 4.3). Within the Atlantic, the Caribbean and Brazil clustered first, quickly joined by the Brazilian islands, then by Azores-Madeira, and finally by the East Atlantic region that appeared as a relatively distinct region. Floeter et al. (2008) performed a biogeographical analysis of the entire Atlantic reef fish fauna. The regions they defined are broadly similar to those found using only butterflyfish, but with some differences. For example, the Brazilian coast and the islands off Brazil were clustered into a unique region and not two regions as in the present analysis. This could arise in our analysis due to the presence of an endemic species (*Prognathodes obliquus*) and lower diversity in the oceanic islands. Another difference is the closer distance of Azores-Madeira to the Caribbean and Brazilian regions than to the other eastern Atlantic sites in the present study. This is presumably because of the presence of a single vagrant species from the western Atlantic in the Azores-Madeira region that may have been

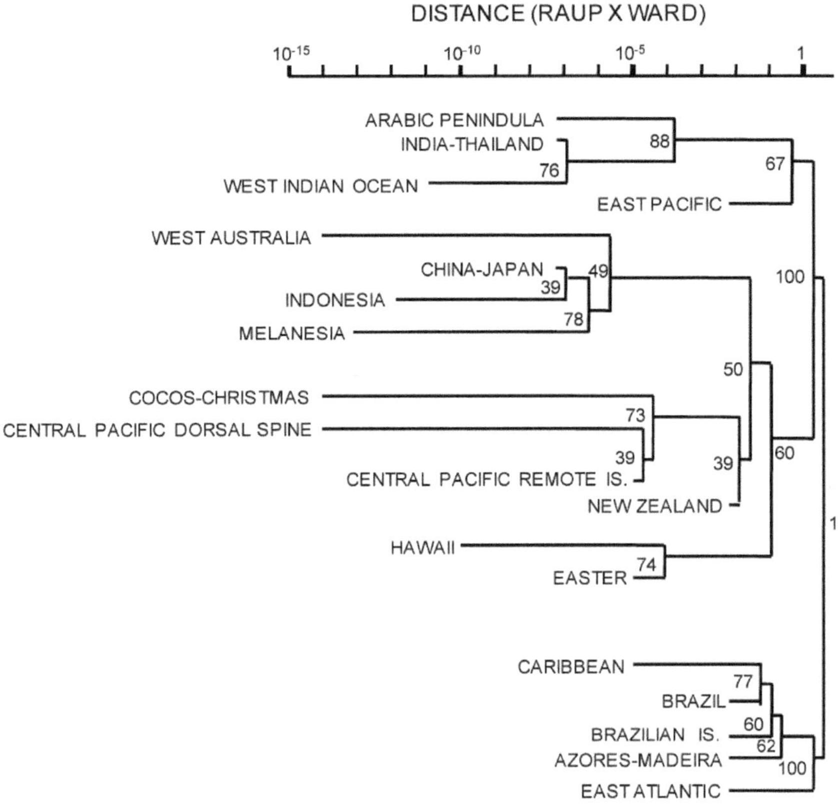

Fig. 4.3 Consensus tree of the 19 biogeographic regions. Values at nodes represent bootstrap support (% of 1000 replicates).

transported by the Gulf Stream. Conversely, Floeter et al. (2008) separated the Ascension and St Helena islands in the tropical eastern Atlantic could be separated from the eastern Atlantic, but our analysis did not distinguish these regions due to the presence of 2 endemic species. The clustering of these islands with the remaining eastern Atlantic is due to one of these species, *Chaetodon sanctaehelenae*, which has been reported from the Canaries (Brito and Falcon, 1996).

Within the Indo-Pacific, regions of the Indian Ocean clustered together, with the exception of the Cocos-Christmas region that clustered with the central Pacific regions. Similarly, regions of the Pacific Ocean clustered together, with the exception of the eastern Pacific region, which joined the Indian Ocean cluster. Many analyses have investigated the definition of biogeographic regions of reef fishes or have mapped the distribution of reef fish diversity in the Indo-Pacific (e.g., Springer, 1982; Kemp, 1998; Mora et al., 2003; Santini and Winterbottom, 2002; Allen, 2007; Kulbicki, 2007).

Springer (1982) was the first to suggest the importance of the tectonic plate margins in the distribution of reef fishes in the Indo-Pacific, in particular the very sharp drop in diversity at the western margin of the Pacific Plate. Subsequently, the first definition of major biogeographic regions for reef fishes was proposed that matched the limits of these plates. Following on from this work Thresher (1991) attempted a worldwide classification of 21 reef fish sites based on species composition. Indo-Pacific sites did not cluster according to their geographic affinities but more according to their physical environments, with island size, island type, and connectivity seemingly important factors. Santini and Winterbottom (2002) arbitrarily defined 18 regions in the Indo-Pacific, mainly based on geological features, but their study did not encompass the central Pacific (i.e., Polynesia, Hawaii, and the Line Islands). More recently, Kulbicki (2007) examined the reef fish fauna of 65 sites in the west and central Pacific and identified nine biogeographical regions that displayed a remarkable overlap with geological features such as tectonic plate margins.

In the present study, we found strong congruence with the previous classifications of biogeographic regions for both corals and reef fishes. There was, however, some variation in the relatedness of the various regions. For example, in the Indian Ocean, India-Thailand and the West Indian Ocean regions were most ecologically similar and were later joined by the Arabian Peninsula region. In contrast, Santini and Winterbottom (2002) found that the Arabian Peninsula first clustered with eastern Africa, then the Mascarene and Seychelles islands, and only after that clustered with India. Similarly, Kemp (1998) found on a smaller scale that the Socotra fauna (in the Gulf of Aden) was linked first to the Red Sea, then Somalia, and then to western India. Veron (2000) found that corals of the Arabian Peninsula clustered with the Mascarene and the eastern Indian Ocean regions on a similar level, which is similar to the present study on butterflyfish.

Multi-dimensional scaling of the 19 biogeographic regions further outlined existing gradients in butterflyfish biodiversity across the world (Fig. 4.4). The analysis revealed four independent provinces, the Atlantic, East Atlantic, Indo-Pacific, and East Pacific. The presence of two centres of diversity, one in Indonesia and the other in the Caribbean, raises the question of the importance of these centres in the distribution of species among regions, and the relationships between regions (Briggs, 2005). These centres appear to neither be centres of endemism (Hughes et al., 2002; Allen, 2007; but different opinions expressed by Roberts et al., 2002, Mora et al., 2003) nor centres of origin (Santini and Winterbottom, 2002) and the factors contributing to these peaks in diversity are complex as will be discussed below.

In the Atlantic the total number of species within the Chaetodontidae is low (15 species) and it is therefore difficult to draw strong conclusions.

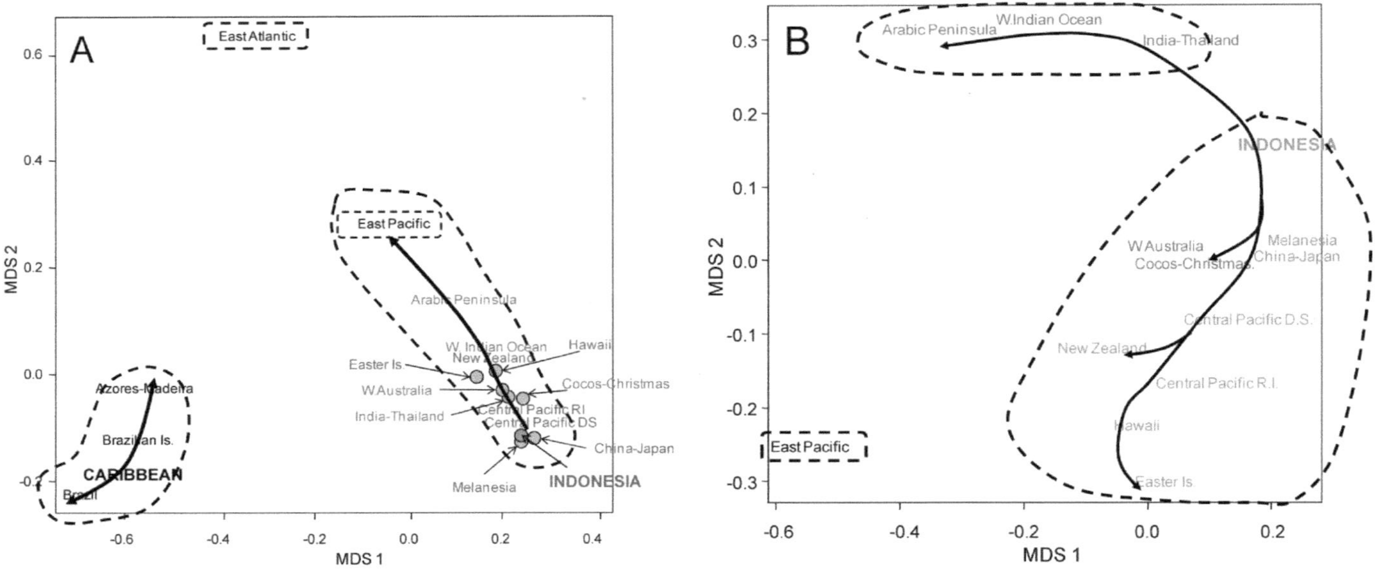

Fig. 4.4 Multidimensional scaling **(A)** all 19 biogeographic regions, **(B)** only Indo-Pacific regions. Regions from the Indian Ocean are in red; regions from the west and central Pacific are in blue. MDS1 and MDS2 are axes of the multidimensional scaling. The arrows indicate the shortest pathways from one region to another.

Colour image of this figure appears in the colour plate section at the end of the book.

However the eastern Atlantic has a unique fauna, its five species being not found anywhere else in the Atlantic, which explains its isolated position on Fig. 4.4A. Brazil has the most species in common with the Caribbean, and while there is a continental continuum between these two regions the Amazon and Orinoco Rivers' fresh water flows represent barriers to species dispersal. Recently however, it has been proposed that some species are able to use deep-water reefs beneath these river plumes as stepping-stones between the two regions (Rocha, 2003). More surprising is the paucity of butterflyfish species on the islands off Brazil. These islands are small and isolated which could in part explain this paucity but other factors such as current patterns and coral diversity may have also contributed to the observed patterns.

In the Indo-Pacific (including East Pacific) two features are remarkable (Fig. 4.4). First, the Indian Ocean and west Pacific faunas show a high degree of similarity with Indonesia and this similarity decreases regularly in both oceans with increasing distance from Indonesia. Second, the eastern Pacific shares very little in common with the rest of the Indo-Pacific, reflecting its isolated position (Fig. 4.4B). In the Indo-Pacific, ecological similarity to Indonesia may decrease with geographic distance for different reasons in the two oceans. In the Pacific, the lowest similarity is found with small and isolated island systems (Easter Island, Hawaii) whereas in the Indian Ocean the Arabian Peninsula, which presents the lowest similarity with Indonesia, has a continental continuum with Indonesia (Fig. 4.4B). This continuum is however disturbed by three physical barriers (two river systems and one upwelling) that have generated differences in similarity higher than those created by isolation or island size in that ocean.

The peculiar situation of the East Pacific region as an independent region located at the extreme east of the eastward Pacific gradient implies that this relatively independent region has recently started to link with the rest of the Pacific. Detailed data analysis indicates that the stepping stone islands are probably the Galapagos as all seven butterflyfish species found in the East Pacific are present in the Galapagos whereas there are two species in the Galapagos that are present in the Indo-Pacific but absent in the East-Pacific.

Alpha Diversity

"Absolute" alpha (α) diversity

The number of butterflyfish species at a site varies from one (*Chaetodon sedentarius,* a vagrant species in the islands of Azores in the Atlantic) to 55 species (Bali–Indonesia) with strong differences between and within regions (Fig. 4.1). In particular there are far less species in the Atlantic and

East Pacific than anywhere in the Indo-Pacific. A number of studies on reef fishes indicate very similar geographical patterns with the distribution of butterflyfishes (e.g., Randall, 1998; Roberts et al., 2002; Bellwood et al., 2005; Allen, 2000, 2007). One notices also the very abrupt gradient at the limits of the geographic range of butterflyfishes in many areas of the Indo-Pacific; in some cases the number of species drops from over 20 species to 0 with little or no transition. Within the Indo-Pacific diversity decreases as the distance from the Indo-Australian archipelago increases, with these decreases coinciding with increasing latitude, decreasing island size, and increased isolation. Similar patterns have been shown for scleractinian corals (Veron, 2000) and given the high dependence of many butterflyfishes on corals for food and/or habitat (See Cole and Pratchett, Chapter 5) links may exist between butterflyfish and coral diversity on a large scale. Here the joint effect of five large-scale factors (i.e., latitude, distance to biodiversity centres, island size, connectivity, and regional coral diversity) on the distribution of α diversity of butterflyfishes was analysed on a worldwide scale using GLMs. Biodiversity centres were defined for each ocean as the areas where the coral reef fish diversity is the highest: the Coral Triangle for the Indo-Pacific (Carpenter and Springer, 2005; Allen, 2007), the southern Caribbean for the Atlantic (Briggs, 2005; Floeter et al., 2008), the Panama isthmus for the Eastern Pacific (Mora and Robertson, 2005; Robertson and Cramer, 2009). Connectivity was defined on a semi-quantitative scale: low (islands more than 400 km from the nearest island), medium (200–400 km), high (100–200 km), very high (< 100 km). Regional scale coral diversity was extracted from the maps published by Veron (2000).

Regional coral diversity was the factor that best explained the distribution of butterflyfishes (Table 4.1). With the exception of the East Pacific, the number of Chaetodontidae species increased significantly as the regional diversity of coral increased (Fig. 4.5). Bellwood and Hughes (2001) found a significant correlation between coral diversity and reef fish diversity (represented by 13 families) in the Indo-Pacific (including the East Pacific; $r = 0.65$, $N = 113$ sites, $p < 0.001$). This correlation proved even stronger using our worldwide butterflyfish dataset (non-parametric Spearman rank order correlation $r = 0.87$, $N = 178$ sites, $p < 0.001$). However to our knowledge, no one has demonstrated if this is a result of similar environmental requirements for both butterflyfishes (or reef fishes in general) and corals, or if there is co-evolution of the two groups. If these congruent patterns were a result of co-evolution we may expect the proportion of obligate coral-feeding butterflyfishes to be highest where coral diversity is highest (assuming there is a correlation between coral diversity and coral abundance). However, the proportion of both coral feeding species and coral specialists tended to be constant (slope of regression not significantly different from 0) as coral

Table 4.1 Correlation between butterflyfish diversity and large-scale factors. N: number of sites; NS: not significant; *: $p < 0.05$; **: $p < 0.01$; ***: $p < 0.001$. The numbers represent the correlation coefficient of a single regression, except in the last column. In the last column the numbers represent the total r^2 for GLMs involving all the previous factors. Factors are symbolised by C: connectivity; D: distance to biodiversity centres; L: latitude. Factors are ranked by increasing importance.

	Island Size	**Latitude**	**Connectivity**	**Distance to Biodiversity Centres**	**Regional Coral Diversity**	**Multiple regression R^2; factors**
Atlantic (N=34)	0.19 NS	0.17 NS	0.57***	–0.68***	0.70***	0.62; D
East Pacific (N=15)	–0.36 NS	–0.21 NS	–0.61*	–0.13 NS	0.22 NS	0.33; C
Indian (N=31)	–0.21 NS	–0.53**	0.01 NS	–0.61***	0.58***	0.41; D, L
Pacific (N= 91)	0.34**	–0.27*	0.69***	–0.68***	0.78***	0.52; C, D

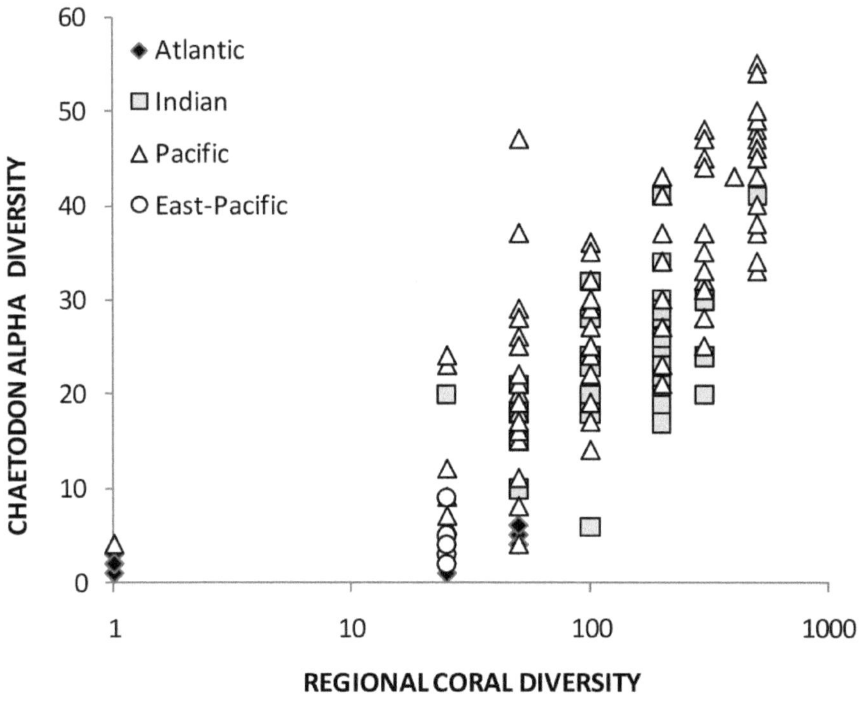

Fig. 4.5 Relationship between the alpha diversity of Chaetodontidae and regional coral diversity.

diversity increased (Fig. 4.6). While not conclusive, this suggests that these congruent patterns are likely to reflect similar environmental requirements as opposed to co-evolution.

The four other large-scale factors tested had varying effects on butterflyfish diversity among the different ocean basins (Table 4.1). The most consistent factor was the distance to the biodiversity centre with a significant decrease in diversity as this distance increases. This occurred in all oceans except the East Pacific. Such an effect has been demonstrated for both reef fish and corals in the Indo-Pacific, including the East Pacific (Bellwood and Hughes, 2001). Connolly et al. (2003) investigated if such patterns could be attributed to the "mid-domain" effect (MDE), which predicts a peak or a plateau in species richness near the centre of ranges within a bounded domain (reviewed by Zapata et al., 2003; Colwell et al., 2004; Pimm and Brown, 2004). They found that the diversity of reef fish and corals was higher along the eastern coast of Africa and in the Indo-Australian archipelago and lower in the Pacific, (especially the East Pacific) than predicted by the MDE. Their major explanation was a link with oceanic circulation. Bellwood et al. (2005) extended this to investigate the relative importance of MDE,

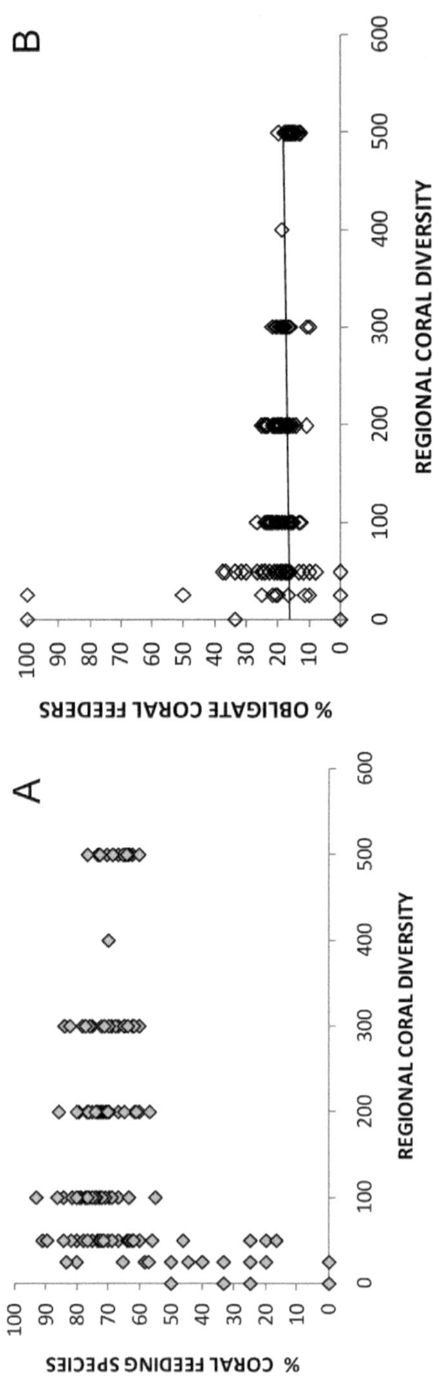

Fig. 4.6 Proportion of **(A)** coral feeding butterflyfish species and **(B)** coral specialist (obligate corallivore) butterflyfish species as regional coral diversity increases.

energy and habitat area and found that reef area and MDE best explained the distribution of diversity, and that energy (sea surface temperature and solar irradiance) had a relatively minor effect. Our analysis of butterflyfishes may provide further insights. In particular, connectivity, latitude, and island size had different effects in the Indian Ocean compared to the Pacific Ocean. In the latter, diversity was highly correlated to connectivity and island size. The diversity of butterflyfishes decreased as islands became smaller and more isolated, whereas in the Indian Ocean these factors had no significant effect—butterflyfish diversity decreased only with latitude. There are striking differences in the distribution of continents and islands in these oceans and there are different types of physical barriers to the geographical distribution of coral reef fishes, which may explain the observed distributions. In the Indian Ocean there are few islands but a continuous coastline from Indonesia to South Africa. Three natural barriers interrupt the progression of diversity along this coastline: the delta system in the Bay of Bengal, the delta system in Pakistan, and an upwelling area in Oman. Diversity drops westwards of these barriers suggesting that they are effective dispersal barriers for these fishes. However, diversity increases again some distance from these barriers suggesting that these barriers may not have been effective all the time or that other mechanisms are at play. The island system formed by the Lakshadweep, Maldives, and Chagos groups may have represented a stepping-stone towards the Mascarene group and the Seychelles, and then towards Madagascar and South Africa. If this view were correct, one would expect an accumulation in the southern end of eastern Africa as species could come from two directions, north from the Arabian Peninsula and east via the aforementioned islands. In such a case neither connectivity nor island size would be significant factors, as observed. In contrast, in the western Pacific there was a very strong overlap between the gradient of island size, island isolation, and decrease in diversity with increasing distance from the biodiversity centre (as indicated by the exclusion of island size as a significant factor in the multiple regression—Table 4.1). In such a case it is clear that diversity could be explained by the dispersal and colonising capacities of the various species. In particular, the lowest diversities found in the most remote areas such as the Hawaiian Archipelago or Easter Island, are supported by this hypothesis. Latitude was also shown to play a minor but significant role, suggesting that the temperature tolerance of these fishes may be important, especially at the southern and northern extremes of their distributions, such as in northern New Zealand or Korea.

In the Atlantic, the effects of large-scale factors were somewhat similar to what was observed in the west Pacific, a decrease in diversity with distance to the biodiversity centre (Caribbean Sea–Briggs, 2005; Floeter et al., 2008) and with isolation (remote islands supported less diversity). Such a pattern has already been described indirectly by Floeter et al. (2008) for reef fishes in general in the Atlantic; in particular, these authors showed the role of barriers such as the Amazon, the Benguela current, and the width of the Atlantic.

In the East Pacific the distribution of butterflyfish diversity was counter-intuitive, with diversity being negatively correlated with connectivity. The highest diversity is indeed found in the Galapagos with an accumulation of both species from the central Pacific and East Pacific endemics. This suggests that this archipelago is a sink for these two sources. For this reason, there is no real biodiversity centre for butterflyfishes in that region and therefore, logically, no correlation with such a factor. There was very little gradient in island size in this region, with either small islands or a continent and the higher diversity observed on islands was not significantly different (t-tests, $p > 0.05$) from the diversity on the continent. Mora and Robertson (2005) investigated the factors behind the latitudinal variations in shore fish diversity in the East Pacific. They found that diversity varied according to the geographic range of species, the diversity of species with wide ranges being mainly explained by MDE and the diversity of restricted range species being explained by habitat variables. Energy was not a significant variable as there was a confounding effect with MDE. No latitudinal variations for butterflyfishes were found in that region, which suggests that they do not follow the general distribution of shorefishes in this region. One potential reason is that these butterflyfish have specific dispersal and colonising capacities that are different for species coming from the Indo-Pacific and those endemic to the East Pacific.

"Relative" α diversity

Butterflyfishes are only one of many families of reef fishes. The relative importance of this family amongst reef fishes may fluctuate according to identifiable factors. An increase in the proportion of butterflyfish according to one factor would suggest that these fish are better adapted than the 'average' reef fish to the conditions imposed by this factor. For instance, if their relative α diversity increases as islands become smaller and more isolated, then this would be an indication of an above average dispersal and colonising capacity. Bellwood and Hughes (2001) analysed the proportion of 13 families of coral and reef fishes in the Indo-Pacific (including the East Pacific). One of their major findings was that the proportion of these families tended towards a constant as total diversity increased. We adopted a similar approach to analyse the butterflyfishes on a worldwide scale, each ocean being analysed separately as there may be specific effects in each ocean as observed for "absolute" α diversity. One of the problems in defining the proportion of butterflyfishes is to properly assess the total reef fish fauna (Mora et al., 2008). Many families, in particular small and/or cryptic species such as Muraenidae, Scorpaenidae, Apogonidae, Blenniidae, or Gobiidae, are very difficult to census completely. Therefore it was decided to use only families that are easily sampled (as Bellwood and Hughes, 2001; Roberts et al., 2002) to represent the total reef fish fauna (see Appendix I for the list of selected families). The proportion of butterflyfishes was clearly not constant within an ocean (Fig. 4.7). Within each ocean there was a decreasing trend of this proportion as total reef fish diversity increased; sites with low % of butterflyfishes showed a large variance in total species richness. The next focus of this chapter, then, is to understand the factors that generated this variance.

The distribution of the sites in the Indo-Pacific may be divided into sites with low total reef fish diversity (< 200 species) and those with higher diversity (following Bellwood and Hughes, 2001). A correlation analysis between the proportion of butterflyfishes (within the reef fish species pool) and the distance to the biodiversity centre, island size, connectivity, and latitude showed a striking difference between the Indian and Pacific oceans. In the Indian Ocean the proportion of butterflyfish species was significantly and negatively correlated ($r = -0.45$; $p = 0.0012$) to the distance

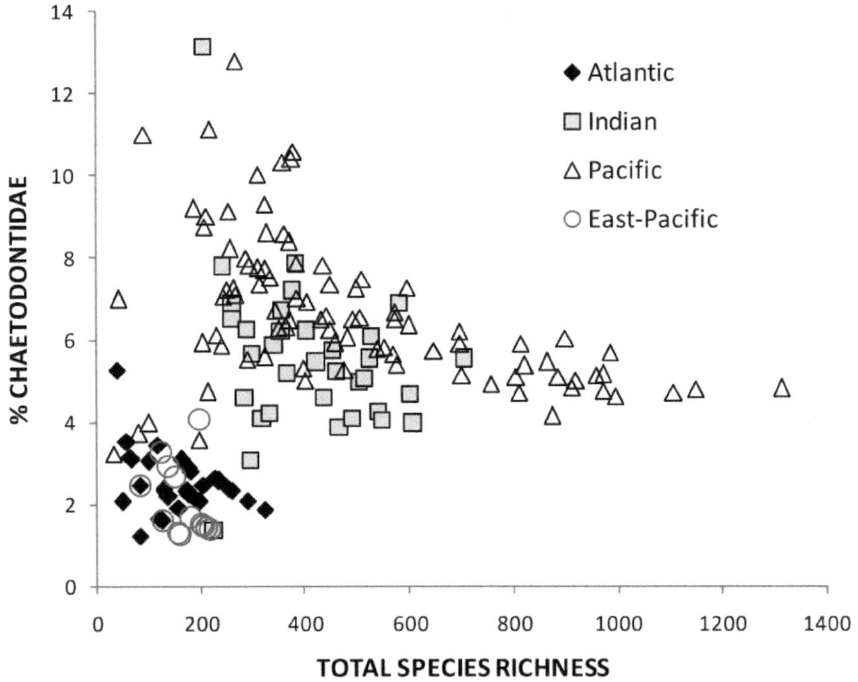

Fig. 4.7 Contribution of butterflyfishes to total reef fish diversity. Each point is one of 174 sites.

to the biodiversity centre and to island size ($r=-0.37$, $p=0.02$). Latitude and connectivity were also significant factors, but in a GLM ($r^2=0.42$, $N=33$) the component of variance they explained was no longer significant once distance to the biodiversity centre was taken into account. In contrast, in the western Pacific none of the previous factors had a significant effect (GLM $r^2<10^{-5}$, $N=81$). In addition there was no relationship between the proportion of butterflyfish species and the average values of their life-history traits in the western Pacific whereas in the Indian Ocean several traits were significant (see Appendix III for the distribution of life-history traits). In particular, on sites where the proportion of butterflyfish species was high, these species tended to be larger, more species fed on non-coral invertebrates, and they tended not to school. In the Atlantic there was a negative correlation (multiple regression $r^2=0.17$, $N=34$) with connectivity and latitude ($p=0.02$ for both factors) and no significant relationship with life-history traits. In the East Pacific connectivity was a significant factor ($r^2=0.84$, $N=15$, $p<10^{-5}$) but there was a very strong correlation with life-history traits. In particular when the proportion of butterflyfish species was high, most species were small and tended to feed on coral or/and plankton, but did not feed on algae. These results suggest that different mechanisms are at play in each ocean but at present it is difficult to identify these mechanisms and their causes.

Beta Diversity

There are many ways of defining and estimating beta diversity (Koleff et al., 2003). Here we used two indices. The first, β_I, is based only on the gradient in butterflyfish species numbers within a region and was estimated as $\beta_I = \gamma/\alpha$ (Whittaker, 1972), where γ is the total number of species within a region (or sub-region) and α the average number of species across sites within the region (or sub-region). The second index, β_{II}, takes into account the butterflyfish species composition and was estimated as the variance of the presence/absence matrix (Legendre et al., 2005).

Beta diversity is difficult to analyse in the context of this chapter because the number of sites per region or sub-region varied considerably and was usually low. Subsequently only sub-regions with at least 3 sites were retained for the analysis. In general, there was little correlation between beta diversity and most factors (Table 4.2). β_I diversity was negatively correlated with coral diversity and the proportion of coral feeding species. This suggests that the presence of a high coral diversity will favour a rather homogeneous distribution of species, whereas low coral diversity will generate a stronger gradient in species diversity. In the Pacific Ocean body size increased with β_I diversity, which suggests that species heterogeneity is highest where species are the largest, which coincides with a lower

Table 4.2 Correlation between beta diversity (estimated at the sub-region level) and life-history traits and large scale factors. Only Spearman correlation coefficients were estimated as the data set was not sufficient for more complex analyses, such as GLMsNS: not significant; *: $p < 0.05$; **: $p < 0.01$; ***: $p < 0.001$. β_I: beta diversity estimated from Whittaker (1972); β_{II}: beta diversity from Legendre et al. (2005). "World": all sub-regions (>2 sites/sub-region). "Pacific": sub-regions in the Pacific Ocean (> 2 sites/sub-region), the Easter Island sub-region being excluded. N: number of sub-regions. Note that the factors, '% algae feeding species', 'number of diet items', 'schooling behaviour', and 'latitude' are not shown because they are never significantly correlated to beta diversity.

	γ Diversity	% Coral feeding Species	% Mobile. Invert. feeding species	Species Size	Average Species Geographic Range	Geological Age	Island Size	Connectivity	Distance to Biodiversity Centres	Coral Diversity
β_I World N= 27	NS	NS	NS	NS	NS	NS	NS	NS	NS	–0.40*
β_I Pacific N=17	NS	–0.51*	NS	0.67**	NS	NS	NS	NS	NS	–0.37*
β_{II} World N= 27	0.77***	0.56**	–0.58**	NS	0.43*	–0.50*	NS	NS	NS	NS
β_{II} Pacific N= 17	0.51*	NS	NS	NS	–0.61**	NS	0.67**	0.61**	–0.62**	NS

proportion of coral feeding species (coral feeders being on average smaller than other species). The level of correlation was much higher for β_{II} diversity that takes into account the difference in species composition amongst sites. High β_{II} diversity, reflecting high spatial heterogeneity in species composition and variance in species numbers, was found in the Pacific on large and/or connected islands (or continents) and was associated with islands near the biodiversity centre where average species geographical areas are small. These relationships suggest that high spatial heterogeneity in the species composition of butterflyfishes is found in areas where α and γ diversities are high. This raises the question of species "saturation" as regional diversity increases (Loreau, 2000; Hillebrand, 2005). In the case of butterflyfishes there is little evidence to suggest saturation of species diversity at such scales, even at the highest regional diversities there was very little variation in the rate of increase of the local diversity (Fig. 4.8). While the asymptotic relationships between α and γ cannot be considered as an indication of local species saturation (Loreau, 2000), the lack of an asymptote can be considered as an absence of local species saturation.

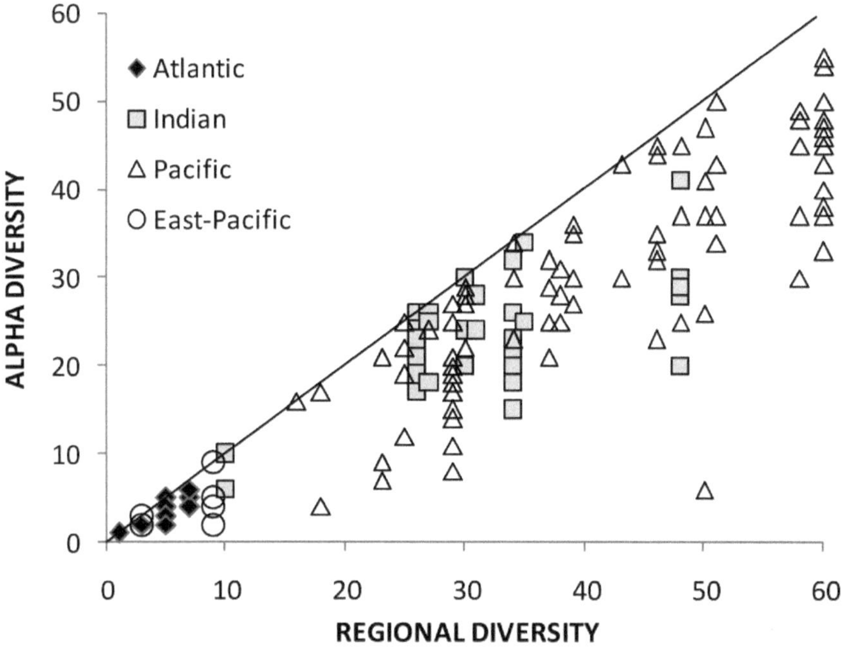

Fig. 4.8 Relationship between local (alpha) and regional diversity. Each point represents a different site. The diagonal line represents the one for one relationship (local = regional diversity).

Geographic Range

In biogeography there is often a special interest oriented towards endemic species, as they may be important to understand the mechanisms behind colonisation. Conversely, there is often little attention directed towards widespread species that may have life-history traits that are the most efficient for dispersal and colonisation. Endemism, by definition, is scale-dependent. As the size of the defined area increases, the rate of endemism will increase (Fig. 4.9). For this chapter, geographic ranges were estimated based on the number of sites where each species is recorded (rather than projections of the total geographic range) to avoid any erroneous extrapolations based on patchy records of occurrence. The number of endemics were then assessed based on whether species are recorded at only one site, within just one sub-region, or a single ocean basin. There are only 6 species of Chaetodontidae that are endemic at the scale of sites, versus 34 endemic species at the level of sub-regions. A high proportion of the butterflyfish species (58% of all species) are endemic to one ocean, and all the Atlantic species are endemic at that scale. Within the Atlantic, the highest endemism is found in the East Atlantic sub-region (5 species, 100% endemic) followed by the Caribbean (42%). In the East Pacific, endemism is high in the islands (22%) but nil along the coast. The Red Sea has by far the highest endemism (23.5%) within the Indian Ocean and the largest number of endemic species (8) across all of the world's oceans. Only 3 other sub-regions in the Indian Ocean have endemic species, Western Australia having the highest proportion (6.7%). The rate of endemism is the lowest in the Pacific; only two sub-regions have an endemism rate above 10% (the Easter Island group: 17.4%; and Hawaii:

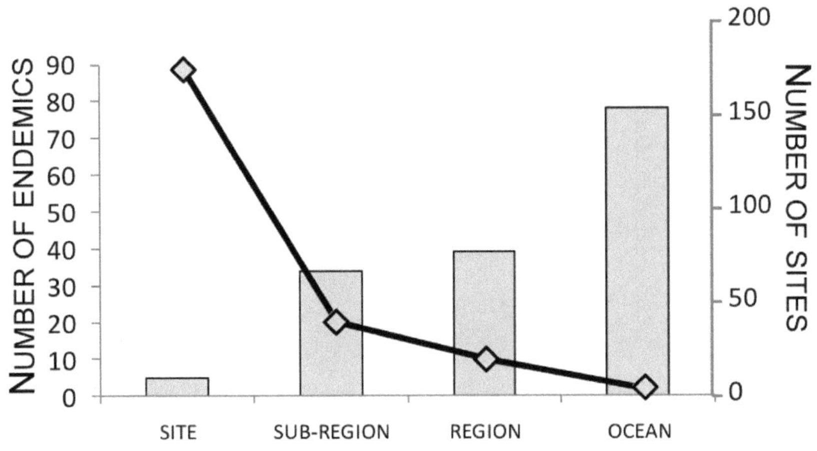

Fig. 4.9 Number of endemic species (bars) according to the area level considered (squares; linked by a line for clarity).

16%) and 16 out of 22 sub-regions have no endemic butterflyfish species. There is, therefore, no relationship between the endemism and diversity of butterflyfishes. This distribution confirms previous findings that there is no relationship between centres of diversity and endemism for reef fishes (Hughes et al., 2002; Allen, 2007; but see Roberts et al., 2002 and Mora et al., 2003). It also shows that in the Indo-Pacific the areas of high endemism are at the limit of the butterflyfish distribution; these areas are characterised by a high level of isolation (remote islands in the case of Easter Island and Hawaii; isolated sea in the case of the Red Sea).

A geographic range index (GRI), based on the average geographic range of the species within a site, was estimated to try to understand the relationship between diversity and geographic range. A high GRI is indicative of a large proportion of wide ranging species and, on the other hand, a low GRI may be associated with large proportions of species with restricted ranges and possibly high endemism. Butterflyfish species with a restricted range are the most numerous (Fig. 4.10). This is the most common type of distribution (Gaston, 2003) and has been documented for several types of reef organisms (Roberts et al., 2002). Butterflyfish diversity was significantly correlated to GRI (r=0.60; N=171) on a global scale, but this relationship was not significant in most oceans except the Pacific (r= –0.35; N=91). The latter negative correlation can be expected as butterflyfish diversity decreases as connectivity decreases in the Pacific, thus favouring species with high dispersal or colonising capacities. The absence of

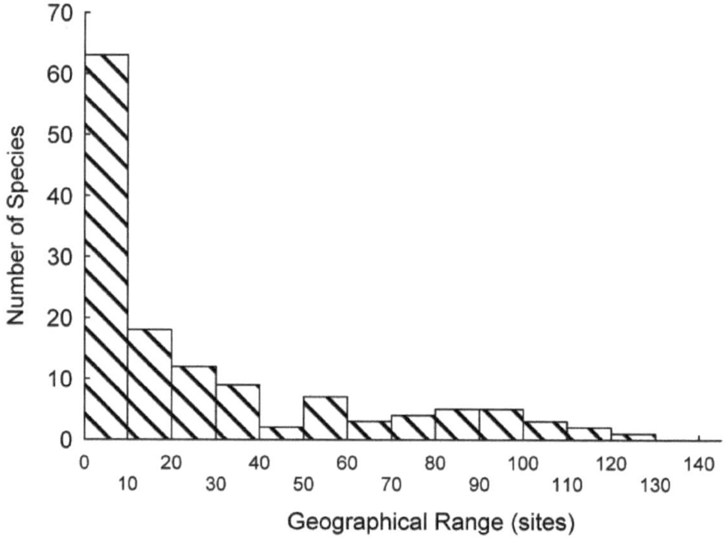

Fig. 4.10 Number of species according to geographical range (measured by the number of sites where the species has been recorded).

correlation in the other oceans is probably linked to the potential dispersal along large landmasses, which does not favour species with high dispersal capacities. This is confirmed by the analysis of the relationship between GRI and several large-scale factors (Table 4.3). In both the Indian and West Pacific Oceans, island size, latitude, and connectivity were significantly negatively related to GRI. In other words, the geographic range of species tended to be greater on smaller, isolated islands at high latitudes. Such relationships may be associated with higher dispersal and colonising abilities. The low level of correlation with the distance to the biodiversity centres is in fact due to a dome shaped curve (Fig. 4.11), indicating that low GRI values in the West Pacific are found either near the biodiversity centre or far from it.

The next question is to determine if specific life-history traits may be related to geographic range (Robertson, 2001). Despite a number of significant correlations (Table 4.4), the only consistent relationship across the four oceans was a negative correlation between geographic range and the proportion of schooling species. In other words, species with the highest schooling capacities have either a low dispersal capacity or a low colonising success. Restricting the analysis to the Indo-Pacific where many species (52) are found in both the Indian and Pacific Oceans reveals a much stronger structure in the Pacific, where almost all life-history traits were very significantly correlated; in the Indian Ocean, only body size reached a high level of correlation. In areas of the Pacific where GRI is low there is a lower proportion of butterflyfish species feeding on coral and algae, a lower proportion of coral specialists, a smaller number of diet items, and a higher proportion of schooling and plankton feeding species. This means that in peripheral regions butterflyfish species will tend to have small geographic ranges and be characterised by smaller body sizes, lower dependency on coral in their diets, and a stronger tendency to school. This may be expected since these regions have low coral diversity. This also suggests that whenever coral diversity is sufficient, butterflyfish species will colonise since most corallivorous species have a wide geographic range and demonstrate a high adaptation to coral.

Table 4.3 Spearman correlation coefficients between the geographic range index (GRI) and large-scale factors, within each of the major geographic locations. N: number of sites; NS: not significant; *: $p < 0.05$; **: $p < 0.01$; ***: $p < 0.001$.

	Island Size	**Latitude**	**Connectivity**	**Distance to Biodiversity Centres**	**Regional Coral Diversity**
Atlantic (N=34)	–0.04 NS	–0.06 NS	0.25 NS	–0.46*	–0.01 NS
East Pacific (N=15)	0.17 NS	0.07 NS	–0.35 NS	0.21 NS	0.25 NS
Indian (N=31)	–0.50**	–0.36*	–0.47**	–0.38*	0.10 NS
Pacific (N= 91)	–0.29**	–0.29**	–0.40***	0.16 NS	–0.30**

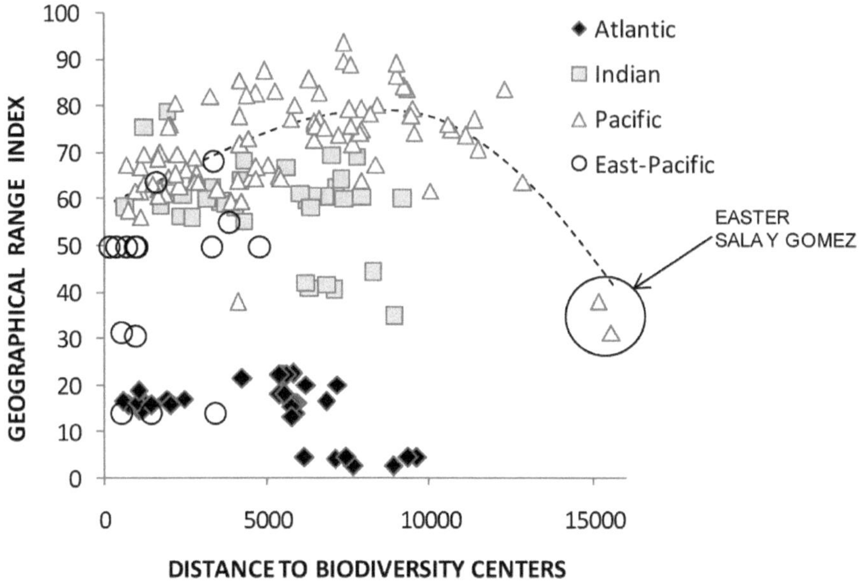

Fig. 4.11 Relationship between the geographical range index (GRI) and the distance (km) to the biodiversity centres for the four oceans. A 2° polynomial regression is significant at $p < 10^{-5}$ for the Pacific Ocean.

Geological Age and Phylogeography

To identify if butterflyfish diversity in an area is linked to the geological history, a first step is to determine if the distribution of the phylogenetic diversity supports the definition of the biogeographical regions based on the species distribution. The phylogeny of the butterflyfishes has been established recently for 71 species among the 134 species of the family (Fessler and Westneat, 2007). This well-resolved and well-supported topology was used to assess the phylogenetic structure of the 19 biogeographic regions defined above (Fig. 4.2). The sequences for 71 chaetodontids from the study of Fessler and Westneat (2007) were downloaded (GenBank accession numbers EF616820 to EF617232), aligned for each data set with Clustal X (Thompson et al., 1993) and concatenated into a single file for subsequent analyses. A total of 3223 nucleotides were aligned and a neighbour-joining tree was built following the bioNJ algorithm (Gascuel, 1997) using the JC model for distance computations as implemented in PAUP* 4.0b10 (Swofford, 2002). The tree and the presence/absence data matrix were used for subsequent analyses of the phylogenetic structure of the communities using Phylocom (Webb et al., 2008).

Table 4.4 Significance level of Spearman correlation coefficients between geographic range and geological age versus dietary composition (e.g., major dietary items), body size, schooling behaviour and depth range. N: number of sites; NS: not significant; *: $p < 0.05$; **: $p < 0.01$; ***: $p < 0.001$.

	Algae	Coral	Other Sessile Invertebrates	Mobile. Invert	Plankton	Coral specialists	Body Size	Solitary	Pairing	School	Depth range
Geographic range											
Atlantic (N=34)	NS	***	***	***	NS	**	_*	***	NS	_***	_*
East Pacific (N=11)	_***	NS	_***	NS	NS	NS	NS	***	*	_**	NS
Indian (N=35)	NS	NS	_**	***	NS	**	***	NS	NS	_***	NS
Pacific (N= 91)	***	***	**	_**	_**	***	***	NS	***	_**	NS
TOTAL (N=175)	***	***	_***	_***	_*	**	**	NS	NS	_*	_*
Geological Age											
Atlantic (N=34)	NS	NS	_**	NS	_**	NS	**	NS	NS	**	NS
East Pacific (N=11)	***	_***	*	*	_*	_*	**	_***	NS	***	*
Indian (N=35)	NS	*	NS	*	_**	NS	**	NS	**	_*	NS
Pacific (N= 91)	NS	_*	_***	*	**	_**	NS	NS	_**	**	NS
TOTAL (N=175)	**	_***	**	***	_*	_***	***	_**	_***	**	*

The distribution of the phylogenetic distances between species from the same biogeographic region or belonging to distinct biogeographic regions provided contrasted patterns (Fig. 4.12). In each case, a bimodal distribution was found, the first mode corresponding to comparisons among species within genera and the second mode among species from distinct genera. The mode at 0.08 within regions and the mode starting at 0.09 amongst regions involved comparisons between *Chaetodon-Prognathodes* and the other genera (*Coradion, Chelmonops, Chelmon, Forcipiger, Hemitaurichthys, Johnrandallia,* and *Heniochus*). Strikingly, phylogenetic distances among species of the same genera (i.e., the first mode) were lower within- than among-regions, indicating a closer phylogenetic relatedness among congeneric species from the same region. This result suggests that the speciation events occurring at the region's boundaries tend to be older on average since phylogenetic distances are larger among region than within region.

The ordination of the biogeographic provinces according to the average phylogenetic distances among species, revealed a poorly resolved cluster (Fig. 4.13). Only some of the Atlantic regions clustered together (Brazil, Azores-Madeira, Brazilian Islands, and the Caribbean), indicating that species from that area are more closely related among them than with those from the Indian and Pacific Oceans. The two most basal *Chaetodon* species (*C. hoefleri* and *C. robustus*) are found in the eastern Atlantic, thus promoting

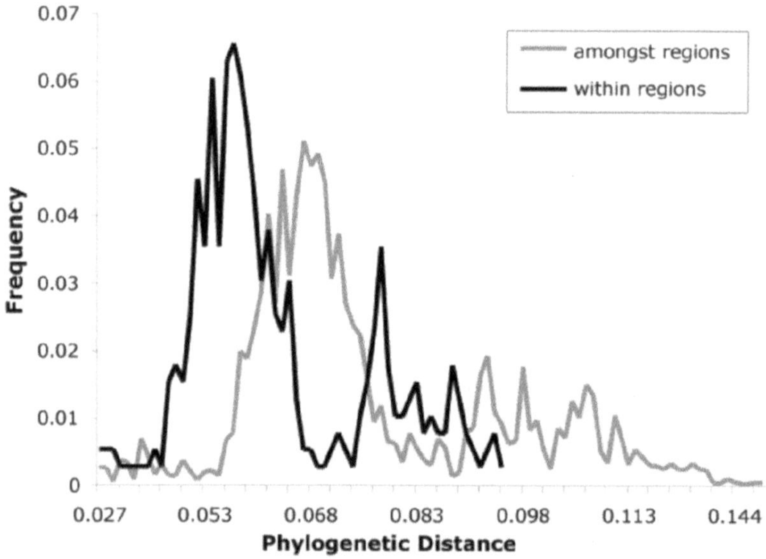

Fig. 4.12 Distribution of phylogenetic distances between species belonging to the same biogeographic region (within regions) and species belonging to distinct biogeographic region (amongst regions). Frequencies are expressed as percent of the total number of distances computed, namely 399 within regions and 7182 amongst regions.

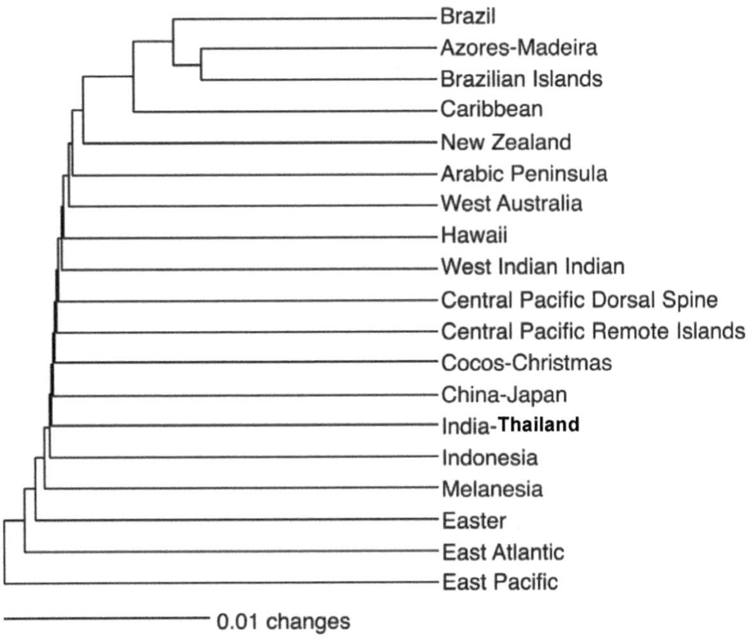

Fig. 4.13 UPGMA dendrogram of the 19 regions computed based on the mean pair wise phylogenetic distance between species communities. UPGMA: Unweighted Pair Group Method with Arithmetic Mean is a clustering algorithm that assesses the dissimilarity between the new cluster and the existing cluster by the distance between the means.

distinctiveness from the western Atlantic. By contrast, phylogenetic communities from the other regions diverged from one another on average by the same phylogenetic distance. This suggests that dispersal has been important enough to promote the spatial expansion of older lineages throughout the biogeographical provinces aforementioned, and hide the signal in the communities' phylogenetic structure of the chronology of differentiation of these regions (Fig. 4.12). Despite the apparent homogeneity of the phylogenetic structure of the biogeographical regions, we further explored the phylogenetic composition of the 19 regions by using several metrics of communities' phylogenetic structure (Fig. 4.14).

We first computed Faith's (1992) index of phylogenetic diversity (PD) for each region. This index is the proportion of total branch length in the phylogeny associated with the taxa in a particular region. The Net Relatedness Index (NRI) and the Nearest Taxon Index (NTI) according to Webb (2000) and Webb et al. (2002) were then computed. These indices describe the differences between the average phylogenetic distances (NRI) and the nearest neighbour distance (NTI) in observed and null communities

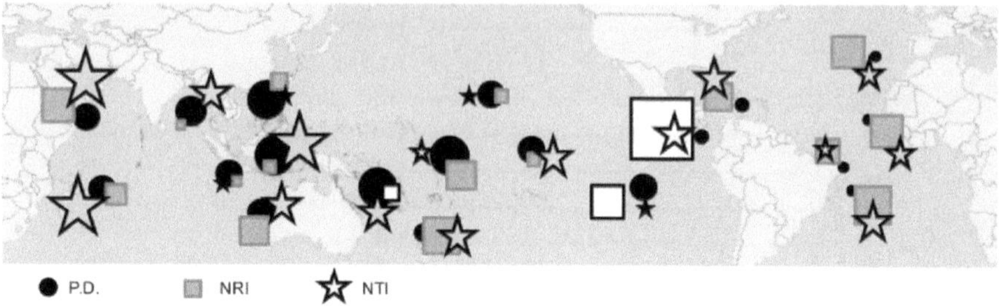

Fig. 4.14 Regional values of PD, NRI and NTI for Chaetodontidae across the world. PD: phylogenetic distance, radius of circle is proportional to value of PD. NRI: Net Relatedness Index, size of square is proportional to NRI value, grey represents >0 and white <0 values. NTI: Nearest Taxon Index, size of star is proportional to NTI value, black represents values <–0.3; +0.3>, grey represents >0.3 and white <–0.3 values.

Colour image of this figure appears in the colour plate section at the end of the book.

(no phylogenetic structure), standardized by the standard deviation of phylogenetic distances in the null communities. High values of NRI indicate that species are more phylogenetically related than expected by a null model, and negative values are found when species are more phylogenetically divergent than expected by chance. High values of NTI indicate that the closest related species are closer than expected and negative values suggest species further than expected.

Phylogenetic diversity (PD) was found to be largely heterogeneous amongst the 19 regions (Fig. 4.14). The highest PD values were observed at the centre of biodiversity (Indonesia) and nearby regions (Melanesia, China-Japan) with very low values in the Atlantic and East Pacific. This means that regions with the highest species diversities had also the highest phylogenetic diversities. This was confirmed by the NRI where values were generally higher in the Atlantic indicating that species were more phylogenetically related than expected under a null community model. This suggests that within the Atlantic regions most species had a common history. The East Pacific and Easter Island regions were characterised by high phylogenetic divergence. For the East Pacific this may be related to the mix of endemic species with species originating from the West Pacific. For the Easter Island region the high phylogenetic divergence could be due to the grouping of Pitcairn-Dulcie (21 species) with species-poor Easter and Sala-Gomez islands (9 and 7 species respectively).

The Nearest Taxon Index (NTI) was higher in the Indian Ocean compared to random values. In other words, there tends to be more phylogenetic homogeneity in that ocean than elsewhere. Associated with the higher-than-expected species diversity (Connolly et al., 2003) in the western part of the Indian Ocean, this could be the reflection of a common dispersal history. However, the highest NTI value was found in Indonesia, indicating that species in that region harbour a very low divergence to their closest relative compared to other regions. This is in contrast with the surrounding regions where NTI is low. This may be an indication that speciation may have also occurred in the area. Together, these phylogenetic structure results indicate that the highest phylogenetic diversity was found in the biodiversity centre (Indonesia) but also that frequently, species from this region are closely related. This apparent discrepancy suggests that this biodiversity centre hosts very divergent lineages that increase the phylogenetic diversity values, but also hosts very speciose lineages with a high level of closely related species. This result converges with the findings of Barber and Bellwood (2005) that suggested *Halichoeres* (Labridae) communities in the coral triangle resulted from a mixed influence of old speciation events from the middle Miocene at 12 million years ago (Ma) and recent speciation events during the Pliocene around 3 Ma.

The distribution of species divergence suggests that speciation events are older amongst regions than within regions (Fig. 4.12). The pattern of relationships amongst regions based on average phylogenetic divergence, however, remains unclear (Fig. 4.13) suggesting that dispersal frequently occurred during the diversification of the Chaetodontidae. Thus, ecological factors that are likely to affect dispersal and establishment of species in new areas may be related to historical parameters. The recent phylogenetic study of the family (Fessler and Westneat, 2007) provided a very detailed timeframe for the diversification of this group of fish. This set covers a majority of species (71 species of ~134 species), but in some areas the number of species covered may be low compared to the total number of species. We computed the average species age within islands according to the species age estimates from Fessler and Westneat (2007) as an indicator of the relative contribution of recent and older speciation events to extant communities. Generally, the average species age was related to several life-history traits (Table 4.4). In particular sites with a high average species age supported species that tended not to feed on coral, were above average body size, and tended to school. Average species age was also negatively related to regional coral diversity and distance from biodiversity centres, and positively related to island size. If average species age is ranked according to longitude and ocean, the homogeneity of this age for the Indo-Pacific is evident as well as the superior ages of the Atlantic and East Pacific and the peculiar position of the East Atlantic (Fig. 4.15).

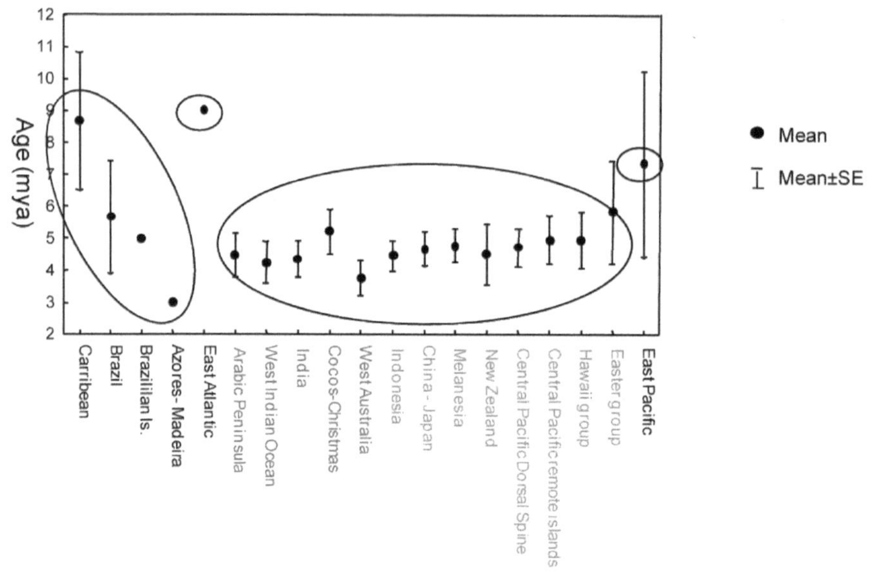

Fig. 4.15 Average age (mya) of species in each of the 19 biogeographic regions. Regions are ordered according to longitude. Indian Ocean regions are in red, west Pacific regions are in blue.

Colour image of this figure appears in the colour plate section at the end of the book.

CONCLUSION

Butterflyfishes are a conspicuous component on coral reefs around the world. Even though they do not necessarily represent a large proportion of the total reef fish diversity or abundance, their link to coral makes them an ideal model to understand the relationships between coral reef fishes and coral on a global basis. Based on checklists from 175 sites across the world, the geographical distribution of 134 species of butterflyfishes and their major life-history traits were analysed, with specific attention upon the potential effects of large-scale factors such as island size and connectivity, distance to centres of biodiversity, latitude, and regional coral diversity. The strength of butterflyfishes as a model for macro-ecological analyses of coral reef fishes is confirmed by the general convergence of the findings for butterflyfishes with previous findings from analyses based on larger sets of species. This strongly suggests that a number of findings based only on butterflyfishes presented in this chapter may be representative of coral reef fish in general, but of course this will need to be validated as our knowledge on the spatial distribution, biology, ecology, and phylogeny of other coral reef fish progresses.

Based upon butterflyfish species composition, 19 regions and 39 sub-regions were defined. This is the first attempt to define biogeographical regions for coral reef fishes on a worldwide basis, and revealed congruence with most previous findings across smaller geographical scales. Butterflyfishes therefore, could represent an interesting model for large-scale reef fish biogeography, except in the Atlantic and eastern Pacific where the diversity of these fishes is low. The highest diversities of butterflyfishes were found in the Indonesian-Philippine region for the Indo-Pacific Ocean and in the Caribbean in the Atlantic Ocean. The rate of endemism of these fishes was highest in the Atlantic, but there were several spots of high endemism in the Indo-Pacific, in particular in the Red Sea, Hawaii, and Easter Island. The distribution of these endemic species confirms the absence of overlap between biodiversity centres and diversity hotspots for reef fishes. The diversity of butterflyfishes was related to several large-scale factors. Regional coral diversity was the best factor explaining the geographical distribution of these fishes, with the distance to biodiversity centres being the second best factor, except for the East Pacific Ocean where this factor played no role. Island size, connectivity, and latitude had similar influences on butterflyfish diversity in both the Atlantic and Pacific Oceans, but only latitude was significant in the Indian Ocean.

Most butterflyfish species tend to have a rather restricted geographic range. There is little correlation between the average range of species on a site and the life-history traits of the species, except in the Pacific where diet, size, and schooling behaviour change significantly with geographic range (Pratchett, Chapter 9).

Phylogenetic analyses of the 71 chaetodontids from the study of Fessler and Westneat (2007) suggested that Chaetodontidae might have dispersed at a fast rate in the Indo-Pacific with little local speciation, whereas in the Atlantic there was a clear separation between East and West. On the other hand, the highest phylogenetic divergence was found in the East Pacific. Phylogeny pointed also at a probable species accumulation rather than speciation in the Indonesian-Philippine biodiversity centre, which could be related to a mid-domain effect. In the Indian Ocean, phylogeny suggested that dispersal has been rather homogeneous compared to the Pacific. Geological ages are higher and more heterogeneous in the Atlantic and East Pacific than in the Indo-Pacific. In addition there was a negative correlation between average species age and coral feeding or coral dependence, which suggests that the dependence of butterflyfishes to coral is a relatively recent event (see also Bellwood and Pratchett, Chapter 1).

REFERENCES

Allen, G.R. 2000. Indo-Pacific coral-reef fishes as indicators of conservation hotspots. *Proceedings of the 9th International Coral Reef Symposium, Bali*, pp. 921–926.

Allen, G.R. 2007. Conservation hotspots of biodiversity and endemism for Indo-Pacific coral reef fishes. *Aquatic Conservation: Marine and Freshwater Ecosystems* 18: 541–556.

Barber, P.H. and D.R. Bellwood. 2005. Biodiversity hotspots: evolutionary origins of biodiversity in wrasses (*Halichoeres*: Labridae) in the Indo-Pacific and new world tropics. *Molecular Phylogenetics and Evolution* 35: 235–253.

Bellwood, D.R. and T.P. Hughes. 2001. Regional-Scale Assembly Rules and Biodiversity of Coral Reefs. *Science* 292: 1532–1535.

Bellwood, D.R., T.P. Hughes, S.R. Connolly and J. Tanner. 2005. Environmental and geometric constraints on Indo-Pacific coral reef biodiversity. *Ecology Letters* 8: 643–651.

Bellwood, D.R., P.C. Wainwright, C.J. Fulton and A.S. Hoey. 2006. Functional versatility supports coral reef biodiversity. *Proceedings of the Royal Society B* 273: 101–107.

Briggs, J.C. 2005. Coral reefs: conserving the evolutionary sources. *Biological Conservation* 126: 297–305.

Brito, A. and J.M. Falcon. 1996. Capture of the St. Helena butterflyfish, *Chaetodon sanctaehelenae* (Chaetodontidae) in the Canary Islands. *Cybium* 20: 99–100.

Carpenter, K.E. and V.G. Springer. 2005. The center of the center of marine shore fish biodiversity: the Philippine Islands. *Environmental Biology of Fishes* 72: 467–480.

Colwell, R.K., C. Rahbek and N.J. Gotelli. 2004. The Mid-Domain Effect and Species Richness Patterns: What Have We Learned So Far? *American Naturalist* 163: E1–E23.

Connolly, S.R., D.R. Bellwood and T.P. Hughes. 2003. Indo-Pacific biodiversity of coral reefs: deviations from a mid-domain model. *Ecology* 84: 2178–2190.

Faith, D.P. 1992.Conservation evaluation and phylogenetic diversity. *Biological Conservation* 61: 1–10.

Fessler, J.L. and M.W. Westneat. 2007. Molecular phylogenetics of the butterflyfishes (Chaetodontidae): Taxonomy and biogeography of a global coral reef fish family. *Molecular Phylogenetics and Evolution* 45: 50–68.

Floeter, S.R., L.A. Rocha, D.R. Robertson, J.C. Joyeux, W.F. Smith-Vaniz, P. Wirtz, A.J. Edwards, J.P. Barreiros, C.E.L. Ferreira, J.L. Gasparini, A. Brito, J.M. Falcón, B.W. Bowen and G. Bernardi. 2008. Atlantic reef fish biogeography and evolution. *Journal of Biogeography* 35: 22–47.

Gascuel, O. 1997. BIONJ: an improved version of the NJ algorithm based on a simple model of sequence data. *Molecular Biology and Evolution* 14: 685–695.

Gaston, K.J. 2000. Global patterns in biodiversity. *Nature* 405: 220–227.

Gaston, K.J. 2003. *The structure and dynamics of geographic ranges*, Oxford University Press, Oxford, UK.

Hillebrand, H. 2005. Regressions of local on regional diversity do not reflect the importance of local interactions or saturation of local diversity. *Oikos* 110: 195–198.

Hughes, T.P., D.R. Bellwood and S.R. Connolly. 2002. Biodiversity hotspots, centres of endemicity, and the conservation of coral reefs. *Ecology Letters* 5: 775–784.

Kemp, J. 1998. Zoogeography of the coral reef fishes of the Socotra Archipelago. *Journal of Biogeography* 25: 919–933.

Kent, M. 2005. Biogeography and macroecology. *Progress in Physical Geography* 29: 256–264.

Kent, M. 2007. Biogeography and macroecology: now a significant component of physical geography. *Progress in Physical Geography* 31: 643–657.

Koleff, P., K.J. Gaston and J.J. Lennon. 2003. Measuring beta diversity for presence-absence data. *Journal of Animal Ecology* 72: 367–382.

Kulbicki, M. 2007. Biogeography of reef fishes of the French territories in the South Pacific. *Cybium* 31: 275–288.

Kulbicki, M., Y. Bozec and A. Green. 2005. Implications of biogeography in the use of butterflyfishes (Chaetodontidae) as indicators for Western and Central Pacific areas. *Aquatic Conservation: Marine and Freshwater Ecosystems* 15: S109–S126.

Legendre, P. and L. Legendre. 1998. *Numerical ecology*. 2nd ed., Elsevier, Amsterdam, The Netherlands.

Legendre, P., D. Borcard and P.R. Peres-Neto. 2005. Analyzing beta diversity: partitioning the spatial variation of community composition data. *Ecological Monographs* 75: 435–450.

Loreau, M. 2000. Are communities saturated? On the relationship between α, ß and γ diversity. *Ecology Letters* 3: 73–76.

MacArthur, R.H. and E.O. Wilson. 1967. *The theory of island biogeography*, Princeton University Press, Princeton, USA.

Magurran, A.E. 2004. *Measuring biological diversity*, Wiley-Blackwell, Malden, USA.

Moilanen, A. and M. Nieminen. 2002. Simple connectivity measures in spatial ecology. *Ecology* 83: 1131–1145.

Mora, C. and D.R. Robertson. 2005. Causes of latitudinal gradients in species richness: a test with fishes of the Tropical Eastern Pacific. *Ecology* 86: 1771–1782.

Mora, C., P.M. Chittaro, P.F. Sale, J.P. Kritzer and S.A. Ludsin. 2003. Patterns and processes in reef fish diversity. *Nature* 421: 933–936.

Mora, C., D.P. Tittensor and R.A. Myers. 2008. The completeness of taxonomic inventories for describing the global diversity and distribution of marine fishes. *Proceedings of the Royal Society B: Biological Sciences* 275: 149–155.

Pimm, S.L. and J.H. Brown. 2004. Domains of diversity. *Science* 304: 831–833.

Randall, J.E. 1995. *Coastal fishes of Oman*, Crawford House Publishing, Bathurst, Australia.

Randall, J.E. 1998. Zoogeography of Shore Fishes of the Indo-Pacific Region. *Zoological Studies* 37: 227–268.

Raup, D.M. and R.E. Crick. 1979. Measurement of faunal similarity in paleontology. *Journal of Paleontology* 53: 1213–1227.

Ricklefs, R.E. 2004. A comprehensive framework for global patterns in biodiversity. *Ecology Letters* 7: 1–15.

Roberts, C.M., C.J. McClean, J.E.N. Veron, J.P. Hawkins, G.R. Allen, D.E. McAllister, C.G. Mittermeier, F.W. Schueler, M. Spalding and F. Wells. 2002. Marine biodiversity hotspots and conservation priorities for tropical reefs. *Science* 295: 1280–1284.

Robertson, D.R. 2001. Population maintenance among tropical reef fishes: inferences from small-island endemics. *Proceedings of the National Academy of Sciences* 98: 5667–5670.

Robertson, D.R. and K.L. Cramer. 2009. Shore fishes and biogeographic subdivisions of the Tropical Eastern Pacific. *Marine Ecology Progress Series* 380: 1–17.

Rocha, L.A. 2003. Patterns of distribution and processes of speciation in Brazilian reef fishes. *Journal of Biogeography* 30: 1161–1171.

Santini, F. and R. Winterbottom. 2002. Historical biogeography of Indo-western Pacific coral reef biota: is the Indonesian region a centre of origin? *Journal of Biogeography* 29: 189–205.

Springer, V.G. 1982. Pacific plate biogeography, with special reference to shore fishes. *Smithsonian Contributions to Zoology* 367: 1–182.

Springer, V.G. and J.T. Williams. 1990. Widely distributed Pacific plate endemics and lowered sea-level. *Bulletin of Marine Science* 47: 631–640.

Swofford, D.L. 2002. *Phylogenetic analysis using parsimony (* and other methods)*, Sinauer Associates, Sunderland, USA.

Thompson, J.D., D.G. Higgins and T.J. Gibson. 1994. CLUSTAL W: improving the sensitivity of progressive multiple sequence alignment through sequence weighting, position-specific gap penalties and weight matrix choice. *Nucleic Acids Research* 22: 4673–4680.

Thresher, R.E. 1991. Geographic variability in the ecology of coral reef fishes: evidence, evolution and possible implications. In *The Ecology of Fishes on Coral Reefs*, P.F. Sale (ed.). Academic Press, Inc., San Diego, USA, pp. 401–435.

Vàzquez, D.P. and R.D. Stevens. 2004. The latitudinal gradient in niche breadth: concepts and evidence. *American Naturalist* 164: E1–E19.

Veron, J.E.N. 2000. *Corals of the World*, Australian Institute of Marine Science, Townsville, Australia.
Ward Jr., J.H. 1963. Hierarchical grouping to optimize an objective function. *Journal of the American Statistical Association* 58: 236–244.
Webb, C.O. 2000. Exploring the phylogenetic structure of ecological communities: an example for rain forest trees. *American Naturalist* 156: 145–155.
Webb, C.O., D.D. Ackerly, M.A. McPeek and M.J. Donoghue. 2002. Phylogenies and community ecology. *Annual Review of Ecology Evolution and Systematics* 33: 475–505.
Webb, C.O., D.D. Ackerly and S.W. Kembel. 2008. Phylocom: software for the analysis of phylogenetic community structure and trait evolution. *Bioinformatics* 24: 2098–2100.
Whittaker, R.H. 1972. Evolution and measurement of species diversity. *Taxon* 21: 213–251.
Whittaker, R.J., K.J. Willis and R. Field. 2001. Scale and species richness: towards a general, hierarchical theory of species diversity. *Journal of Biogeography* 28: 453–470.
Willig, M.R., D.M. Kaufman and R.D. Stevens. 2003. Latitudinal gradients of biodiversity: pattern, process, scale, and synthesis. *Annual Review of Ecology Evolution and Systematics* 34: 273–309.
Zapata, F., K. Gaston and S. Chown. 2003. Mid-domain models of species richness gradients: assumptions, methods and evidence. *Journal of Animal Ecology* 72: 677–690.

CHAPTER 5

Diversity in Diet and Feeding Behaviour of Butterflyfishes: Reliance on Reef Corals versus Reef Habitats

Andrew J. Cole[a] and *Morgan S. Pratchett*[b,*]

INTRODUCTION

Considerable research attention has focused on the feeding habits of coral reef butterflyfishes, especially *Chaetodon* butterflyfishes (reviewed by Cole et al., 2008), largely due to their propensity to feed on corals (more specifically, reef-building scleractinian corals). Coral feeding infers a very high dependence upon living corals, which has led to suggestions that changes in the abundance (or behaviour) of butterflyfishes may provide an effective indicator for declines in coral cover (e.g., Khalaf and Crosby, 2005; Gochfeld, 2006), or even declining condition of entire coral reef ecosystems (Reese, 1977; Hourigan et al., 1988; Crosby and Reese, 1996, 2005). Butterflyfishes, especially highly specialised coral-feeding species, are certainly sensitive to changes in the availability of scleractinian corals (e.g., Crosby and Reese, 2005; Samways, 2005; Pratchett et al., 2006; Graham, 2007; Wilson et al., Chapter 14). However, responses of butterflyfishes to coral loss are very complex, highly species-specific, and often delayed, which

ARC Centre of Excellence for Coral Reef Studies, James Cook University, Townsville, QLD 4811, Australia.
[a]Email: andrew.cole3@jcu.edu.au
[b]Email: morgan.pratchett@jcu.edu.au
*Corresponding author

greatly reduces their effectiveness as indicators of coral health (Crosby et al., Chapter 10). Variation in dietary breadth and composition among butterflyfishes, which determines coral-dependence, is nonetheless very important in understanding inherent vulnerabilities of butterflyfishes to sustained and ongoing coral loss associated with the worldwide degradation of coral reef ecosystems (Pratchett et al., 2008a, Pratchett, Chapter 6).

Few trophic studies of coral reef fish communities ascribe corallivores a significant role in reef processes (e.g., Connell, 1973; Hatcher, 1988). Despite this, corallivores represent a functional group that is one of the most specialised on coral reefs (Randall, 1974). These fishes have co-evolved with and are inextricably linked to the corals on which they feed (Reese, 1977, 1981; Bellwood and Pratchett, Chapter 1). Coral-feeding provides the most direct trophic link between reef-building scleractinian corals and higher consumers (Glynn, 2004; Cole et al., 2008). Scleractinian corals are responsible for producing a significant component of the exogenous carbon on coral reefs (Reaka-Kudla, 1997; Veron, 2000) making corals an important energy source for those species capable of accessing it. Energy fixed by corals may enter trophic webs incidentally by ingestion and assimilation of coral mucous that settles on reef substrates (Coffroth, 1990; Wild et al., 2004), but corallivorous fishes make this carbon immediately available to higher consumers. Glynn (2004) found coral feeding organisms to be an important link in the food web of Eastern Pacific Reefs with over 287 links between corals, coral predators and top level carnivores.

Earlier studies on coral predation (e.g., Robertson, 1970; Connell, 1973) suggested that there were relatively few coral reef fishes capable of feeding on corals, because it requires extreme levels of morphological and physiological adaptations to account for the anti-predatory defences of corals. However, more recent evaluation of the literature reveals coral-feeding is relatively widespread, having been reported for 128 species of coral reef fishes from 11 families (Cole et al., 2008), though coral-feeding is most prevalent and best-known among butterflyfishes (Reese 1977, Harmelin-Vivien and Bouchon-Navaro, 1983; Roberts and Ormond, 1992; Cox, 1994; Rotjan and Lewis, 2008). Butterflyfishes (family Chaetodontidae) account for 54% (69/128 species) of all fishes that are reported to feed on corals (Fig. 5.1), and 61% (25/41 species) of fishes that feed entirely on coral (Cole et al., 2008). Among coral reef butterflyfishes at least 50% of species feed at least in part on hard (scleractinians) or soft (alcyonarians) corals (Fig. 5.2). The remaining species are mostly benthic carnivores, which feed predominantly on small discrete invertebrates (e.g., polychaetes) or zooplankton (Harmelin-Vivien and Bouchon-Navaro, 1983; Sano, 1989).

Butterflyfishes are a diverse family of acanthopterygian fishes comprising 125 nominal species, of which 79 are generally associated with coral reefs (Allen et al., 1998; Froese and Pauly, 2008). Butterflyfishes are

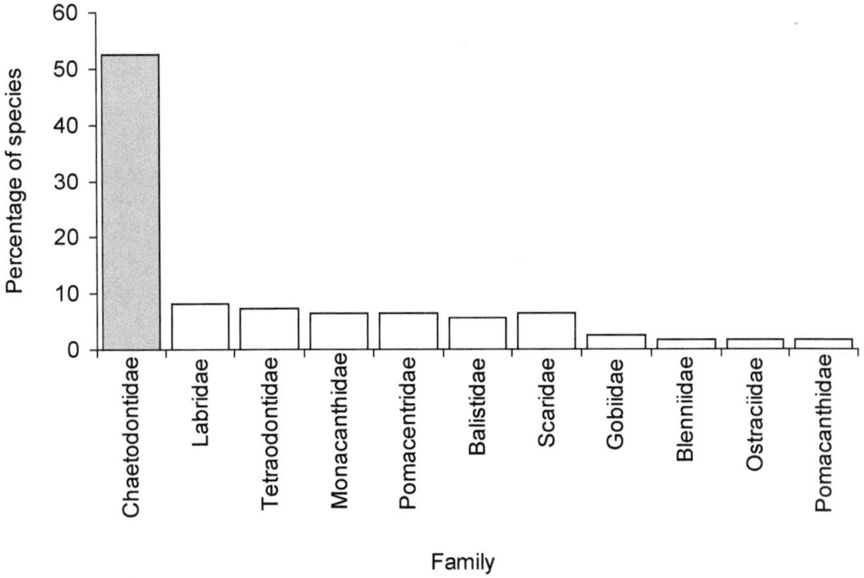

Fig. 5.1 Percentage of coral-feeding fishes (128 species) in each family, showing the disproportionate number of corallivores within the family Chaetodontidae, based on data presented in Cole et al. (2008). Corallivorous fishes comprise all species that have been reported to feed directly on live corals, including a few species (5 species, e.g., *C. vagabundus*) that only very occasionally and incidentally feed on the surface of live corals.

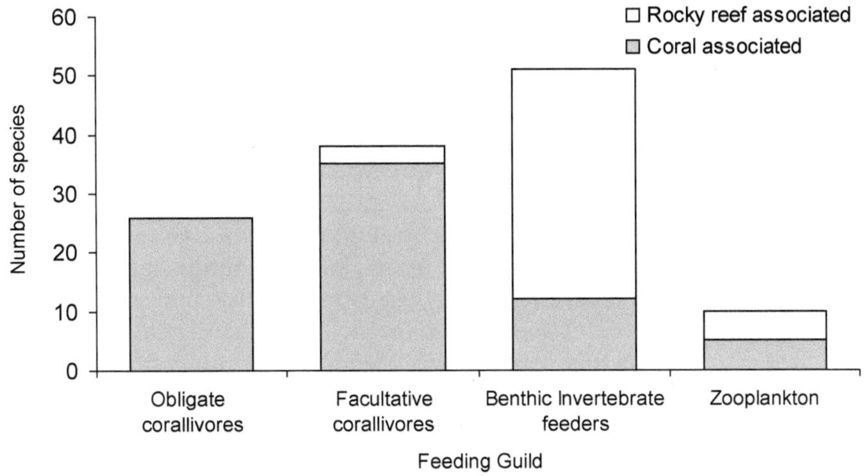

Fig. 5.2 Number of butterflyfishes associated with coral versus rocky (temperate or deep-water) reefs within each of four major feeding guilds. Obligate corallivores represent species that rely on coral for >80% of material ingested, whereas facultative corallivores eat 20–80% coral. Each species was classified based on descriptions in identification books and, where applicable, peer-reviewed journals. When there was doubt about the extent to which coral-feeding occurred, species were classified as facultative.

characterised by small terminal mouths, strong jaws, and bristle-like teeth, which are well-suited to scraping live tissue from the surface of scleractinian corals (Motta, 1988). Morphological specialisations of feeding apparatus provide exceptional diversity in foraging modes and diets within the family (Konow et al., Chapter 3). For example, *Forcipiger longirostris* has a highly modified elongate jaw with entirely novel suspensorial joints, which affords considerable control in capturing its elusive prey, the caridean shrimp, from within complex reef substrates (Ferry-Graham et al., 2001a,b). However, the most derived group of butterflyfishes are probably the species of *Chaetodon* that feed exclusively on scleractinian corals (Blum, 1989; Ferry-Graham et al., 2001b; Harmelin-Vivien, 2002). Even among coral-feeding species, there is significant variation in the nature and extent of corallivory (Pratchett, 2005; Cole et al., 2008). Most notably, it is now apparent that there are at least two fundamentally different strategies used to exploit live corals by butterflyfishes, whereby some butterflyfishes feed ostensibly on coral mucous (e.g., *Chaetodon ornatissimus*) while most other coral-feeding butterflyfishes consume live coral tissue (Hobson, 1974; Cole et al., 2008; Bellwood et al., 2009). There are also differences in the degree of dietary specialisation exhibited by coral-feeding butterflyfishes (Pratchett 2005, 2007; Berumen and Pratchett, 2008) that relates to further variation in feeding strategies and behaviour (e.g., territorial aggression, Berumen and Pratchett, 2006).

Among specialist corallivores (e.g., *Chaetodon trifascialis*) there is a tendency for butterflyfishes to maintain non-overlapping feeding territories from which they exclude conspecifics and con-generics (especially coral feeding species) to ensure exclusive access to specific resources (Sutton, 1985; Tricas, 1989b; Wrathall et al., 1992; Righton et al., 1998; Berumen et al., 2005; Berumen and Pratchett, 2006; Gochfeld, 2006; Righton and Mills, 2006). In contrast, most non-corallivorous butterflyfishes are often wide roaming and rarely exhibit any aggression towards conspecifics or con-generics (Roberts and Ormond, 1992; Berumen and Pratchett, 2006). Additionally, most corallivorous species occur as individuals or in pairs (Roberts and Ormond, 1992; Cox, 1994; Pratchett et al., 2006) although several species are gregarious and form feeding aggregations which roam across the reef (e.g., *Chaetodon lunula, C. daedalma*) (Hobson, 1974; Allen et al., 1998).

The purpose of this review is to document the diversity in feeding behaviour and diets of chaetodontid fishes. Foremost, we distinguish three major groups of butterflyfishes based on their predominant feeding behaviour: corallivores, non-coral benthic invertebrate feeders, and planktivores. We assign all nominal species to one of these three groups based on best available evidence from a multitude of sources (Table 5.1). There is however, relatively limited information available on the dietary composition for many species of butterflyfishes. For example, *Heniochus singularis* is often reported to feed exclusively on scleractinian corals

Table 5.1 Major dietary habits of coral reef butterflyfishes (I—benthic invertebrate feeder, O—obligate corallivore, F—Facultative corallivores, and P—planktivore). Evidence for trophic assignment was either anecdotal (Anec) when reported in field guides, but there is no supporting published data, based on field observation of feeding behaviour (Obs) or gut-content analysis (Gut). Where there was no information available for a species, the diet was assumed based phylogenetic associations (Phyl), whereby most monophyletic groupings appear to have similar dietary habits (Bellwood et al., 2009).

Species	Distribution	Diet	Evidence	Reference
Amphichaetodon				
A. howensis	SW Pacific	I	Anec	Michael (2004)
A. melbae	San Felix Is.	I	Anec	Michael (2004)
Chaetodon				
C. adiergastos	West Pacific	F	Anec	Allen et al. (1998); Michaels (2004); Froese and Pauly (2008)
C. andamanenisis	East Indian	O	Obs	Pratchett, Chapter 6
C. argentatus	West Pacific	F	Gut	Sano (1989)
C. assarius	West Australia	P	Anec	Allen et al. (1998); Michael (2004)
C. aureofasciatus	Northern Australia	O	Obs	Pratchett (2005)
C. auriga	Indo-Pacific	F	Gut	Sano et al. (1984); Harmelin-Vivien and Bouchon-Navaro (1981, 1983); Gregson et al. (2008)
C. auripes	Western Pacific	F	Gut	Sano et al. (1984)
C. austriacus	Red Sea	O	Gut	Harmelin-Vivien and Bouchon-Navaro (1981); Wrathall et al. (1992); Righton et al. (1998); Alwany et al. (2003)
C. baronessa	Pacific	O	Gut	Anderson et al. (1981); Sano (1989); Pratchet (2005); Berumen et al. (2005)
C. bennetti	Indo-Pacific	O	Gut	Harmelin-Vivien and Bouchon-Navaro (1983); Sano (1989)
C. blackburnii	Western Indian	F	Anec	Harmelin-Vivien (1989)
C. burgessi	West Pacific	I	Anec	Michael (2004)
C. capistratus	Caribbean, West Indies	F	Gut	Birkeland and Neudecker (1981); Gore (1984); Lasker (1985); Neudecker (1985); Pitts (1991)

Table 5.1 contd....

Table 5.1 contd.

Species	Distribution	Diet	Evidence	Reference
C. citrinellus	Indo-Pacific	F	Gut	Bouchon-Navaro (1986); Harmelin-Vivien (1989); Sano (1989); Pratchett (2005); Gregson et al. (2008)
C. collare	Red Sea, Indian Ocean	F	Anec	Allen et al. (1998); Kuiter (2002); Froese and Pauly (2008)
C. daedalma	NW Pacific	F	Gut	Sano (1989)
C. declivis	Marquesas, Line Is.	I	Anec	Michael (2004)
C. decussatus	North Indian	F	Anec	Michael (2004)
C. dialeucos	East Africa	I	Anec	Michael (2004)
C. dolosus	East Africa	I	Anec	Michael (2004)
C. ephippium	Indo-Pacific	F	Gut	Harmelin-Vivien and Bouchon-Navaro (1983); Sano et al. (1984)
C. falcula	Indian	F	Obs	McClanahan et al. (2005)
C. fasciatus	Red Sea	F	Gut	Harmelin-Vivien and Bouchon-Navaro (1981)
C. flavirostris	Pacific	F	Anec	Allen et al. (1998); Kuiter (2002)
C. flavocoronatus	Guam	I	Anec	Michael (2004)
C. fremblii	Hawaiian Islands	I	Anec	Michael (2004)
C. gardineri	NW Indian Ocean	F	Anec	Michael (2004)
C. guentheri	West Pacific	P	Anec	Michael (2004); Froese and Pauly (2008)
C. guttatissimus	Indian	O	Obs	Pratchett et al., 2013
C. hoefleri	West Africa	I	Anec	Michael (2004)
C. humeralis	Eastern Pacific	I	Anec	Froese and Pauly (2008)
C. interruptus	Indian	F	Anec	Allen et al. (1998); Michael (2004)
C. kleinii	Indo-Pacific	F	Gut	Sano (1989); Pratchett (2005)
C. larvatus	Red Sea	O	Gut	Zekeria et al. (2002)
C. leucopleura	West Indian	F	Anec	Michael (2004)
C. lineolatus	Indo-Pacific	I	Gut	Sano (1989); Pratchett (2005); Gregson et al. (2008)
C. litus	Easter Is.	P	Anec	Michael (2004)

Table 5.1 contd....

Table 5.1 contd.

Species	Distribution	Diet	Evidence	Reference
C. lunula	Indo-Pacific	F	Gut	Harmelin-Vivien and Bouchon-Navaro (1983); Harmelin-Vivien (1989); Pratchett (2005); Gregson et al. (2008)
C. lunulatus	Western-Pacific	O	Gut	Sano (1989); Cox (1994); Pratchett et al. (2004); Berumen et al. (2005); Gregson et al. (2008)
C. marleyi	Southern Africa	I	Anec	Michael (2004)
C. melapterus	Southern Red Sea, Arabian Gulf	O	Anec	Michael (2004); Froese and Pauly (2008)
C. melannotus	Red Sea, Indo-Pacific	O	Gut	Alino et al. (1988, 1992); Sano (1989); Pratchett (2005)
C. mertensii	Indo-Pacific	F	Gut	Harmelin-Vivien (1989)
C. mesoleucos	Red Sea	F	Gut	Zekeria et al. (2002)
C. meyeri	Indo-Pacific	O	Gut	Sano et al. (1984); Sano (1989)
C. miliaris	Hawaii	P	Gut	Ralston (1981)
C. mitratus	Indian	I	Obs	Michael (2004)
C. multicinctus	Hawaii	O	Gut	Tricas (1985, 1989a,b); Cox (1994); Aeby (2002); Gochfeld (2004)
C. nigropunctatus	Arabian Gulf	O	Anec	Shokri et al. (2005)
C. nippon	Pacific	F	Gut	Sano (1989)
C. ocellatus	West Atlantic	F	Anec	Pitts (1991); Michael (2004)
C. ocellicaudus	West Pacific	O	Anec	Allen et al. (1998); Froese and Pauly (2008)
C. octofasciatus	Indo-Pacific	O	Obs	Pratchett, Chapter 6
C. ornatissimus	Indo-Pacific	O	Gut	Reese (1977); Harmelin-Vivien and Bouchon-Navaro (1983); Sano (1989); Cox (1994)
C. oxycephalus	Indo-west Pacific	I	Anec	Allen et al. (1998); Michael (2004)
C. paucifasciatus	Red Sea	F	Gut	Harmelin-Vivien and Bouchon-Navaro (1981)
C. pelewensis	Pacific	O	Gut	Harmelin-Vivien and Bouchon-Navaro (1983)
C. plebeius	Indo-Pacific	O	Gut	Sano (1989); Pratchett (2005); Gregson et al. (2008)

Table 5.1 contd....

Table 5.1 contd.

Species	Distribution	Diet	Evidence	Reference
C. punctatofasciatus	Indo-Pacific	O	Gut	Sano (1989); Nagelkerken et al. (2009)
C. quadrimaculatus	Pacific	F	Gut	Harmelin-Vivien and Bouchon-Navaro (1983); Hourigan et al. (1988)
C. rafflesi	Indo-Pacific	F	Gut	Sano (1989); Pratchett (2005); Gregson et al. (2008)
C. rainfordi	West Pacific	O	Obs	Pratchett (2005); Gregson et al. (2008)
C. reticulatus	Pacific	O	Gut	Harmelin-Vivien and Bouchon-Navaro (1983)
C. robustus	West Africa	I	Phyl	
C. sanctaehelene	St. Helena, Ascension Is.	I	Anec	Allen et al. (1998)
C. sedentarius	Tropical West Atlantic	I	Gut	Randall (1967)
C. selene	Western Pacific	I	Anec	Michael (2004)
C. semion	Indo-Pacific	I	Gut	Sano (1989); Pratchett (2005)
C. semilarvatus	Red Sea	O	Gut	Zekeria et al. (2002)
C. smithi	Rapa, Pitcairn, Illots de Bass Is.	P	Anec	Allen et al. (1998); Michael (2004)
C. speculum	Indo-Pacific	F	Gut	Sano (1989); Pratchett (2005); Gregson et al. (2008)
C. striatus	Tropical West Atlantic	I	Gut	Randall (1967); Sazima and Sazima (2001); Bonaldo et al. (2005)
C. tinkeri	Central Pacific	I	Anec	Michael (2004)
C. triangulum	Indian	O	Anec	Allen et al. (1998); Froese and Pauly (2008)
C. trichrous	Society Is.	F	Obs	Randall (2005)
C. tricinctus	Lord Howe	F	Obs	Kuiter (1996)
C. trifascialis	Red Sea, Indo-Pacific	O	Gut	Reese (1981); Harmelin-Vivien and Bouchon-Navaro (1981, 1983); Harmelin-Vivien (1989); Irons (1989); Sano (1989); Alwany et al. (2003); Pratchett (2005); Samways (2005); Gregson et al. (2008)
C. trifasciatus	Indian	O	Gut	Harmelin-Vivien (1989)

Table 5.1 contd....

Table 5.1 contd.

Species	**Distribution**	**Diet**	**Evidence**	**Reference**
C. ulietensis	Indo-Pacific	F	Gut	Sano et al. (1984); Harmelin-Vivien and Bouchon-Navaro (1983); Sano (1989); Pratchett (2005); Gregson et al. (2008)
C. unimaculatus	Indo-Pacific	O	Gut	Harmelin-Vivien and Bouchon-Navaro (1983); Cox (1986, 1994); Sano (1989); Wylie and Paul (1989); Pratchett (2005); Gregson et al. (2008)
C. vagabundus	Indo-Pacific	I	Gut	Sano et al. (1984); Pratchett (2005); Gregson et al. (2008)
C. wiebeli	NW Pacific rim	F	Anec	Michael (2004)
C. xanthocephalus	Indian	F	Gut	Harmelin-Vivien (1989)
C. xanthurus	West Pacific	F	Anec	Michael (2004); Froese and Pauly (2008)
C. zanzibarensis	West Indian	O	Obs	Samways (2005); Pratchett et al. (2013)
Chelmon marginalis	Northwestern Ausralia	I	Anec	Michael (2004)
C. muelleri	North Australia	I	Anec	Allen et al. (1998); Froese and Pauly (2008)
C. rostratus	Indo-Pacific	I	Anec	Allen et al. (1998); Michael (2004)
Chelmonops				
C. curiosus	Southwestern Australia	I	Anec	Allen et al. (1998); Froese and Pauly (2008)
C. truncatus	Southeastern Australia	I	Anec	Allen et al. (1998); Michael (2004); Froese and Pauly (2008)
Coradion				
C. altivelis	Indo-Pacific	I	Gut	Sano (1989)
C. chrysozonus	Indo-Pacific	I	Anec	Allen et al. 1998
C. melanopus	Southeast Asia	I	Anec	Allen et al. (1998); Michael (2004)
Forcipiger				
F. flavissimus	Indo-Pacific, Red Sea	I	Gut	Harmelin-Vivien and Bouchon-Navaro (1983); Motto (1988)
F. longirostris	Indo-Pacific	I	Gut	Hobson (1974); Sano (1989); Motto (1988)

Table 5.1 contd....

Table 5.1 contd.

Species	Distribution	Diet	Evidence	Reference
Hemitaurichthys				
H. multispinosus	Pitcairn Is.	P	Anec	Froese and Pauly (2008)
H. polylepis	Pacific	P	Gut	Hobson (1974); Sano (1989)
H. thompsoni	Central Pacific	P	Anec	Michael (2004)
H. zoster	Indian	P	Anec	Michael (2004)
Heniochus				
H. acuminatus	Indo-Pacific	I	Anec	Michael (2004)
H. chrysostomus	Indo-Pacific	F	Gut	Harmelin-Vivien and Bouchon-Navaro (1983); Sano (1989)
H. diphreutes	Red Sea, Indo-Pacific	P	Anec	Michael (2004); Khalaf and Crosby (2005)
H. intermedius	Red Sea	F	Gut	Harmelin-Vivien and Bouchon-Navaro (1981)
H. monoceros	Indo-Pacific	I	Gut	Sano (1989), Allen et al. (1998)
H. pleurotaenia	NE Indian	I	Anec	Michael (2004)
H. singularius	Indo-Pacific	O	Gut	Sano (1989)
H. varius	Pacific	F	Gut	Sano (1989)
Johnrandallia				
J. nigrirostris	Eastern Pacific	I	Anec	Allen et al. (1998), Michael (2004), Froese and Pauly (2008)
Parachaetodon				
P. ocellatus	Indo-west Pacific	I	Anec	Allen et al. (1998), Michael (2004)
Prognathodes				
P. aculeatus	Caribbean	I	Gut	Randall (1967)
P. aya	Gulf of Mexico	I	Anec	Michael (2004)
P. brasilliensis	Brazil	I	Anec	Michael (2004)
P. dichrous	St Helena, Ascension Is.	I	Anec	Michael (2004)
P. falcifer	Eastern Pacific	I	Anec	Michael (2004)
P. guezei	Reunion, Mauritius and Comoro Is.	I	Phyl	
P. guyanensis	West Indies	I	Anec	Froese and Pauly (2008)
P. guyotenis	Maldives and NW Pacific	I	Phyl	

Table 5.1 contd....

Table 1 contd.

Species	Distribution	Diet	Evidence	Reference
P. marcellae	West Africa	I	Anec	Michael (2004)
P. obliquus	St Pauls Rocks	I	Phyl	
Roa				
R. australis	NW Australia	I	Phyl	
R. excelsa	Hawaiian Is., Guam	I	Anec	Allen et al. (1998)
R. jayakari	NW Indian Ocean	I	Phyl	
R. modestus	Western Pacific	I	Anec	Michael (2004)

(Table 5.1), though this conclusion stems from preliminary analyses of gut contents from only two individual fishes (Sano, 1989), and field observations suggest this species rarely, if ever, feeds from the surface of live corals (Berumen and Pratchett, unpublished data). By compiling all available data, this chapter serves to identify major gaps in our understanding of the feeding behaviour for coral reef butterflyfishes, which is critical to the future conservation of these species. Establishing which species are dependant upon corals for food is particularly important in the context of the increasing threats to coral reef biodiversity, including climate-induced coral bleaching, increasing incidence of coral disease, and direct anthropogenic stresses (Hughes, 1994; Sebens, 1994; Chadwick-Furman, 1996). Notably, corallivores are particularly sensitive to the declining abundance of certain prey corals, and are typically among the first and worst affected fishes during extensive coral loss (Wilson et al., 2006; Baker et al., 2008; Pratchett et al., 2008a). However, some butterflyfishes exhibit geographic variability in feeding mode and cannot easily be ascribed to only one feeding guild (Cole et al., 2008). Additionally, within each guild there is a gradient between specialist feeders and generalists, whereby the latter are capable of greater flexibility in the range of prey consumed, and may therefore withstand major changes in prey availability (Pratchett et al., 2004; Pratchett, Chapter 6).

2. MEASURING DIETARY COMPOSITION

Dietary composition of butterflyfishes has been documented predominantly through gut content analyses (e.g., Bouchon-Navaro, 1986), or *in situ* feeding observations (e.g., Pratchett, 2005). Gut content analyses are fundamental in establishing the proportional consumption of small cryptic invertebrates (e.g., polychaete worms, and crustaceans) that are often eaten by benthic feeding butterflyfishes, but have only limited application in relation to coral feeding. Gut content is important in confirming that butterflyfishes actually ingest coral tissue, whereby the presence of recognisable polyps, zooxanthellae and/or nematocysts in the gut contents of butterflyfishes provides incontrovertible evidence of coral feeding (e.g., Reese, 1977; Zekeria et al., 2002). However, it is very difficult to estimate proportional consumption of corals (versus other potential prey items) from gut content analyses, because gut contents of putative coral-feeding butterflyfishes are generally dominated by indistinct mucous (Sano et al., 1984). The brown mucous found in the stomachs of putative corallivores is often assumed to represent partially digested coral tissue (Harmelin-Vivien and Bouchon-Navaro, 1981, 1983; Sano, 1989; Zekeria et al., 2002), but gastric juices almost certainly contribute to increasing the volume of mucous in gut contents, making it difficult to estimate the relative volumetric quantity of coral mucous. Moreover, different dietary items may be digested and assimilated at different rates, such that the relative occurrence of identifiable dietary components in gut contents of fishes may have little relation to the relative consumption of different prey items (MacDonald et al., 1982; Polis and Strong, 1996; Mariani et al., 2002).

Recent studies (e.g., Berumen et al., 2005; Pratchett, 2005) have tended to estimate proportional consumption of corals by butterflyfishes based on the proportional number of bites taken from the surface of live corals within a specified period. There are many assumptions implicit in using feeding observations to estimate proportional coral feeding (Pratchett, 2005). Foremost, when butterflyfishes bite on the surface of a coral it is generally assumed that they are feeding on live coral tissue, while they may actually be feeding on coral mucous (e.g., Hobson, 1974), or picking micro-invertebrates from the surface of live corals. Moreover, there may be differences in the amount of material consumed with each bite, such that the proportional number of bites on different prey does not accurately reflect the relative volumes of prey types that are actually ingested. However, biases associated with *in situ* feeding observations are probably no worse than for gut content analyses. Further, *in situ* feeding observations are currently the only way to assess the relative consumption of different coral species (Gore, 1984; Cox, 1994; Irons, 1989; Alwany et al., 2003; Pratchett, 2005; Berumen et al., 2005), which is important in quantifying dietary specialisation of butterflyfishes and cannot be distinguished based on gut content analysis.

Two further techniques that might prove useful in resolving some of the inherent problems associated with gut content analysis and *in situ* feeding observations are stable isotope analysis (e.g., Nagelkerken et al., 2009) and the use of lipid biomarkers (e.g., Wilson et al., 2001). These techniques have great potential in distinguishing corallivorous and non-corallivorous species especially for more cryptic feeders (e.g., *Heniochus* species); however it is unlikely that either technique will be able to identify relative consumption of individual coral species. Lipid bio-markers compare the ratio of fatty acids inside fish tissues with that found in available dietary items; from this it is possible to get an indication of which fatty acids and dietary items are being assimilated. Similarly, stable isotope analysis utilises natural variation in isotopic ratios (mainly ^{13}C and ^{15}N) of potential prey items, which are preserved in the tissues of consumers (Fry, 2006). Both techniques have several advantages over short-term feeding observations and gut content analyses. Most importantly they provide a history of food assimilated in terms of weeks to months, depending on the predator's growth rate (Gearing, 1991; Hesslein et al., 1993). It is therefore less subject to short-term variations in diet (which contributes to considerable bias for gut content analyses and *in situ* feeding observations) and provides an extended record of typical feeding behaviour. This may be particularly important if butterflyfishes supplement their largely coral diet by occasionally targeting prey items (e.g., algae) that are enriched in ^{13}C or have distinct lipid signatures. Preliminary data from two studies (Nagelkerken et al., 2009; Mill, 2007) shows apparent separation in ^{13}C and $d^{15}N$ of obligate corallivores (mean = –12.5 and 8.8, respectively) versus non-corallivores (–16.0 and 10.2, respectively). While considerable work is required to improve resolution between live coral tissues and other potential prey items as well as specifically quantify fractionation of coral prey (e.g., Mill et al., 2007), these data do show that stable isotopes may prove useful in identifying coral-feeders and establishing the relative importance of coral versus non-coral prey (Nagelkerken et al., 2009).

3. CORALLIVORES

Butterflyfishes exhibit a great diversity of feeding behaviour, from corallivory to herbivory to planktivory (e.g., Hiatt and Strassburg, 1960; Hobson, 1974; Reese, 1981; Harmelin-Vivien and Bouchon-Navaro, 1981, 1983; Sano, 1989; Roberts and Ormond, 1992). However, scleractinian corals constitute the primary food source for most butterflyfishes and so butterflyfishes are often considered synonymous with coral-feeding (Hourigan et al., 1988; Roberts and Ormond, 1992; Cole et al., 2008). Among butterflyfishes that are found living exclusively on coral reefs, 78% (61/78 species) feed at least in part on live corals, and 46% (28/61) of these species feed almost exclusively on hard and/or soft corals (Fig. 5.2). Butterflyfishes are the most well studied group of corallivorous fishes and there is a wealth of information on their general ecology, including dietary composition (e.g., Reese, 1977, 1981; Tricas, 1985, 1989a; Harmelin-Vivien and Bouchon-Navaro, 1983; Harmelin-Vivien, 1989; Irons, 1989; Sano, 1989; Alwany et al., 2003; Pratchett, 2005), foraging strategies (Tricas, 1985, 1989a,b; Reese, 1989; Gochfeld, 2004), feeding rates (Irons, 1989; Berumen et al., 2005; Zekeria et al., 2002; Gregson et al., 2008), and responses to coral depletion (e.g., Williams, 1986; Bouchon-Navaro and

Bouchon, 1989; Harmelin-Vivien, 1989; Zekeria and Videler, 2000; Bozec et al., 2005; Pratchett et al., 2006). Corallivorous butterflyfishes are found on coral reefs throughout the world, though they are significantly under-represented in the eastern Pacific and Caribbean. In the Indo-Pacific, 56% (62/111 species) of butterflyfishes feed on corals, whereas only 14% (2/14 species) of butterflyfishes in the Caribbean utilize coral as a food source, possibly due to the limited abundance of *Acropora* and *Pocillopora* in the Caribbean, which are the preferred prey of most corallivores (Randall, 1974; Hourigan et al., 1988; Pratchett, 2005; 2007). *Chaetodon capistratus* is the only Caribbean species that consumes coral in any significant proportion, with up to 75% of its diet derived from coral, and it feeds mainly on *Madracis* and *Agaricia* corals (Birkeland and Neudecker, 1981; Gore, 1984; Pitts, 1991).

Corals provide a unique food source for those species capable of exploiting them. Unlike motile prey that are often hidden within the reef interstices and must be located and captured prior to feeding, coral represents a stationary, non-cryptic, renewable food source that, in addition to being widely available, also provides ample protection from predators (Birkeland and Neudecker, 1981; Tricas, 1985). However, coral is defended in other ways (e.g., Gochfeld, 2004) and is widely considered to be nutritionally depauperate, compared to other major prey types (e.g., Birkeland and Neudecker, 1981; Tricas, 1989a; Gregson et al., 2008). In support of this theory, corallivorous species spend a considerable proportion of their time actively foraging (e.g., Birkeland and Neudecker, 1981; Tricas, 1989a, b; Gregson et al., 2008). Tricas (1989b) spent 3,000 hours observing *Chaetodon multicinctus* and found that this species spends 90% of its time feeding. Moreover, most corallivores feed at very rapid rates, taking upwards of 600 bites per hour (Irons, 1989; Berumen et al., 2005; Gregson et al., 2008). These data suggest that inefficiencies in ingesting coral tissues pose a major limitation to energy budgets of coral feed butterflyfishes (Tricas, 1989a), and may explain why many coral-feeders supplement their diet with other non-coral prey (Birkeland and Neudecker, 1981). Most corallivorous butterflyfishes feed only partly on live coral tissue (Fig. 5.2), supplementing their diet with motile invertebrates, sedentary polychaetes, sponges, algae, or other benthic prey (Birkeland and Neudecker, 1981; Harmelin-Vivien and Bouchon-Navaro, 1983; Sano, 1989; Pratchett, 2005). Similarly, among corallivores in other reef fish families (all except the Labridae), most corallivorous species are facultative corallivores, feeding only partly on scleractinian corals and supplementing their diet with other benthic prey (Cole et al., 2008).

Corallivorous fishes span the full spectrum of coral dependence, from species that feed almost exclusively on coral (obligate or exclusive coral feeder, Harmelin-Vivien and Bouchon-Navaro, 1983) to those in which coral is only a minor component (facultative coral feeders, Harmelin-Vivien and Bouchon-Navaro, 1983). Additionally, reliance on coral shown by corallivorous species can vary among individuals and among locations, making it difficult to assess the level of coral dependence for many apparent corallivores (Cole et al., 2008). *Chaetodon auriga*, for example, takes only a very small proportion of bites (<4%) on corals at Lizard Island, in the northern Great Barrier Reef (GBR), Australia (Pratchett, 2005; Gregson et al., 2008), but is recorded to feed extensively (taking up to 61% of bites) on live coral in other geographic regions (e.g., French Polynesia, Harmelin-Vivien and Bouchon-Navaro, 1983; Japan, Sano et al., 1984; Red Sea, Harmelin-Vivien and Bouchon-Navaro, 1981; Marshall Islands, Hiatt and Strasburg, 1960; Chagos Archipelago, central Indian Ocean, Graham, 2007).

The majority of coral-feeding butterflyfishes feed primarily on scleractinian corals, although a small number of species (e.g., *Chaetodon melannotus, C. ulientensis*, and *C. unimaculatus*) also feed heavily on soft corals (Alcyonarians) (Pratchett, 2005; Berumen, unpublished data). *C. melannotus* is the best known of the soft-coral feeders, and consistently feeds on soft corals in higher proportions than hard-corals (Harmelin-Vivien and Bouchon-Navaro, 1981; Sano, 1989; Alino et al., 1992; Pratchett, 2005), while several other butterflyfishes occasionally target soft-corals, mainly in areas where soft-corals are particularly abundant. On the Great Barrier Reef, Pratchett (2005) observed *C. unimaculatus* to consume large amounts of soft coral, but elsewhere in the Pacific it is a specialist hard coral feeder (Cox, 1986, 1994; Hourigan et al., 1988; Berumen and Pratchett, 2006). In the Red Sea, where there is high abundance of soft coral and low densities of other soft-coral feeders (e.g., *C. melannotus*), *C. auriga* has been observed to take >50% of bites from soft corals (Berumen, unpublished data). Soft corals are generally avoided by coral predators, possibly due to their extensive physical and chemical defences; soft coral tissues are heavily embedded with calcium carbonate sclerites, and typically contain high concentrations of toxic secondary metabolites (Sammarco and Coll, 1992; Wylie and Paul, 1989). Butterflyfishes are also the only corallivores found to consume soft coral; all other corallivorous families feed on scleractinian corals and avoid soft corals (Cole et al., 2008).

3.1 Obligate Coral-feeders

Obligate coral-feeders, in the strictest sense, are butterflyfishes that have a specific reliance on corals for food and are adversely affected by significant limitations in the abundance of live coral (Pratchett et al., 2008a). As it is difficult to assess actual reliance on corals, the term obligate coral-feeders has come to refer to those species that feed almost exclusively on corals, (e.g., Harmelin-Vivien and Bouchon-Navaro, 1983; Sano, 1989; Berumen et al.,

2005; Pratchett, 2005; Graham, 2007) implicitly assuming that exclusive coral feeders are virtually incapable of exploiting non-coral prey. An operational criterion to categorise obligate coral-feeding butterflyfishes, proposed by Pratchett (2005), includes all butterflyfishes for which more than 80% of their diet comprises live coral. Corallivorous fishes that have diets containing more than 80% coral exhibit a disproportionate decline following coral loss, which indicates that these species are highly dependant on coral for their survival (Pratchett et al., 2008a). By this definition, a relatively high proportion of butterflyfishes are obligate hard coral feeders. For example just over one-third (25 species) of coral feeding butterflyfishes have been shown to feed almost entirely on scleractinian coral (Fig. 5.2, Table 5.1). In most locations throughout the Indo-Pacific, obligate corallivores tend to be more speciose and more abundant than facultative or non-coral feeders (Fig. 5.3). The only exception is in locations where large schools of planktivorous butterflyfishes are encountered (e.g., Hawaii), which are numerically dominant to benthic-feeding species (Brock et al., 1979). Moreover, the actual number of obligate coral-feeding butterflyfishes may be under-estimated, due to limited data on the feeding ecology of many species that are currently regarded as facultative coral-feeders (Table 5.1). There are at least 2 species of butterflyfish (e.g., *Chaetodon adiergastos*,

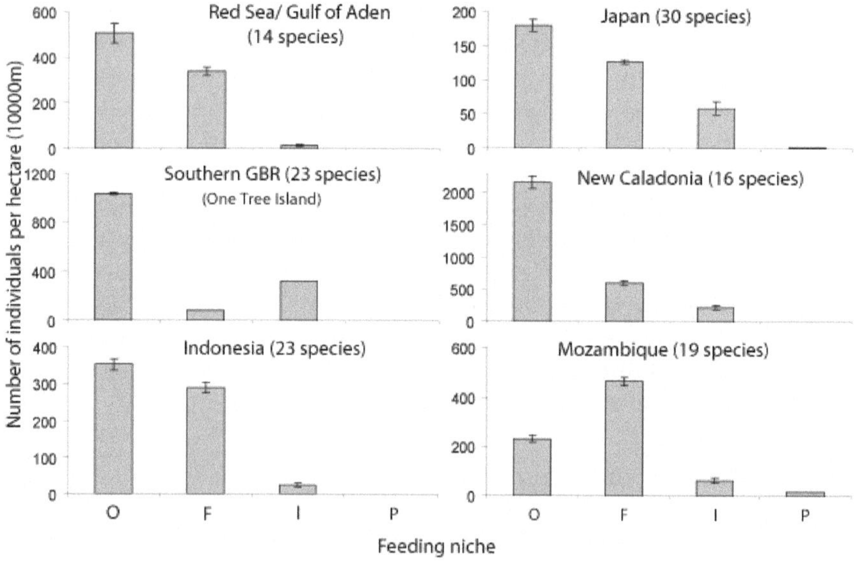

Fig. 5.3 Relative abundance of obligate (O) and facultative (F) corallivorous, benthic invertebrate feeders (I) and planktivorous (P) butterflyfish on coral reefs throughout the Indo-Pacific. Red Sea/Gulf of Aden (Khalaf and Abdallah, 2005), Japan (Cadoret et al., 1999), One Tree Island (Fowler, 1990), New Caledonia (Kulbicki and Sarramégna, 1999), Indonesia (Adrim and Hutomo, 1989) and Mozambique (Pereira and Videira, 2005).

and *C. interruptus*), which have been reported to feed on corals, but have not received detailed study and so the most conservative strategy was to categorise these species as facultative corallivores. It is possible however, that corals may comprise >80% of the diet for at least one of these species.

Obligate coral-feeding butterflyfishes are often regarded to be among the most evolutionary advanced fishes inhabiting coral reefs (Randall, 1974; Reese, 1977), and recent phylogenetic analyses confirm that these species largely constitute a single monophyletic group of highly derived butterflyfishes (Bellwood et al., 2009; Bellwood and Pratchett, Chapter 1). While it is very difficult to characterise obligate coral-feeding butterflyfishes based on morphological traits (Motta, 1989), there are several morphological adaptations that correspond with broad differences in diet. Most notably, corallivorous butterflyfishes have longer and narrower intestines compared to non-coral feeding butterflyfishes (Motta, 1988; Harmelin-Vivien and Bouchon-Navaro, 1983; Elliot and Bellwood, 2003; Berumen et al., 2011). These long intestines are indicative of diets with low nutritional value and are thought to enable high gut retention times and enable the digestion of complex molecules such as wax esters like cetyl palmitate (Benson and Muscatine, 1974; Kung and Ciereszko, 1985).

Furthermore, differences in the dentition of butterflyfishes tends to reflect their dietary habits, whereby species which grasp or tear their prey have numerous rows of curved teeth that encircle the mouth allowing them to hold onto their vagile prey (Motta, 1988, 1989). Coral feeding species show several adaptations in accessing coral polyps. Firstly, coral-scrapers, which includes species that irritate corals to stimulate production of excess mucous, use their lower jaw to scrape multiple polyps per bite. These species (e.g., *Chaetodon ornatissimus*) generally have relatively large mouths, wider than they are tall; jaw protrusion is limited and results in a shovel-like mouth with an anteriorly massed pad of teeth that is used to comb the surface of the coral (Motta, 1988, 1989). The other method of feeding on coral is to nip off individual polyps one at a time (e.g., *C. trifascialis*). Polyp feeders have small jaws and only a few rows of teeth that combine to form forcep-like mouths, which efficiently remove single polyps (Motta, 1989). Lastly, *C. unimaculatus* and its sister-species *C. interruptus* are unique amongst butterflyfishes. These two species have extremely robust jaws and teeth that are used to bite deep into hard corals and remove a large proportion of carbonate skeleton with each bite (Hobson, 1974; Cox, 1986; Motta, 1988; Sano, 1989).

Exclusive coral feeding confers major reliance on abundant coral growth and obligate coral-feeders are foremost among coral reef fishes in their association with live coral (Halford et al., 2004) and are extremely vulnerable to any declines in coral cover (reviewed by Wilson et al., 2006; Pratchett et al., 2008a; Wilson et al., Chapter 9). However, positive relationships between the abundance of obligate corallivores and live coral cover are not always apparent. Limited availability of coral clearly limits the abundance of obligate corallivores, especially following declines in coral cover caused by outbreaks of the coral-feeding sea star *Acanthaster planci* (Williams, 1986; Sano et al., 1987) or climate-induced coral bleaching (Sano, 2004; Pratchett et al., 2006), but obligate corallivores are not always abundant in areas with high coral cover (e.g., Bell et al., 1985; Fowler, 1990; Kulbicki et al., 2005). Population regulation of corallivorous butterflyfishes may be differentially attributable to resource-limitation or recruitment-limitation in different locations and habitats, though it is known that recruitment rates for butterflyfishes are very low (Pratchett et al., 2008b). Further, obligate corallivores may not be limited by the abundance of live coral *per se*, but by availability of specific prey corals (e.g., Berumen and Pratchett, 2008) owing to inter-specific differences in specific prey preferences.

Obligate coral-feeding butterflyfishes differ greatly in the range of different corals that they consume as well as their proportional consumption of different corals (e.g., Pratchett, 2005) and variability in the dietary habitats is likely to influence their responses to different types of disturbances (Pratchett et al., 2008a). *Chaetodon trifascialis*, for example, specialises on *Acropora hyacinthus* (Reese, 1981; Irons, 1989; Harmelin-Vivien, 1989; Alwany et al., 2003; Pratchett, 2005, 2007) and when *A. hyacinthus* is scarce, *C. trifascialis* is rarely found (Reese, 1981; Berumen and Pratchett, 2006). *C. trifascialis* is among the most specialised of all obligate coral-feeding butterflyfishes (Irons, 1989; Pratchett, 2005) and is often the first and worst affected of all butterflyfishes during major disturbances (e.g., climate-induced coral bleaching) that cause extensive coral depletion (Wilson et al., 2006; Pratchett, Chapter 6; Wilson et al., Chapter 9).

In contrast to specialist coral feeders, highly generalist obligate corallivores can access a greater diversity of prey and may partially compensate for initial depletion of preferred prey resources by increasing intake of alternate prey types (i.e., prey switching) and/or increasing the area over which they forage (Tricas, 1989b; Pratchett et al., 2004; Samways, 2005). Some generalists consume prey items in similar proportions to their availability (Birkeland and Neudecker, 1981), and are capable of changing their diet based on local environmental conditions such as coral cover (Neudecker, 1985; Pratchett et al., 2004; Berumen et al., 2005) or the presence of superior competitors (Cox, 1994; Pratchett, 2005). As a consequence of this flexibility, these generalists are expected to be more resilient to subtle changes in the make up of the coral community. For example, *Chaetodon lunulatus* is highly adaptable and can feed on a wide diversity of different coral species (Berumen et al., 2005; Pratchett, 2005) and largely escaped significant negative effects associated with extensive coral bleaching in the central GBR by feeding increasingly on coral species (e.g., *Favia* spp.) that are less susceptible to coral bleaching (Pratchett et al., 2004). Alternatively, some generalists have specialised patterns of prey use, but these preferences vary greatly between reef systems. For example, in Hawaii, *Chaetodon unimaculatus* is an obligate hard coral feeder that specialises on *Montipora verrucosa* (Cox, 1986, 1994), however in Japan and on the Great Barrier Reef, *C. unimaculatus* feeds primarily on soft coral, and hard coral is a relatively minor dietary component (Sano, 1989; Pratchett, 2005, 2007). Further studies on the feeding biology of obligate corallivores across extensive geographic ranges (e.g., *C. trifascialis*) may reveal population-level specialisation but high versatility at the species-level (*sensu* Fox and Morrow, 1981).

Despite significant inter-specific variation in dietary specialisation, it is worth noting that even the most generalist coral-feeding butterflyfishes (e.g., *C. lunulatus*) are still quite specialised and typically consume less than 30% of available coral species (e.g., Pratchett, 2005). These pronounced and often broadly over-lapping prey preferences of obligate coral-feeding butterflyfishes lead to a concentration of feeding effort on a few distinct coral species, which may have significant effects on community structure of coral assemblages (Neudecker, 1977, 1979; Cox, Chapter 7). When preferred prey are scarce, selective feeding by corallivorous butterflyfishes may cause localised extinction of these corals. Neudecker (1977) showed that the coral, *Pocillopora damicornis*, which was abundant in shallow lagoons of Guam did not persist when transplanted to exposed reef slopes (15 and 30m) with high densities of piscine corallivores, although caged controls grew well at these depths. The impact of corallivory may also be more pronounced under conditions where corals are stressed or depleted owing to other disturbances (e.g., Glynn, 1985, 1988; Guzman and Robertson, 1989; Bellwood et al., 2006; Rotjan et al., 2006). Rotjan et al. (2006) found that corallivores exacerbated the effects of climate-induced coral bleaching for colonies of *Montastrea* spp. as heavily grazed colonies exhibited slower recovery rates compared to ungrazed colonies. The stress induced by long recovery times has direct implications for colony fitness and fecundity (Baird and Marshall, 2002). Corallivorous grazing and coral bleaching appear to interact and exacerbate one another.

3.2 Facultative Corallivores

If obligate corallivores are species that absolutely rely on coral for food, facultative coral-feeders would represent species that opportunistically exploit corals, but could equally survive without any coral in their diet (Pratchett et al., 2008a). In reality, facultative corallivores represent species that often feed heavily, but not exclusively on corals (e.g., Harmelin-

Vivien and Bouchon-Navaro, 1983; Sano, 1989; Pratchett, 2005; Graham, 2007), though little is known about their actual reliance on coral for food. Studies that have examined consequences of coral depletion on coral-feeding butterflyfishes, almost invariably show that obligate corallivores are disproportionately affected (Bouchon-Navaro et al., 1985; Williams, 1986; Pratchett et al., 2006), but some facultative corallivores are adversely affected by extensive coral loss (Sano et al., 1987). Pratchett (2001) revealed that two facultative coral-feeding butterflyfishes (*Chaetodon citrinellus* and *C. kleinii*) exhibited significant declines in abundance following coral depletion by localised infestation of *Acanthaster planci* in the northern GBR. These findings suggest that the proportion of scleractinian corals in the diet of butterflyfish is not necessarily a true indication of their actual reliance on corals. While obligate coral feeding butterflyfish must have access to at least some coral species, facultative corallivores may be equally reliant on scleractinian corals to provide essential nutrients and/or provide a mixed diet to maximise assimilation efficiency (Birkeland and Neudecker, 1981). In the extreme, both facultative and obligate corallivores do not persist in areas devoid of all scleractinian corals. For example, Sano et al. (1987) reported a complete absence of all coral-feeding butterflyfish (facultative and obligate corallivores) on a 'rubble reef' in southern Japan, decimated by extensive outbreaks of *A. planci*. It is likely that a greater range of different butterflyfish species, including those species with only a partial reliance on scleractinian corals, will be affected with increasing severity of coral depletion and subsequent loss of structural complexity.

Specific reliance on coral prey by facultative corallivores is apparent based on their high degree of dietary specificity (Pratchett, 2005). If facultative corallivores opportunistically feed on corals, simply because it provides a readily accessible food source to supplement the capture of other discrete prey items, then patterns of coral-feeding might be expected to correspond with the abundance or accessibility of different coral species. Alternatively, patterns of coral-feeding by facultative corallivores might be expected to complement coral preferences of locally abundant obligate coral-feeders, to minimise direct competition (Anderson et al., 1981). Surprisingly though, facultative corallivores are highly selective, and often exhibit very similar patterns of coral preference to specialist obligate coral feeders (Hourigan et al., 1988; Pratchett, 2005, 2007). In the northern GBR, for example, Pratchett (2005, 2007) showed that virtually all facultative and obligate coral-feeding butterflyfishes preferred either *Acropora hyacinthus* or *Pocillopora damicornis*.

The proportional consumption of live coral (both hard and soft corals combined) by facultative corallivores ranges from 6% for *Chaetodon xanthocephalus* (Harmelin-Vivien 1989) and *C. mesoleucos* (Zekeria et al., 2002) up to 70% for *C. ulientensis* (Harmelin-Vivien and Bouchon-Navaro,

1983). There are several species that only very occasionally feed on corals, but we consider these to be primarily non-coral feeders. For example, Pratchett (2005) found that *Chaetodon vagabundus* took 1.51% (32/2, 116 bites) of bites from scleractinian coral, which probably represents incidental grazing on the surface of live corals, rather than specific bites targeting live coral. Therefore, facultative corallivores are categorised as those species that consume >5% but <80% live coral, though for this chapter all butterflyfishes that consume <10% live coral were carefully reviewed and considered on a case by case basis. For example, Harmelin-Vivien and Bouchon-Navaro (1983) report that *Forcipiger flavissimus* consumes 6% coral (by volume), however we conclude that coral consumption is entirely incidental, due to extensive feeding on motile invertebrates from among the branches of live corals. In most cases when feeding data was limited, butterflyfish were only considered to be facultative (e.g., *C. flavirostris, C. gardineri*) if convincing anecdotal evidence indicated that coral browsing occurred (Table 5.1). It is worth noting however, that for some of these facultative corallivores (e.g., *C. flavirostris, C. leucopleura, C. gardineri*) the extent to which coral feeding occurs is largely unknown and coral is possibly only a very minor component (<5%) of their diet.

Aside from coral, butterflyfishes consume mostly algae and/or polychaetes (see section 4). The extent to which corallivorous butterflyfishes are also herbivorous is potentially important, but controversial. Irrespective of consumption of distinct algal material, which is often found in gut contents of facultative corallivores (e.g., *Chaetodon citrinellus*), Harmelin-Vivien and Bouchon-Navaro (1983) consider that consumption of symbiotic algae (zooxanthellae) within the tissues of corals confers some degree of herbivory, which may be reflected in their elongated guts (Elliot and Bellwood, 2003). However, Parker (1984) showed that zooxanthellae pass through the guts of butterflyfishes (and other corallivorous fishes) completely undigested; fecal material from these corallivorous fishes comprises high densities of zooxanthellae, which are photosynthetically active and capable of re-establishing symbioses with corals. Rather than deriving any nutrition from incidental consumption of zooxanthellae, coral-feeders may actually be fundamental in dispersal of zooxanthellae among aposymbiotic corals (Parker, 1984). Similarly, extensive consumption of distinct algal material may be linked to the consumption of material (bacteria and/or small motile organisms) that are associated with the epilithal algal matrix (*sensu* Wilson et al., 2003). Even so, algal material can comprise up to 50% (by volume) of the diet of some butterflyfishes (e.g., *Chaetodon daedalma*, Sano, 1989), suggesting that these species may be important contributors to herbivory, which is considered to be the most important functional role for fishes on coral reefs (e.g., Bellwood et al., 2003).

4. NON-CORAL FEEDERS

Scientific interest in the feeding ecology of butterflyfishes is clearly centred on corallivory, leading to a disproportionate number of studies that consider corallivorous species (e.g., Reese, 1977; Harmelin-Vivien and Bouchon-Navaro, 1981, 1983; Gore, 1984; Neudecker, 1985; Alino et al., 1988; Tricas, 1989a, b; Irons, 1989; Zekeria et al., 2002, 2005; Berumen et al., 2005). Despite the large number of butterflyfishes that do not feed on corals (61/125 species) and the potentially interesting questions relating to competitive coexistence and resource portioning among these species (Pratchett, 2005), there are very few studies that have specifically investigated the feeding habitats and dietary composition of non-coral feeders (e.g., Birkeland and Neudecker, 1981). Almost all knowledge of non-corallivorous butterflyfishes comes from studies of feeding behaviour extending across entire assemblages of butterflyfishes (e.g., Anderson et al., 1981; Harmelin-Vivien and Bouchon-Navaro, 1983; Sano, 1989; Pratchett, 2005) or even more extensive studies of the trophic habitats of coral reef fishes (Hiatt and Strasburg, 1960; Randall, 1967; Hobson, 1974; Sano et al., 1984). Many of these non-coralline invertebrate feeding butterflyfishes also live in very deep and largely inaccessible habitats, which has greatly limited studies on their biology and ecology. For example, almost nothing is known about the biology or ecology of most *Prognathodes* and *Roa* species, because these butterflyfishes typically occur at depths >80m and several of these species (e.g., *P. guezi*, and *P. guyotensis*) are known only from a few distinct specimens collected using traps (Allen et al., 1998; Pauly and Froese, 2008).

Non-corallivorous butterflyfishes are obviously less constrained by the abundance of live coral compared to strict corallivores (Pratchett et al., 2008a; Wilson et al., Chapter 9) and thus, are not necessarily associated with coral rich habitats. This is apparent both in the geographical distribution of many non-corallivores which extend into temperate environments with no coral reefs (Allen et al., 1998), but also, within coral reef environs, non-corallivores tend to exhibit more even densities across a broad range of different habitats (Pratchett and Berumen, 2008; Pratchett et al., 2008b). While corallivorous butterflyfishes are concentrated in areas with rich coral growth (e.g., Birkeland and Neudecker, 1981; Bouchon-Navaro et al., 1985; Findley and Findley, 1985; Bouchon-Navaro and Bouchon, 1989; Pitts, 1991; Cadoret et al., 1999; Pratchett and Berumen, 2008), Pratchett and Berumen (2008) found no significant relations between abundance of non-corallivorous butterflyfishes and live coral cover among 16 study sites at Lizard Island, in the northern GBR. However, many non-corallivorous butterflyfishes (e.g., *Forcipiger flavissimus*) are found exclusively on coral reefs (Fig. 5.2), and like corallivorous butterflyfishes these species sometimes exhibit significant declines in abundance following extensive coral depletion (Pratchett, 2001; Bouchon-Navaro et al., 1985). Hourigan (1989) suggested that close associations between *Chaetodon* butterflyfish and scleractinian corals are not entirely based on feeding requirements, but rather scleractinian corals may represent essential habitat for newly settled recruits. Pratchett et al. (2008b) specifically examined settlement preferences of corallivorous versus non-corallivorous butterflyfishes and did not find any non-corallivorous species that settled to live coral. However, it is also well established that coral rich habitats moderate competition and/ or predation among coral reef fishes, and can enhance local abundance of fishes that do not feed, shelter or recruit on live corals (Almany, 2004; Hixon and Jones, 2005). Non-corallivorous butterflyfishes may also rely on specific non-coral prey that associate with rich coral growth (Motta, 1988). Moreover, highly generalist benthic feeding butterflyfishes may be relatively unaffected by changes in the biological structure of benthic reef habitats, but may be adversely affected (like nearly all small to medium sized fishes) by any changes in the topographic structure and complexity of reef habitats (Wilson et al., 2006; Pratchett et al., 2008a).

4.1 Benthic Invertebrate Feeders

Benthic invertebrate feeders are generally considered to be the least specialised of all butterflyfishes, as these species are presumed to feed opportunistically on whatever prey they can find (Birkeland and Neudecker, 1981). For instance, many of these species will abandon their usual benthic-feeding habitat to take advantage of ephemeral pulses of high quality food particles in the water column (Pratchett et al., 2001; Sazima and Sazima, 2001). Unlike coral feeders, which browse continuously across many coral colonies and rarely inflict mortality, benthic invertebrate feeders are direct predators that capture and consume discrete prey items. The greatest limitation to feeding for most carnivorous species is the time taken to detect and capture individual prey items (Holling, 1959), and so these species feed at a much lower rate than their corallivorous counterparts (e.g., Gregson et al., 2008) and are likely to consume any appropriate prey items (e.g., within a given size range) that they encounter (Birkeland and Neudecker, 1981). However, studies which have sufficient resolution in describing the specific dietary composition of benthic invertebrate feeders (based mainly on extensive analysis of gut contents) tend to show that the diets of these seemingly ecologically equivalent species are sometimes very distinct and often complementary (Harmelin-Vivien and Bouchon-Navaro, 1983; Sano, 1989), reflective of either individualistic feeding preferences or active partitioning of prey resources (*sensu* Pratchett, 2005).

Most non-coral feeders feed mainly on errant or sedentary polychaetes (e.g., Pitts, 1991; Zekeria et al., 2002; Pratchett, 2005), which is perhaps reflective of their abundance in reef environments. There are also some butterflyfishes that specialise on very specific motile invertebrates. *Forcipiger longirostris* is one such species, which feeds entirely on elusive caridean shrimp (Hobson, 1974; Sano, 1989). This species has an extremely elongate jaw and possesses modifications to the suspensorium, which allows both the upper and lower jaws to be protruded anteriorly (Ferry-Graham et al., 2001a). A morphological adaptation which allows this species to feed from narrow crevices, whereas its sister-species, *F. flavissimus*, which does not have this modification has a much more generalised diet (Harmelin-Vivien and Bouchon-Navaro, 1983).

4.2 Planktivores

Planktivorous feeding may at the outset appear to be a major departure from the normal benthic-feeding mode of butterflyfishes, but Motta (1988) suggested that the small, moderately protrusible and laterally enclosed mouth of most butterflyfishes is equally effective in picking planktonic particles from the water column as it is at capturing motile invertebrates on and around coral substrates, and corresponds closely with mouth and jaw morphology of other specialised planktivores (Coughlin and Strickler, 1990; Hobson, 1991). The dietary composition of planktivorous butterflyfishes is also highly consistent with other planktivorous reef fishes, feeding mainly on zooplankton and in particular calonoid copepods and larvaceans (Hobson, 1974; Ralston, 1981; Harmelin-Vivien and Bouchon-Navaro, 1983; Sano, 1989). Among butterflyfishes, there are 10 species that are nominal planktivores (e.g., *Hemitaurichthys*; Table 5.1), while a further 7 species (*Chaetodon kleinii, C. lunula, C. striatus, C. trichrous, Heniochus acuminata, H. intermedius*) will occasionally or in specific locations exhibit planktivorous feeding behaviour (Roberts and Ormond, 1992). Planktivorous butterflyfishes are often found in very large aggregations, concentrated in open water along reef edges adjacent to deep-water (Brock and Chamberlain, 1968; Hobson, 1974; Ralston, 1981), which is a common strategy among planktivorous reef fishes to both maximise accessibility to planktonic prey and reduce the risk of predation (Hobson and Chess, 1978; Thresher, 1983; Hobson, 1991). Obligate planktivorous butterflyfish are much more gregarious than facultative planktivores, forming aggregations of up to 1000 individuals (Stimson, 1990; Kuiter, 2002; Michael, 2004), whereas facultative planktivores typically forage in very small groups or join mixed species feeding aggregations. In most locations, planktivorous species account for a very minor component of either the abundance or diversity of butterflyfishes (Fig. 5.3). However, butterflyfish assemblages in Hawaii are dominated by the planktivore, *Chaetodon miliaris* which is the most commonly encountered species in the Hawaiian Islands and reaches densities of over 3000 individuals per hectare (Brock et al., 1979; Ralston, 1981; Stimson, 1990).

In addition to traditional planktivory, several butterflyfish also act as mutualistic cleaners of other reef fish. Cleaning behaviour is a common feature of juvenile stages of several species in the genera *Chaetodon*, *Heniochus* and *Johnrandallia* (e.g., Côté, 2000; Konow and Ferry, Chapter 2) and it has also recently been reported for *Forcipiger flavissimus* (Craig, 2007). This behaviour appears to be largely opportunistic and is most commonly observed among juvenile individuals, although some species do maintain this behaviour as adults (e.g., *Heniochus diphreutes*, *Chaetodon guentheri* and *C.striatus*; Sazima and Sazima, 2001; Michael, 2004). Juveniles that clean, benefit directly with a reduced risk of predation (Côté, 2000). Additionally, exploiting alternative feeding strategies (e.g., cleaning) may provide increased nutrients that enhance growth rates and subsequently overall fitness.

5. CONCLUSIONS

Energy acquisition is fundamental to all aspects of the biology and ecology of living organisms, and so studies of diet and feeding behaviour are critical in understanding that factors that structure population dynamics, species characteristics, and community structure (Krebs, 1998; Hughes, 1989). It is imperative therefore, that future research aims to address limited knowledge of the feeding behaviour for many butterflyfishes, especially those species that are already rare, geographically isolated or potentially reliant on corals for food. In this chapter (and throughout this book) it is apparent that the range and proportional use of different prey types that butterflyfishes consume (e.g., coral versus non-coral prey) has important implications for population size, geographic range, and patterns of habitat-use, as well as their vulnerability to disturbances and changes in resource availability. Most importantly, it seems that highly specialized coral feeding butterflyfishes are extremely vulnerable to disturbances, and are increasingly being replaced by generalist species in degraded reef environments (Bellwood et al., 2006; Pratchett et al., 2008a). Obligate coral feeding butterflyfishes are among the most specialized and highly derived fishes, and should typically dominate butterflyfish assemblages. However, their abundance is critically dependant on rich coral growth (e.g., Birkeland and Neudecker, 1981; Bouchon-Navaro et al., 1985; Findley and Findley, 1985; Bouchon-Navaro and Bouchon, 1989; Pitts, 1991; Cadoret et al., 1999; Berumen and Pratchett, 2006; Pratchett and Berumen, 2008).

Aside from feeding on corals, butterflyfishes play a potentially important role in the trophodynamics and ecosystem function of coral reef ecosystems as herbivores, planktivores, benthic carnivores, and ectoparasite feeders (cleaners). Only half of coral reef butterflyfish species feed and recruit to corals, but remaining species are nonetheless often dependant on coral reef habitats, as they are critically dependant on the habitat complexity or specific non-coral prey provided by rich coral growth. Overall, butterflyfishes exhibit considerable diversity in dietary composition and feeding strategies, which often belie their apparent similarity in morphology among species (Motta, 1988). It is exactly these traits (including diverse feeding habitats, contrasting levels of specialization, similarities in size and gross morphology) that make butterflyfishes such a useful subject to study the causes and consequences of inter-specific differences in feeding and diet.

6. REFERENCES

Adrim, M. and M. Hutomo. 1989. Species composition, distribution and abundance of Chaetodontidae along reef transects in the Flores Sea. *Netherlands Journal of Sea Research* 23: 85–93.

Aeby, G.S. 2002. Trade-offs for the butterflyfish, *Chaetodon multicinctus*, when feeding on coral prey infected with trematode metacercariae. *Behavioral Ecology and Sociobiology* 52: 158–165.

Alino, P.M., P.W. Sammarco and J.C. Coll. 1988. Studies of the feeding preferences of *Chaetodon melannotus* (Pisces) for soft corals (Coelenterata; Octocorallia). *Proceeding of the Sixth International Coral Reef Symposium, Townsville, Australia* 31–36.

Alino, P.M., J.C. Coll and P.W. Sammarco. 1992. Toxic prey discrimination in a highly specialised predator *Chaetodon melannotus* (Block et Schneider): visual vs. chemical cues. *Journal of Experimental Marine Biology and Ecology* 104: 209–220.

Allen, G.R., R. Steene and M. Allen. 1998. *A Guide to Angelfishes & Butterflyfishes*. Odyssey Publishing, Perth.

Almany, G.R. 2004. Differential effects of habitat complexity, predators and competitors on abundance of juvenile and adult coral reef fishes. *Oecologia* 141: 105–113.

Alwany, M., E. Thaler and M. Stachowitsch. 2003. Food selection in two corallivorous butterflyfishes, *Chaetodon austriacus* and *Chaetodon trifascialis*, in the northern Red Sea. *Marine Ecology* 24: 165–177.

Anderson, G.R.V., A.H. Ehrlich, P.R. Ehrlich, J.D. Roughgarden, B.C. Russell and F.H. Talbot. 1981. The community structure of coral reef fishes.*American Naturalist* 117: 476–495.

Baird, A.H. and P.A. Marshall. 2002. Mortality, growth and reproduction in scleractinian corals following bleaching on the Great Barrier Reef. *Marine Ecology-Progress Series* 237: 133–141.

Baker, A.C., P.W. Glynn and B. Riegl. 2008. Climate change and coral reef bleaching: An ecological assessment of long-term impacts, recovery trends and future outlook. *Estuarine, Coastal and Shelf Science* 80: 435–471.

Bell, J.D., M.L. Harmelin-Vivien and R. Galzin. 1985. Large scale spatial variation in abundance in butterflyfishes (Chaetodontidae) on Polynesian reefs. *Proceedings of the Fifth International Coral Reef Symposium, Tahiti, French Polynesia* 5: 421–426.

Bellwood, D.R., A.S. Hoey and J.H. Choat. 2003. Limited functional redundancy in high diversity systems: resilience and ecosystem function on coral reefs. *Ecology Letters* 6: 281–285.

Bellwood, D.R., A.S. Hoey, J.L. Ackerman and M. Depczynski. 2006. Coral bleaching, reef fish community phase shifts and the resilience of coral reefs. *Global Change Biology* 12: 1587–1594.

Bellwood, D.R., S. Klanten, M.S. Pratchett, N. Konow and L. van Herwerden. 2009. Evolutionary history of the butterflyfishes (f: Chaetodontidae) and the rise of coral feeding fishes. *Journal of Evolutionary Biology* 23: 335–349.

Benson, A. and L. Muscatine. 1974. Wax in coral mucus: energy transfer from corals to reef fishes. *Limnology and Oceanography* 19: 810–814.

Berumen, M.L. and M.S. Pratchett. 2006. Effects of resource availability on the competitive behaviour of butterflyfishes (Chaetodontidae). *Proceedings of the Tenth International Coral Reef Symposium, Okinawa, Japan* 644–650.

Berumen, M.L. and M.S. Pratchett. 2008. Trade-offs associated with dietary specialization in corallivorous butterflyfish (Chaetodontidae: *Chaetodon*). *Behavioral Ecology and Sociobiology* 62: 989–994.

Berumen, M.L., M.S. Pratchett and M.I. McCormick. 2005. Within-reef differences in diet and body condition of coral-feeding butterflyfishes (Chaetodontidae). *Marine Ecology-Progress Series* 287: 217–227.

Berumen, M.L., M.S. Pratchett and B.A. Goodman. 2011. Relative gut lengths of coral reef butterflyfishes (Pisces: Chaetodontidae). *Coral Reefs* 30: 1005–1010.

Blum, S.D. 1989. Biogeography of the Chaetodontidae: an analysis of allopatry among closely related species. *Environmental Biology of Fishes* 25: 9–31.

Birkeland, C. and S. Neudecker. 1981. Foraging behaviour of two Caribbean chaetodontids: *Chaetodon capistratus* and *C. aculeatus*. *Copeia* 1: 169–178.

Bonaldo, R.M., J.P. Krajewski and I. Sazima. 2005. Meals for two: foraging activity of the butterflyfish *Chaetodon striatus* (Perciformes) in southeast Brazil. *Brazilian Journal of Biology* 65: 211–215.

Bouchon-Navaro, Y. 1986. Partitioning of food and space resources by chaetodontid fishes on coral reefs. *Journal of Experimental Marine Biology and Ecology* 103: 21–40.

Bouchon-Navaro, Y. and C. Bouchon. 1989. Correlations between chaetodontid fishes and coral communities of the Gulf of Aquaba (Red Sea). *Environmental Biology of Fishes* 25: 47–60.

Bouchon-Navaro, Y., C. Bouchon and M.L. Harmelin-Vivien. 1985. Impact of coral degradation on a chaetodontid fish assemblage (Moorea, French Polynesia). *Proceedings of the Fifth International Coral Reef Symposium, Tahiti, French Polynesia* 5: 427–432.

Bozec, Y.M., S. Doledec and M. Kulbicki. 2005. An analysis of fish-habitat associations on disturbed coral reefs: chaetodontid fishes in New Caledonia. *Journal of Fish Biology* 66: 966–982.

Brock, V.E. and T.C. Chamberlain. 1968. A geological and ecological reconnaissance of western Oahu, Hawaii, principally by means of the research submarine 'Asherah'. *Pacific Science* 22: 373–394.

Brock, R.E., C. Lewis and R.C. Wass. 1979. Stability and structure of a fish community on a coral patch reef in Hawaii. *Marine Biology* 54: 381–292.

Cadoret, L., M. Aderjoud and M. Tsuchiya. 1999. Spatial distribution of chaetodontid fish in coral reefs of the Ryukyu Islands, southern Japan. *Journal of the Marine Biological Asscoiation of the United Kingdom* 79: 725–735.

Chadwick-Furman, N.E. 1996. Reef coral diversity and global change. *Global Change Biology* 2: 559–568.

Coffroth, M.A. 1990. Mucous sheet formation on poritid corals: an evaluation of coral mucus as a nutrient source on reefs. *Marine Biology* 105: 39–49.

Cole, A.J., M.S. Pratchett and G.P. Jones. 2008. Diversity and functional importance of coral-feeding fishes on tropical coral reefs. *Fish and Fisheries* 9: 286–307.

Connell, J.H. 1973. Population ecology of reef-building corals. *In:* O.A. Jones and R. Endean, editors. *Biology and geology of coral reefs*, II, Biology 1. Academic Press, New York, USA pp. 247–268.

Côté, I.M. 2000. Evolution and ecology of cleaning symbiosis in the sea. *Oceanography and Marine Biology Annual Review* 38: 311–355.

Coughlin, D.J. and J.R. Strickler. 1990. Zooplankton capture by a coral reef fish: an adaptive response to evasive prey. *Environmental Biology of Fishes* 29: 35–42.

Cox, E.F. 1986. The effects of a selective corallivore on growth rates and competition for space between two species of Hawaiian corals. *Journal of Experimental Marine Biology and Ecology* 101: 161–174.

Cox, E.F. 1994. Resource use by corallivorous butterflyfishes (Family Chaetodontidae) in Hawaii. *Bulletin of Marine Science* 54: 535–545.

Craig, M.T. 2007. Facultative cleaning by the forcepsfish, *Forcipiger flavissimus* (Chaetodontidae). *Copeia* 459–461.

Crosby, M.P. and E.S. Reese. 1996. *A Manual for Monitoring Coral Reefs with Indicator Species: Butterflyfishes as Indicators of Change on Indo-Pacific Reefs*. Office of Ocean and Coastal Resource Management, National Oceanic and Atmospheric Administration, Silver Spring, MD.

Crosby, M.P. and E.S. Reese. 2005. Relationship of habitat stability and intra-specific population dynamics of an obligate corallivore butterflyfish. *Aquatic Conservation: Marine and Freshwater Ecosystems* 15: S13–S25.

Elliott, J.P. and D.R. Bellwood. 2003. Alimentary tract morphology and diet in three coral reef fish families. *Journal of Fish Biology* 63: 1598–1609.

Ferry-Graham, L.A., P.C. Wainwright, C.D. Hulsey and D.R. Bellwood. 2001a. Evolution and mechanics of long jaws in butterflyfishes (family Chaetodontidae). *Journal of Morphology* 248: 120–143.

Ferry-Graham, L.A., P.C. Wainwright and D.R. Bellwood. 2001b. Prey capture in long-jawed butterflyfishes (Chaetodontidae): the functional basis of novel feeding habits. *Journal of Experimental Marine Biology and Ecology* 256: 167–184.

Findley, J.S. and M.T. Findley. 1985. A search for pattern in butterfly fish communities. *American Naturalist* 126: 800–816.

Fowler, A.J. 1990. Spatial and temporal patterns of distribution and abundance of chaetodontid fishes at One Tree Reef, Southern GBR. *Marine Ecology-Progress Series* 64: 39–53.

Fox, L.R. and P.A. Morrow. 1981. Specialization: species property or local phenomenon. *Science* 211: 887–893.

Froese, R. and D. Pauly. Editors. 2008. FishBase. World Wide Web electronic publication. www.fishbase.org, version (04/2008).

Fry, B. 2006. *Stable isotope ecology*. Springer, USA.

Gearing, J.N. 1991. The study of diet and trophic relationships through natural abundance ^{13}C. *In: Carbon Isotope Techniques*, D.C. Coleman and B. Fry (ed.). Academic Press, San Diego pp. 201–218.

Glynn, P.W. 1985. Corallivore population sizes and feeding effects following El Nino (1982–1983) associated coral mortality in Panama. *Proceedings of the Fifth International Coral Reef Symposium, Tahiti, French Polynesia* 4: 183–188.

Glynn, P.W. 1988. Predation on coral reefs: some key processes, concepts and research directions. *Proceedings of the Sixth International Coral Reef Symposium,Townsville, Australia* 2: 51–62.

Glynn, P.W. 2004. High complexity food webs in low-diversity Eastern Pacific Reef-Coral Communities. *Ecosystems* 7: 358–367.

Gochfeld, D.J. 2004. Predation-induced morphological and behavioral defenses in a hard coral: implications for foraging behavior of coral-feeding butterflyfishes. *Marine Ecology-Progress Series* 267: 145–158.

Gochfeld, D.J. 2006. Use and misuse of butterflyfishes as indicators of coral reef health. *Marine Environmental Research* 62: S387–S387.

Gore, M.A. 1984. Factors affecting the feeding behavior of a coral reef fish, *Chaetodoncapistratus*. *Bulletin of Marine Science* 35: 211–220.

Graham, N.A.J. 2007. Ecological versatility and the decline of coral feeding fishes following climate driven coral mortality. *Marine Biology* 153: 119–127.

Gregson, M.A., M.S. Pratchett, M.L. Berumen and B.A. Goodman. 2008. Relationship between butterflyfish (Chaetodontidae) feeding rate and coral consumption on the Great Barrier Reef. *Coral Reefs* 27: 455–715.

Guzman, H.M. and D.R. Robertson. 1989. Population and feeding responses of the corallivorous pufferfish *Arothron meleagris* to coral mortality in the Eastern Pacific. *Marine Ecology-Progress Series* 55: 121–131.

Halford, A., A.J. Cheal, D. Ryan and D.M. Williams. 2004. Resilience to large-scale disturbance in coral and fish assemblages on the Great Barrier Reef. *Ecology* 85: 1892–1905.

Harmelin-Vivien, M.L. 1989. Implications of feeding specializations on the recruitment processes and community structure of butterflyfishes. *Environmental Biology of Fishes* 25: 101–110.

Harmelin-Vivien, M.L. 2002. Energetics and fish diversity on coral reefs. *In:* P.F. Sale, editor. *Coral Reef Fishes, Dynamics and Diversity in a Complex Ecosystem.* Academic Press, Elsevier Science, USA pp. 265–274.

Harmelin-Vivien, M.L. and Y. Bouchon-Navaro. 1981. Trophic relationships among Chaetodontid fishes in the Gulf of Aquaba (Red Sea). *Proceedings of the Fourth International Coral Reef Symposium, Manila, Phillipines* 2: 119–127.

Harmelin-Vivien, M.L. and Y. Bouchon-Navaro. 1983. Feeding diets and significance of coral feeding among chaetodontid fishes in Moorea (French Polynesia). *Coral Reefs* 2: 119–127.

Hatcher, B.G. 1988. Coral reef primary productivity: A beggar's banquet. *Trends in Ecology and Evolution* 3: 106–111.

Hesslein, R.H., K.A. Hallard and P. Ramlal. 1993. Replacement of sulfur, carbon, and nitrogen in tissue of growing broad whitefish (*Coregonus nasus*) in response to a change in diet traced by $\delta^{34}S$, $\delta^{13}C$, and $\delta^{15}N$. *Canadian Journal of Fisheries and Aquatic Sciences* 50: 2071–2076.

Hiatt, R.W. and D.W. Strasburg. 1960. Ecological relationships of the fish fauna on coral reefs of the Marshall Islands. *Ecological Monographs* 30: 65–127.

Hixon, M.A. and G.P. Jones. 2005. Competition, predation, and density-dependant mortality in demersal marine fishes. *Ecology* 86: 2847–2859.

Hobson, E.S. 1974. Feeding relationships of teleostean fishes on coral reefs in Kona, Hawaii. *Fisheries Bulletin* 72: 915–1031.

Hobson, E.S. 1991. Trophic relationships of fishes specialized to feed on zooplankters above coral reefs. *In:* P.F. Sale, editor.*The Ecology of Fishes on Coral Reefs*. Academic Press pp. 69–93.

Hobson, E.S. and J.R. Chess. 1978. Trophic relationships among fishes and plankton in the lagoon at Enewetak Atoll, Marshall Islands. *Fisheries Bulletin* 76: 133–153.

Holling, C.S. 1959. Some characteristics of simple types of predation and parasitism. *Canadian Entomology* 91: 385–399.

Hourigan, T.F. 1989. Environmental determinants of butterflyfish social-systems. *Environmental Biology of Fishes* 25: 61–78.

Hourigan, T.F., T.C. Tricas and E.S. Reese. 1988. Coral reef fishes as indicators of environmental stress in coral reefs. *In:* D.F. Soule and G.S. Kleppel, editors. *Marine Organisms as Indicators.* Springer Verlag, New York pp. 107–135.

Hughes, T.P. 1989. Community structure and diversity of coral reefs: the role of history. *Ecology* 70: 275–279.

Hughes, T.P. 1994. Catastrophes, phase shifts and large-scale degradation of a Caribbean coral reef. *Science* 265: 1547–1551.

Irons, D.K. 1989. Temporal and areal feeding behavior of the butterflyfish, *Chaetodon trifascialis,* at Johnston-Atoll. *Environmental Biology of Fishes* 25: 186–193.

Khalaf, M.A. and M. Abdallah. 2005. Community structure of butterflyfishes in the Red Sea and Gulf of Aden. *Aquatic Conservation: Marine and Freshwater Ecosystems* 15: S77–S89.

Khalaf, M.A. and M.P. Crosby. 2005. Assemblage structure of butterflyfishes and their use as indicators of Gulf of Aqaba benthic habitat in Jordan. *Aquatic Conservation: Marine and Freshwater Ecosystems* 15: S27–S43.

Krebs, C.J. 1998. *Ecological methodology*. Addison Wesley Longman, New York.

Kuiter, R.H. 1996. *Guide to sea fishes of Australia*. New Holland, Sydney.

Kuiter, R.H. 2002. *Butterflyfishes, Bannerfishes and their relatives*. TMC Publishing, Chorleywood, UK.

Kulbicki, M. and S. Sarramegna. 1999. Comparison of density estimates derived from strip transect and distance sampling for underwater visual censuses: a case study of Chaetodontidae and Pomacanthidae. *Aquatic Living Resources* 12: 315–325.

Kulbicki, M., Y.M. Bozec and A. Green. 2005. Implications of biogeography in the use of butterflyfishes (Chaetodontidae) as indicators for western and central Pacific areas. *Aquatic Conservation-Marine and Freshwater Ecosystems* 15: S109–S126.

Kung, S.S. and L.S.Ciereszko. 1985. Occurrence of the wax cetyl palmitate in stomachs of the corallivorous butterfly fish *Chaetodon trifascialis*. *Coral Reefs* 4: 45–46.

Lasker, H.R. 1985. Prey preferences and browsing pressure of the butterflyfish *Chaetodon capistratus* on Caribbean gorgonians. *Marine Ecology-Progress Series* 21: 213–220.

MacDonald, J.S., K.G. Waiwood and R.H. Green. 1982. Rates of different digestion of different prey in atlantic cod (*Gadus morhua*), ocean pout (*Macrozoarces americanus*), winter flounder (*Pseudopleuronectes americanus*), and American plaice (*Hippoglossoides platessoides*). *Canadian Journal of Fisheries and Aquatic Sciences* 39: 651–659.

Mariani, S., A. Maccaroni, F. Massa, M. Rampacci and L. Tancioni. 2002. Lack of consistency between trophic interrelationships of five sparid species in two adjacent central Mediterranean coastal lagoons. *Journal of Fish Biology* 61: 138–147.

McClanahan, T., J. Maina, C.J. Starger, P. Herron-Perez and E. Dusek. 2005. Detriments to post-bleaching recovery of corals. *Coral Reefs* 24: 230–246.

Michael, S.W. 2004. Angelfishes and Butterflyfishes. T.F.H. Publications, Neptune City.

Mill, A.C., J.K. Pinnegar and N.V.C. Polunin. 2007. Explaining isotope trophic-step fractionation: why herbivorous fish are different. *Functional Ecology* 21: 1137–1145.

Mill, A.C. 2007. *Stable isotope data as reef food-web descriptors in a dynamic tropical environment*. PhD dissertation,University of Newcastle upon Tyne, UK.

Motta, P.J. 1988. Functional morphology of the feeding apparatus of ten species of Pacific butterflyfishes (Perciformes, Chaetodontidae): an ecomorphological approach. *Environmental Biology of Fishes* 22: 39–67.

Motta, P.J. 1989. Dentition patterns among Pacific and western Atlantic butterflyfishes (Perciformes, Chaetodontidae): relationship to feeding ecology and evolutionary history. *Environmental Biology of Fishes* 25: 159–170.

Nagelkerken, I., G. van der Velde, S.L.J. Wartenbergh, M.M. Nugues and M.S. Pratchett. 2009. Cryptic dietary components reduce dietary overlap among sympatric butterflyfishes (Chaetodontidae). *Journal of Fish Biology* 75: 1123–1143.

Neudecker, S. 1977. Transplant experiments to test the effect of fish grazing on coral distribution. *Proceedings of the 3rd International Coral Reef Symposium*, Miami, Florida 317–323.

Neudecker, S. 1979. Effects of grazing and browsing fishes on the zonation of corals in Guam. *Ecology* 60: 666–672.

Neudecker, S. 1985. Foraging patterns of Chaetodontid and Pomacanthid fishes at St. Croix (Iles vierges, U.S.A.). *Proceedings of the Fifth International Coral Reef Symposium, Tahiti, French Polynesia* 5: 415–414.

Parker, G.M. 1984. Dispersal of zooxanthellae on coral reefs by predators on cnidarians. *Biological Bulletin* 167: 159–167.

Pereira, M.A.M. and E.J.S. Videira. 2005. Distribution and community structure of butterflyfishes (Pisces: Chaetodontidae) in southern Mozambique. *Western Indian Ocean Journal of Marine Science* 4: 39–46.

Pitts, P.A. 1991. Comparative use of food and space by three Bahamian butterflyfishes. *Bulletin of Marine Science* 48: 749–756.

Polis, G. and D.R. Strong. 1996. Food web complexity and community dynamics. *American Naturalist* 147: 813–846.

Pratchett, M.S. 2001. Dynamics of outbreak populations of crown-of-thorns starfish (*Acanthaster planci* L.), and their effects on coral reef ecosystems. PhD dissertation, James Cook University, Townsville, Australia.

Pratchett, M.S. 2005. Dietary overlap among coral-feeding butterflyfishes (Chaetodontidae) at Lizard Island, northern Great Barrier Reef. *Marine Biology* 148: 373–382.

Pratchett, M.S. 2007. Dietary selection by coral-feeding butterflyfishes (Chaetodontidae) on the Great Barrier Reef, Australia. *Raffles Bulletin of Zoology* 171–176.

Pratchett, M.S. and M.L. Berumen. 2008. Interspecific variation in distributions and diets of coral reef butterflyfishes (Teleostei: Chaetodontidae). *Journal of Fish Biology* 73: 1730–1747.

Pratchett, M.S., N. Gust, G. Goby and S.O. Klanten. 2001. Consumption of coral propagules represents a significant trophic link between corals and reef fish. *Coral Reefs* 20: 13–17.

Pratchett, M.S., S.K. Wilson, M.L. Berumen and M.I. McCormick. 2004. Sublethal effects of coral bleaching on an obligate coral feeding butterflyfish. *Coral Reefs* 23: 352–356.

Pratchett, M.S., S.K. Wilson and A.H. Baird. 2006. Long-term monitoring of the Great Barrier Reef. *Journal of Fish Biology* 69: 1269–1280.

Pratchett, M.S., P.L. Munday, S.K. Wilson, N.A.J. Graham, J.E. Cinner and D.R. Bellwood. 2008a. Effects of climate-induced coral bleaching on coral-reef fishes- ecological and economic consequences. *Oceanography and Marine Biology: An Annual Review* 46: 251–296.

Pratchett, M.S., M.L. Berumen, M.J. Marnane, J.V. Eagle and D.J. Pratchett. 2008b. Habitat associations of juvenile versus adult butterflyfishes. *Coral Reefs* 27: 541–551.

Pratchett, M.S., N.A.J. Graham and A.J. Cole. 2012. Specialist corallivores dominate butterflyfish assemblages in coral-dominated reef habitats. *Journal of Fish Biology* 82: 1177–1191.

Ralston, S. 1981. Aspects of the reproductive biology and feeding ecology of *Chaetodon miliaris*, a Hawaiian endemic butterflyfish. *Environmental Biology of Fishes* 6: 167–176.

Randall, J.E. 1967. Food habits of reef fishes of the West Indies. *Studies in Tropical Oceanography* 5: 665–847.

Randall, J.E. 1974. The effect of fishes on coral reefs. *Proceedings of the Second International Coral Reef Symposium, Brisbane, Australia* 1: 159–166

Reaka-Kudla, M. 1997. The global biodiversity of coral reefs: A comparison with rain forests. *In:* M.L. Reaka-Kudla, D.E. Wilson and E.O. Wilson, editors. *Biodiversity II: Understanding and protecting our natural resources*. Joseph Henry/National Academy Press, Washington, D.C. pp. 83–108

Reese, E.S. 1977. Coevolution of corals and coral feeding fishes of the family Chaetodontidae. *Proceedings of the Third International Coral Reef Symposium, Miami,* USA 1: 268–274.

Reese, E.S. 1981. Predation on corals by fishes of the family Chaetodontidae: implications for conservation and management of coral reef ecosystems. *Bulletin of Marine Science* 31: 594–604.

Reese, E.S. 1989. Orientation behaviour of butterflyfishes (Family Chaetodontidae) on coral reefs: spatial learning of route specific landmarks and cognitive maps. *Environmental Biology of Fishes* 25: 79–86.

Righton, D. and C. Mills. 2006. Application of GIS to investigate the use of space in coral reef fish: a comparison of territorial behaviour in two Red Sea butterflyfishes. *International Journal of Geographical Information Science* 20: 215–232.

Righton, D., M. Miller and R. Ormond. 1998. Correlates of territory size in the butterflyfish *Chaetodon austriacus* (Ruppell). *Journal of Experimental Marine Biology and Ecology* 226: 183–193.

Roberts, C.M. and R.F.G. Ormond. 1992. Butterflyfish social behaviour, with special reference to the incidence of territoriality: a review. *Environmental Biology of Fishes* 34: 79–93.

Robertson, D.R. 1970. Review of the predators and parasites of stony corals with special reference to symbiotic proso-branch gastropods. *Pacific Science* 24: 43–54.

Rotjan, R.D. and S.M. Lewis. 2008. Impact of coral predators on tropical reefs. *Marine Ecology-Progress Series* 367: 73–91.

Rotjan, R.D., J.L. Dimond, D.J. Thornhill, J.J. Leichter, B. Helmuth, D.W. Kemp and S.M. Lewis. 2006. Chronic parrotfish grazing impedes coral recovery after bleaching. *Coral Reefs* 25: 361–368.

Sammarco, P.W. and J.C. Coll. 1992. Chemical adaptations in the Octocorallia: evolutionary considerations. *Marine Ecology-Progress Series* 88: 93–104.

Samways, M.J. 2005. Breakdown of butterflyfish (Chaetodontidae) territories associated with the onset of a mass coral bleaching event. *Aquatic Conservation: Marine and Freshwater Ecosystems* 15: S101–S107.
Sano, M. 1989. Feeding habits of Japanese butterflyfishes (Chaetodontidae). *Environmental Biology of Fishes* 25: 195–203.
Sano, M. 2004. Short-term effects of a mass coral bleaching event on a reef fish assemblage at Iriomote Island, Japan. *Fisheries Science* 70: 41–46.
Sano, M., M. Shimizu and Y. Nose. 1984. Food habits of teleostean reef fishes in Okinawa Islands, southern Japan. *University Museum University of Tokyo Bulletin* 25: 1–128.
Sano, M., M. Shimizu and Y. Nose. 1987. Long-term effects of destruction of hermatypic corals by *Acanthaster planci* infestation on reef fish communities at Iromote Island, Japan. *Marine Ecology-Progress Series* 37: 191–199.
Sazima, C. and I. Sazima. 2001. Plankton-feeding aggregation and occasional cleaning by adult butterflyfish, *Chaetodon striatus* (Chaetodontidae), in southwestern Atlantic. *Cybium* 25: 145–151.
Sebens, K.P. 1994. Biodiversity of corals: what are we losing and why? *American Zoologist* 34: 115–133.
Shokri, M.R., S.M.R. Fatemi and M.P. Crosby. 2005. The status of butterflyfishes (Chaetodontidae) in the northern Persian Gulf, I.R. Iran. *Aquatic Conservation: Marine and Freshwater Ecosystems* 15: S91–S99.
Stimson, J.S. 1990. Density dependent recruitment in the reef fish *Chaetodon Miliaris*. *Environmental Biology of Fishes* 29: 1–13.
Sutton, M. 1985. Patterns of spacing in a coral reef fish in two habitats on the Great Barrier Reef. *Animal Behaviour* 33: 1332–1337.
Thresher, R.E. 1983. Environmental correlates of the distribution of planktivorous fishes in the One Tree Reef lagoon. *Marine Ecology-Progress Series* 10: 137–145.
Tricas, T.C. 1985. The economics of foraging in coral feeding butterflyfishes in Hawaii. *Proceedings of the Fifth International Coral Reef Congress, Tahiti, French Polynesia* 5: 409–414.
Tricas, T.C. 1989a. Prey selection by coral-feeding butterflyfishes: strategies to maximize profit. *Environmental Biology of Fishes* 25: 171–185.
Tricas, T.C. 1989b. Determinants of feeding territory size in the corallivorous butterflyfish, *Chaetodon multicinctus*. *Animal Behaviour* 37: 830–841.
Veron, J.E.N. 2000. *Corals of the World*. Australian Institute of Marine Science, Townsville, Australia.
Wild, C., M. Huettel, A. Klueter, S.G. Kremb, M.Y.M. Rasheed and B.B. Jergensen. 2004. Coral mucus functions as an energy carrier and particle trap in the reef ecosystem. *Nature* 428: 66–70.
Williams, D.M. 1986. Temporal variation in the structure of reef slope fish communities (central Great Barrier Reef): short term effects of *Acanthaster planci* infestations. *Marine Ecology-Progress Series* 28: 157–164.
Wilson, S.K., K. Burns and S. Codi. 2001. Sources of dietary lipids in the coral reef blenny *Salarias patzneri*. *Marine Ecology-Progress Series* 222: 291–296.
Wilson, S.K., D.R. Bellwood, J.H. Choat and M.J. Furnas. 2003. Detritus in the epilithic algal matrix and its use by coral reef fishes. *Oceanography and Marine Biology Annual Review* 41: 279–309.
Wilson, S.K., N.A. Graham, M.S. Pratchett, G.P. Jones and N.V.C. Polunin. 2006. Multiple disturbances and the global degradation of coral reefs: are reef fishes at risk or resilient? *Global Change Biology* 12: 2220–2234.
Wrathall, T.J., C.M. Roberts and R.F.G. Ormond. 1992. Territoriality in the butterflyfish *Chaetodon austriacus*. *Environmental Biology of Fishes* 34: 305–308.
Wylie, C.R. and V.J. Paul. 1989. Chemical defenses in three species of *Sinularia* (Coelenterata, Alcyonacea): effects against generalist predators and the butterflyfish *Chaetodon unimaculatus* Bloch. *Journal of Experimental Marine Biology and Ecology* 129: 141–160.

Zekeria, Z.A. and J.J. Videler. 2000. Correlation between the abundance of butterflyfishes and coral communities in the southern Red Sea. *Proceedings of the Ninth International Coral Reef Symposium, Bali, Indonesia* 1: 487–492.

Zekeria, Z.A., Y. Dawit, S. Ghebremedhin, M. Naser and J.J. Videler. 2002. Resource partitioning among four butterflyfish species in the Red Sea. *Marine And Freshwater Research* 53: 163–168.

Zekeria, Z.A., Y. Afeworki and J.J. Videler. 2005. The distribution patterns of Red Sea chaetodontid assemblages. *Aquatic Conservation: Marine and Freshwater Ecosystems* 15: S71–S76.

CHAPTER 6

Feeding Preferences and Dietary Specialisation among Obligate Coral-Feeding Butterflyfishes

Morgan S. Pratchett

INTRODUCTION

Coral reef fishes have long been known to exhibit highly specialised patterns of habitat use (e.g., Fautin, 1985; Munday et al., 1997; Gardiner and Jones, 2005), but are generally considered to be highly versatile in their feeding (Ross, 1986; Bellwood et al., 2006c), opportunistically consuming a wide range of different prey. One striking example of this is the range of fishes (including algal farming damselfishes) that abandon their normal feeding habitats to take advantage of ephemeral, lipid-rich prey, feeding on corals eggs during annual mass-spawning by scleractinian corals (Pratchett et al., 2001; McCormick, 2003). There is however, increasing realisation that coral reef fishes can be extremely specialised when it comes to diet. The broad trophic groups once used to categorise reef fishes (e.g., planktivores, herbivores, carnivores and omnivores) are increasingly being sub-divided to capture important differences in ecosystem function (Graham et al., 2006; Hoey and Bellwood, 2011), while some studies simply recognise

ARC Centre of Excellence for Coral Reef Studies, James Cook University, Townsville, Australia.
E-mail: morgan.pratchett@jcu.edu.au

that individual species have unique and important functional roles (e.g., Bellwood et al., 2003, 2006b). Many species have extremely specialised diets or feeding habits, such as *Holacanthus* angelfishes that feed almost exclusively on sponges, and mainly on just one species (Randall and Hartman, 1968) or tube-lip wrasses (*Labroides* and *Labrichthys*) that consume coral mucous from small wounds that they inflict on specific coral types (Cole et al., 2008).

The extent to which species are specialist or generalist in their use of prey has significant ramifications for their biology, ecology and evolution. Highly specialised species, for example, are expected to be much more constrained by the availability of resources and less able to cope with fluctuations in resource availability compared to generalist counterparts, leading to a greatly increased risk of extinction (Vazquez and Simberloff, 2002). Further, several studies on terrestrial animals have revealed strong, often negative correlations between specialisation and range-size (e.g., Eeley and Foley, 1999), attributed to the limited range of locations that can support species with very specific resource requirements (Brown, 1984; McKinney, 1997). These relationships are assumed to apply equally for both marine and terrestrial species, and are increasingly used in assessing extinction vulnerability among coral reef fishes (e.g., Graham et al., 2011). However, there are few studies that have effectively quantified resource specialisation among marine species, let alone specifically testing ecological ramifications of resource specialisation (Munday, 2001, 2004; Pratchett et al., 2012). More generally, there is surprisingly limited data on the variety of prey resources used by most coral reef fishes (Pratchett et al., 2012).

Butterflyfishes (family Chaetodontidae) are among the most specialised coral reef fishes, accounting for >50% of fishes that are capable of feeding on scleractinian corals (Cole and Pratchett, Chapter 5). Dietary preferences and the extent to which coral-feeding butterflyfishes are more or less specialised in their use of available coral prey have been quantified at a range of locations throughout the Indo-Pacific, from the Red Sea (Zekeria, 2003; Alwany et al., 2003) and Persian Gulf (Pratchett et al., 2013b), the central Indian Ocean (Pratchett et al., 2013a), the Indo-Pacific Archipeligo (Nagelkerken et al., 2009), Australia's Great Barrier Reef (Pratchett, 2005, 2007), and east to Hawaii (Cox, 1994) and French Polynesia (Berumen and Pratchett, 2006a). Each of these studies have undertaken *in situ* feeding observations to establish the proportional use of different coral taxa (mostly, species or genera), enabling interspecific comparisons of dietary composition and ecological specialisation. Field-based feeding observations have been criticised because they cannot discern specific prey items that are actually ingested, only recording the benthic surface from which fishes are seen to be biting (Cole and Pratchett, Chapter 5). However, this is the only current way to get necessary resolution in the specific corals that obligate coral-feeding butterflyfishes consume (e.g., Nagelkerken et al., 2009).

Among coral-feeding butterflyfishes, sympatric species exhibit highly contrasting levels of dietary specialization (Pratchett, 2007). The most specialized species, *Chaetodon trifascialis*, feeds on an extremely limited range (<8 species) of different corals, mostly just one species of *Acropora* (Irons, 1989; Pratchett, 2005), though the predominant coral on which it lives and feeds does vary geographically (Reese, 1981). In the northern Great Barrier Reef (GBR), *C. trifascialis* takes 98% of bites from *Acropora hyacinthus* (Pratchett, 2005). In Chagos, *C. trifascialis* was recorded to take 99.5% of bites from *Acropora cytherea* (Pratchett et al., 2013a). Moreover, *C. trifascialis* appears physiologically incapable of assimilating sufficient energy when restricted to feeding on coral prey (e.g., *Porites* spp.) that they do not consume in the wild (Berumen and Pratchett, 2008). In comparison, *C. lunulatus* has been reported to feed on at least 52 different corals species (Pratchett, 2005). *Chaetodon lunulatus* has also been shown alter patterns of prey use in response to changes in resource availability (Pratchett et al., 2004), though this does impose costs in terms of reduced physiological condition for individuals that consume lesser quantities of preferred prey. In all, there are 24 species of *Chaetodon* butterflyfishes (family Chaetodontidae) that are reported to feed predominantly (≥80%), if not exclusively on scleractinian corals (Table 6.1). Most of these species (22/24 species), all except *C. multicinctus* and *C. unimaculatus*, also come from a single distinct phylogenetic clade (Clade 3—Bellwood et al., 2009), providing a unique opportunity to compare prey preferences and dietary specialisation within and among a distinct group of ecologically equivalent fishes, which are also closely related and have very similar basic biology.

Obligate coral feeding butterflyfishes are defined as those species for which scleractinian corals constitute >80% of the diet based on the proportion of bites or proportional weight of stomach contents, following Cole and Pratchett (Chapter 5). This assumes that species which feed almost exclusively (>80%) on corals will be directly dependent on this source of prey (Pratchett et al., 2008). Reliance on coral prey does, however, vary within and among *Chaetodon* butterflyfishes (Cole and Pratchett, Chapter 5), including some instances where species appear to feed exclusively on scleractinian corals at one location, but not others (Table 6.1). *Chaetodon unimaculatus*, for example, feeds mainly (72.9%) on soft corals on the GBR (Pratchett, 2005), whereas Cox (1994) considered this species to be among the most specialised obligate hard coral feeders in Hawaii. In Hawaii, and also French Polynesia, *C. unimaculatus* feeds almost exclusively on scleractinian corals, and mostly *Montipora* (Table 6.1), possibly because there are few soft corals available at these locations (Berumen and Pratchett, 2006a). There are also several other *Chaetodon* species (e.g., *C. nigropunctatus* and *C. vagabundus*) that have been classified as obligate coral feeders, but quantitative analyses of dietary composition have since revealed that scleractinian corals are only a minor

Table 6.1 Proportional use of major coral genera by obligate coral feeding butterflyfishes, based on the proportion of bites taken from different corals at different geographic locations throughout the Indo-Pacific. The predominant coral on which butterflyfishes were recorded to feed (be it *Acropora*, *Montipora*, *Pocillopora*, *Porites* or Others) is shown in bold.

Species	Location	n	Coral use	*Acropora*	*Montipora*	*Pocillopora*	*Porites*	Others	Reference
C. andamanensis	Sumatra	20	91.1%	11.7%	2.4%	35.4%	0.0%	**50.5%**	Hoey and Pratchett (Unpub.)
C. aureofasciatus	Lizard Island, GBR	35	81.8%	34.2%	4.5%	2.3%	10.3%	**48.3%**	Pratchett (2005)
C. austriacus	Aqaba, Egypt		100.0%	**30.9%**	8.6%	24.9%	26.8%	8.8%	Alwany et al. (2003)
C. baronessa	Lizard Island, GBR	152	99.6%	**83.9%**	0.3%	10.3%	0.4%	5.0%	Pratchett (2005)
	Derawan, Borneo	5	100.0%	**38.0%**	24.0%	6.0%	0.0%	32.0%	Nagelkerken et al. (2009)
C. bennetti	Chagos	9	100.0%	2.9%	0.0%	0.0%	7.2%	**89.9%**	Pratchett et al. (2013a)
	Derawan, Borneo	5	94.4%	0.0%	13.0%	0.0%	0.0%	**87.0%**	Nagelkerken et al. (2009)
	Moorea, Fr Polynesia	4	29.4%	0.0%	0.0%	9.1%	0.0%	**90.9%**	Pratchett and Berumen (Unpub.)
C. guttatissimus	Chagos	26	91.6%	**41.7%**	2.4%	1.3%	17.1%	37.4%	Pratchett et al. (2013a)
C. larvatus	Eritrea	15	90.1%		14.7%		15.8%	69.5%	Zekeria (2003)
C. lunulatus	Christmas Island	30	94.0%	13.1%	11.0%	1.4%	**35.0%**	39.5%	Montanari et al. (2011)
	Lizard Island, GBR	146	97.1%	**59.1%**	8.6%	17.4%	3.9%	10.9%	Pratchett (2005)
	Derawan, Borneo	6	95.2%	28.0%	24.0%	6.0%	8.0%	**34.0%**	Nagelkerken et al. (2009)
	Hawaii	4–19	100.0%		12.1%	1.1%	**54.5%**	32.2%	Cox (1994)
	Moorea, French Polynesia	32	86.9%	0.0%	18.1%	8.8%	**56.2%**	16.8%	Berumen and Pratchett (Unpub.)

Table 6.1 contd....

Table 6.1 contd.

Species	**Location**	**n**	**Coral use**	***Acropora***	***Montipora***	***Pocillopora***	***Porites***	**Others**	**Reference**
C. melapterus	Arabian Peninsula	13	100.0%	9.7%	8.1%	23.9%	**31.9%**	26.3%	Pratchett et al. (2013b)
C. meyeri	Chagos	23	96.9%	**47.9%**	6.0%	12.8%	19.4%	14.0%	Pratchett et al. (2013a)
C. multicinctus	Hawaii	4–19	100.0%		14.6%	16.7%	**47.5%**	21.2%	Cox (1994)
C. octofasciatus	Kimbe Bay, PNG	5	88.1%	10.2%	2.4%	6.1%	33.5%	**47.8%**	Pratchett (Unpub.)
C. ornatissimus	Kimbe Bay, PNG	10	100.0%	18.3%	12.7%	12.0%	**44.3%**	12.8%	Pratchett and Berumen (Unpub.)
	Hawaii	4–19	100.0%		14.1%	7.6%	38.9%	**39.4%**	Cox (1994)
	Moorea, French Polynesia	41	99.7%	5.6%	10.6%	**50.5%**	31.3%	2.0%	Berumen and Pratchett (Unpub.)
C. pelewensis	Moorea, French Polynesia	39	97.0%	10.0%	0.0%	**67.5%**	14.0%	8.5%	Berumen and Pratchett (Unpub.)
C. plebeius	Lizard Island, GBR	105	95.7%	**34.8%**	2.7%	31.6%	3.9%	30.1%	Pratchett (2005)
C. punctatofasciatus	Derawan, Borneo	5	86.8%	29.0%	17.0%	13.0%	11.0%	**30.0%**	Nagelkerken et al. (2009)
C. rainfordi	Lizard Island, GBR	33	93.1%	29.5%	0.8%	17.0%	3.3%	**49.4%**	Pratchett (2005)
C. reticulatus	Moorea, French Polynesia	41	100.0%	26.3%	2.8%	**58.7%**	9.9%	2.4%	Berumenand Pratchett (Unpub.)
C. semilarvatus									
C. triangulum									
C. trifascialis	Aqaba, Egypt		100.0%	**86.1%**	0.0%	13.4%	0.0%	0.5%	Alwany et al. (2003)
	Seychelles	16	100.0	**95.1%**	0.0%	2.4%	0.0%	2.4%	Samways (2005)
	Chagos	28	95.8%	**99.5%**	0.0%	0.4%	0.0%	0.1%	Pratchett et al. (2013a)

	Lizard Island, GBR	71	100.0%	**98.6%**	0.0%	1.3%	0.0%	0.0%	Pratchett (2005)
	Moorea, French Polynesia	26	100.0%	**66.1%**	1.7%	32.1%	0.0%	0.0%	Berumen and Pratchett (Unpub.)
C. trifasciatus	Chagos	26	100.0%	**54.8%**	6.3%	1.9%	25.4%	11.6%	Pratchett et al. (2013a)
	Seychelles	18	100.0%	**97.0%**	0.0%	2.5%	0.0%	0.5%	Samways (2005)
	Christmas Island	34	96.4%	4.3%	24.3%	2.6%	**37.5%**	31.3%	Montanari et al. (2012)
C. unimaculatus	Lizard Island, GBR	31	20.6%	**89.0%**	7.0%	1.2%	0.0%	2.8%	Pratchett (2005)
	Hawaii	4–19	100.0%		**78.3%**	0.0%	0.5%	21.2%	Cox (1994)
	Moorea, French Polynesia	41	97.6%	10.4%	**78.6%**	7.0%	1.6%	2.4%	Berumenand Pratchett (Unpub.)
C. zanzibarensis	Chagos	20	99.3%	10.5%	1.4%	13.2%	14.9%	**60.0%**	Pratchett et al. (2013a)
	Seychelles	10	100.0%	0.0%	0.0%	0.0%	**83.3%**	16.6%	Samways (2005)

component of the diet. In the case of *C. nigropunctatus*, Shokri et al. (2005) correlated abundance with live coral cover and implicitly assumed that this species feeds on live coral. Feeding observations in the southern Persian Gulf and Gulf of Oman show however, that *C. nigropunctatus* is a facultative corallivore, which takes <20% of bites from live corals (Pratchett et al., 2013b). This chapter considers only those species that feed on scleractinian (hard) corals, rather than alcyonarian (soft) corals, because of the increased resolution used when distinguishing scleractinian corals (e.g., Pratchett, 2005). This chapter also does not consider *Heniochus singularis*, which Sano (1989) deemed to be an obligate corallivore, though there have not been any rigorous quantitative studies of diet composition.

PREFERRED PREY

Many studies examining patterns of prey use by coral reef butterflyfishes have not distinguished between different types of prey corals (e.g., Harmelin-Vivien and Bouchon-Navaro, 1981, 1983; Bouchon-Navaro, 1986; Sano, 1989; Pitts, 1991; Zekeria et al., 2002), partly because this is simply not possible when determining dietary composition based on stomach contents. Studies of gut contents are important in distinguishing the broad dietary habitats of different butterflyfishes (e.g., obligate versus facultative coral feeders), but obscure key differences in the behaviour, vulnerability, and ecosystem function of these butterflyfishes. Many early studies for example, explicitly assumed that feeding activities of butterflyfishes were equally apportioned across all different corals species leading to erroneous conclusions about the functional importance of coral feeding butterflyfishes (Cole et al., 2012). Hourigan et al. (1988) suggested that coral feeding butterflyfishes have negligible impact on prey corals because they maintain very large territories and feeding is spread across a large number of individual colonies (see also Robertson 1970; Harmelin-Vivien and Bouchon-Navaro, 1983). However, when considering that chronic high levels of feeding are not evenly apportioned among corals (e.g., Neudecker, 1979; Cole and Pratchett, 2011), coral feeding fishes are certainly capable of having significant effects upon the growth, survivorship and community structure of coral assemblages, especially after major disturbance events (Neudecker, 1979; Cox, 1986; Bellwood et al., 2006a; Cole et al., 2012). Limited resolution used in distinguishing the type of corals consumed by coral-feeding butterflyfishes has also led to reports of very high dietary overlap (often 100%) among coral feeding fishes (e.g., Harmelin-Vivien and Bouchon-Navaro, 1983; Pitts, 1991). Studies that have explored specific patterns of prey use of obligate coral-feeding butterflyfishes (e.g., Reese, 1977; Cox, 1994; Irons, 1989; Tricas, 1989; Pratchett, 2005) have shown that coral-feeding butterflyfishes clearly distinguish among different coral species, and generally use only a very small subset of available corals.

The restricted range of prey types consumed by most butterflyfishes may reflect niche diversification and active partitioning of prey resources, which reduces inter specific competition and promotes coexistence among sympatric species (*sensu* Schoener, 1974). Cox (1994) showed that patterns of resource use for butterflyfishes in the field were different from those observed in the laboratory (in the absence of con-generics), suggesting that actual patterns of prey use may be constrained by inter-specific competition (see also Hourigan et al., 1988). Further, Cox (1994) showed that the two most abundant coral-feeding butterflyfishes in Kaneohe Bay (*C. unimaculatus* and *C. lunulatus*) each fed on different coral species (*Montipora verrucosa* and *Porites compressa*, respectively) and thus, had very low levels of dietary overlap (<3.2%). However, at other locations in Hawaii, Cox (1994) showed that pair-wise dietary overlap among sympatric coral-feeding butterflyfishes was often very high (up to 93.9%), and well above levels (40%) expected for groups of species where coexistence is facilitated by resource partitioning (Ross, 1986). Similarly, Pratchett (2005) revealed 60–73% dietary overlap among sympatric coral feeding butterflyfishes at Lizard Island, in the northern GBR, with the greatest overlap being recorded between *C. triascialis* and *C. baronessa*. Both of these species feed predominantly on *A. hyacinthus* and co-occur in exposed reef crest habitats with high cover of this coral (Pratchett, 2005). Moreover, there is conspicuous evidence of direct aggression among coral-feeding butterflyfishes, and the highest incidence of competition was among obligate coral-feeding species with greatest levels of dietary overlap (Berumen and Pratchett, 2006b). These findings suggest that inter-specific competition, while potentially important, has not resulted in strong dietary partitioning among all coral-feeding butterflyfishes. This might be because coral prey is not limiting (Cox, 1994), or other factors, such as spatial or temporal separation of feeding activities, have reduced direct competition or limited encounters between ecologically equivalent species (Anderson et al., 1981; Zekeria et al., 2002).

Resource partitioning has long been viewed as one of the key diversifying processes in ecological communities (Hutchinson, 1959; MacArthur, 1972; Schoener, 1974), leading to contractions in the range of resources used by individual species to minimise resource overlap among potential competitors. The diversity of prey corals and habitat types within most reef environments provides considerable opportunity for resource partitioning among coral-feeding butterflyfishes (Anderson et al., 1981). However, an alternative explanation for differences in the utilisation of available resources relates to inter specific variation in nutritional requirements (e.g., Ebeling and Laur, 1986; Schpigel and Fishelson, 1989), whereby individual species evolve independently to use the minimum set of prey resources that maximise energetic intake and/or satisfy overall nutritional requirements. Specific adaptations may lead to marked differences in feeding preferences among sympatric species, but it is also possible that sympatric species will exhibit highly convergent patterns of resource use (Ebeling and Laur, 1986), especially if certain prey have very high energetic value. Resource partitioning and individualistic responses are not necessarily mutually exclusive, but do provide contrasting predictions about the degree to which patterns of resource use will actually overlap among sympatric species. This can be readily explored given the large body of information about species-specific patterns of coral use by butterflyfishes at a range of different locations (Table 6.1).

Interspecific Differences in Prey use

Despite generally high levels of dietary overlap among sympatric coral-feeding butterflyfishes, detailed studies of dietary composition do reveal important inter specific differences in patterns of prey use (e.g., Pratchett, 2005), based on differences in either i) the preferred prey coral (Pratchett et al., 2013a), or ii) the overall variety or range of prey corals that are used (Pratchett, 2005). At Lizard Island, Pratchett (2005) showed highly convergent patterns of prey preferences for six species of obligate coral feeding butterflyfishes (*C. aureofasciatus*, *C. baronessa*, *C. lunulatus*, *C. plebeius*, *C. rainfordi*, and *C. trifascialis*) with all six species feeding predominantly on *A. hyacinthus* or *Pocillopora damicornis*, yet differences emerged based on the extent to which these butterflyfishes consumed other seemingly less preferred coral prey (Fig. 6.1A). *Chaetodon trifascialis* and *C. baronessa* fed almost exclusively on *A. hyacinthus*, taking 88% and 63% of bites from this one coral species, respectively. When not feeding on *A. hyacinthus*, *C. baronessa* fed mostly on *P. damicornisi*, but also took a significant proportion of bites (~3%) from *Galaxea* and encrusting *Montipora*. *Chaetodon lunulatus*, *C. plebeius*, and *C. rainfordi* also took a significant proportion of bites, and exhibited significant preference for, *P. damicornis* (Pratchett, 2007). *Chaetodon aureofasciatus* was differentiated from other butterflyfishes due to relatively heavy use of *Echinopora* and *Isopora*. In contrast, *C. lunulatus* was the only butterflyfish to take a significant proportion of bites from *Porites*.

Highly convergent patterns of feeding preferences among coral-feeding butterflyfishes reported by Pratchett (2005) are completely at odds with predictions of resource partitioning. However, this does not necessarily mean that competition among coral-feeding butterflyfishes is not important. Pratchett (2005) suggested that the reason why *C. trifascialis* and *C. baronessa* consume such a high proportion of *A. hyacinthus* is because they are competitively dominant and can therefore, partially exclude other

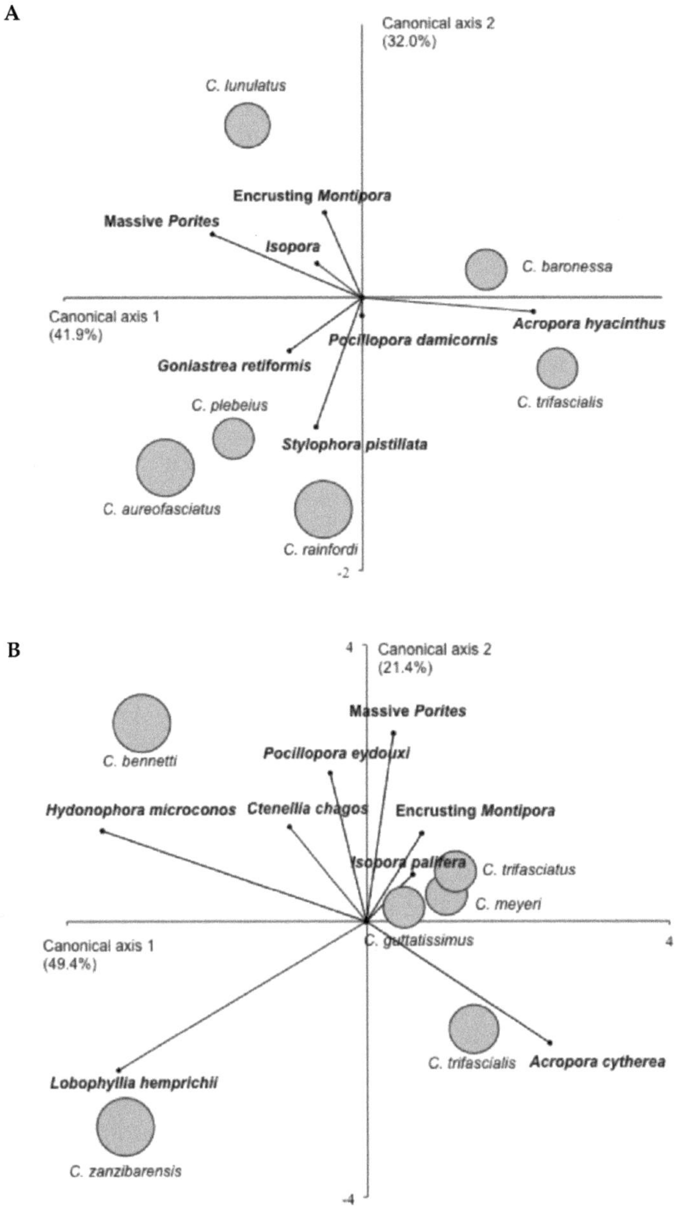

Fig. 6.1 Canonical discriminant analysis (CDA) showing interspecific variation in the dietary composition among sympatric coral-feeding butterflyfishes (6 species) at A) Lizard Island, northern GBR, and B) Chagos, central Indian Ocean. Ellipses represent 95% confidence intervals around the group centroid for each species of butterflyfishes, and vary in size depending on sampling intensity. Structural coefficients, displayed as vectors, are shown for only the most significant response variables.

subordinate species from feeding on this preferred prey. Given limited access to *A. hyacinthus* and also *P. damicornis*, Pratchett (2005) suggested that subordinate butterflyfishes are forced to feed on other suboptimal prey (e.g., *Porites*) to satisfy their nutritional requirements. In support of this hypothesis, marked reductions in the physiological condition of *C. lunulatus* have been documented when their intake of *Acropora* (specifically, *A. hyacinthus*) has been reduced (Pratchett et al., 2004; Berumen et al., 2005). However, observed differences in prey use and physiological condition were largely attributable to differences in overall resource availability, and it is unknown to what extent interspecific competition actually constrains patterns of prey use among these fishes. This would need to be examined using controlled removals of the dominant competitors, testing whether subordinate species (e.g., *C. lunulatus*) increase their intake of *A. hyacinthus* and *P. damicornis* in the absence of *C. trifascialis* and/or *C. baronessa*. An alternative is that these different butterflyfishes are fundamentally dependent upon different types of corals, and that *C. lunulatus* might continue to exploit a wide range of different prey even if it had greater access to *A. hyacinthus*.

A significant demonstration on the importance of competition in moderating prey preferences of coral-feeding butterflyfishes comes from direct comparisons of prey use between laboratory studies (where competitors are absent) versus field conditions. In Hawaii, controlled feeding trials have been undertaken for four species of corallivores under both laboratory and field conditions (Reese, 1977; Hourigan et al., 1988; Cox, 1994). Three (out of 4) species, *C. lunulatus*, *C. multicinctus*, and *C. ornatissimus* exhibit exactly the same order of preferences among common coral species (e.g., *Pocillopora meandrina*>*Montipora verrucosa*>*Porites lobata*), during laboratory trials (Reese, 1977; Hourigan et al., 1988). In the field however, only *C. multicinctus* continued to feed preferentially on *P. meandrina*, while other species (*C. lunulatus* and *C. ornatissimus*) feed mainly on non-preferred corals, presumably to avoid competition with con-generics. *Chaetodon unimaculatus* however, has an altogether different feeding preference to the other sympatric species, feeding preferentially on *Montipora* sp. regardless of the presence of potential competitors (Cox, 1994). These data show that competition may moderate field-based use of different coral prey. Clearly, there are some strong and consistent feeding preferences across a range of different butterflyfishes, but there are also some butterflyfishes (e.g., *C. unimaculatus*) that have feeding preferences that differ markedly from other sympatric species.

In Chagos, four (out of 6) species of obligate corallivores (*C. guttatissimus*, *C. meyeri*, *C. trifascialis*, and *C. trifasciatus*) had similar prey preferences, feeding predominantly (41.7–99.5%) on *Acropora* (Pratchett et al., 2013a). However, two other locally abundant coral-feeding butterflyfishes

(*C. bennetti* and *C. zanzibarensis*) had a markedly different dietary composition (Fig. 6.1B). *Chaetodon zanzibarensis* consumed the widest range of different coral prey (>7genera per 100 bites) of corallivorous butterflyfishes in Chagos, and consumed some coral genera (e.g., *Hydonophora* and *Galaxea*) that were not seen to be eaten by any other corallivores. *Chaetodon bennetti* fed predominantly on *Lobophyllia*, and only rarely consumed *Acropora*, even though it was very abundant at most locations. Similarly in Borneo, Nagelkerken et al. (2009) showed that *C. bennetti* did not consume the widely available coral prey (e.g., *Acropora* and *Pocillopora*) commonly consumed by most other coral-feeding butterflyfishes, but sought out relatively rare, generally massive or encrusting corals, such as *Cyphastrea*, *Favia*, and *Madracis*.

Most of the obligate coral feeding butterflyfishes (14 out of 22 species) for which field-based studies of dietary composition have been undertaken (Table 6.1), feed predominantly (if not preferentially) on *Acropora*, *Montipora*, *Pocillopora*, or *Porites*. This is perhaps not surprising given that one or other of these coral genera tend to be dominant on reefs throughout the Indo-Pacific, which is why it is important to test whether corals are being used disproportionately to their abundance (e.g., Pratchett, 2007). Five (out of 22) species (*C. austriacus*, *C. guttatissimus*, *C. meyeri*, *C. plebeius*, and *C. trifascialis*) fed mainly on *Acropora* corals across the full range of locations where quantitative patterns of prey use have been documented (Table 6.1). Also, *C. triangulum* (the Indian ocean sister species to *C. baronessa*) almost certainly fits within this group, but there is no published data on its actual feeding habitats. Caution must be taken in assuming feeding habitats based purely on phylogenetic affinities, but other recently diverged and almost identical sister species (e.g., *C. austriacus*, *C. trifasciatus*, *C. melapterus* and *C. lunulatus*—Plate 6.1) tend to show very comparable patterns of prey use. At Christmas Island, for example, where *C. lunulatus* and *C. trifasciatus* co-occur and even form mixed species pairs, relative use of different coral genera was virtually identical (Montanari et al., 2012).

Other species (e.g., *C. lunulatus*, *C. trifasciatus* and *C. unimaculatus*) feed predominantly on *Acropora* at some, but not all locations (Table 6.1), presumably because of geographical variation in the availability of *Acropora* or inter-specific competition (discussed later). Another distinct group of butterflyfishes (*C. pelewensis* and *C. reticulatus*) feed predominantly on *Pocillopora*, while the final group (*C. andamanensis*, *C. aureofasciatus*, *C. bennetti*, *C. punctatofasciatus*, *C. rainfordi* and *C. zanzibarensis*) either feed on a wide range of different corals, or target corals other than *Acropora*, *Montipora*, *Pocillopora*, and *Porites* (Table 6.1). In Sumatara, *C. andamanensis* feed predominantly on *Goniastrea*, but did take a significant proportion of bites from *Acropora* and *Pocillopora*, especially given that these corals were generally rare (Hoey and Pratchett, unpublished data).

Plate 6.1 Closely related *Chaetodon* species with strong similarities in their appearance are also similar in terms of dietary composition and specialization. The four species shown are a) *C. lunulatus*, from the Pacific, b) *C. trifasciatus*, from the Indian Ocean, c) *C. austriacus*, found predominantly within the Red Sea, and d) *C. melapterus*, from the north-west Inidian Ocean, including the Red Sea and Persian Gulf. All photos by Luiz Rocha.

Colour image of this figure appears in the colour plate section at the end of the book.

Chaetodon aureofasciatus, *C. punctatofasciatus* and *C. rainfordi* take a high proportion of bites (approximately one third) from *Acropora*, but also consume a wide range of other coral prey, including *Goniastrea*, *Galaxea*, and *Pachyseris*. *Chaetodon bennetti* and *C. zanzibarensis* are unusual in that they appear to be the only coral-feeding butterflyfishes that actively avoid *Acropora* corals (Table 6.1).

Where it is present and abundant, *Acropora* is undoubtedly the most commonly used genus of corals by coral-feeding butterflyfishes. At Lizard Island, for example, 64% of bites recorded during 1,272 independent 3-minute feeding observations across all *Chaetodon* butterflyfishes were from *Acropora* (Fig. 6.2), and the total number of bites taken from *Acropora* (10,184 bites) was 4 times that of the next most frequently consumed coral *Pocillopora* (2,295 bites). Another way to look at this is to actually quantify bite rates on individual colonies of different corals (Cole and Pratchett, 2011), which takes in to account marked differences in their local abundance.

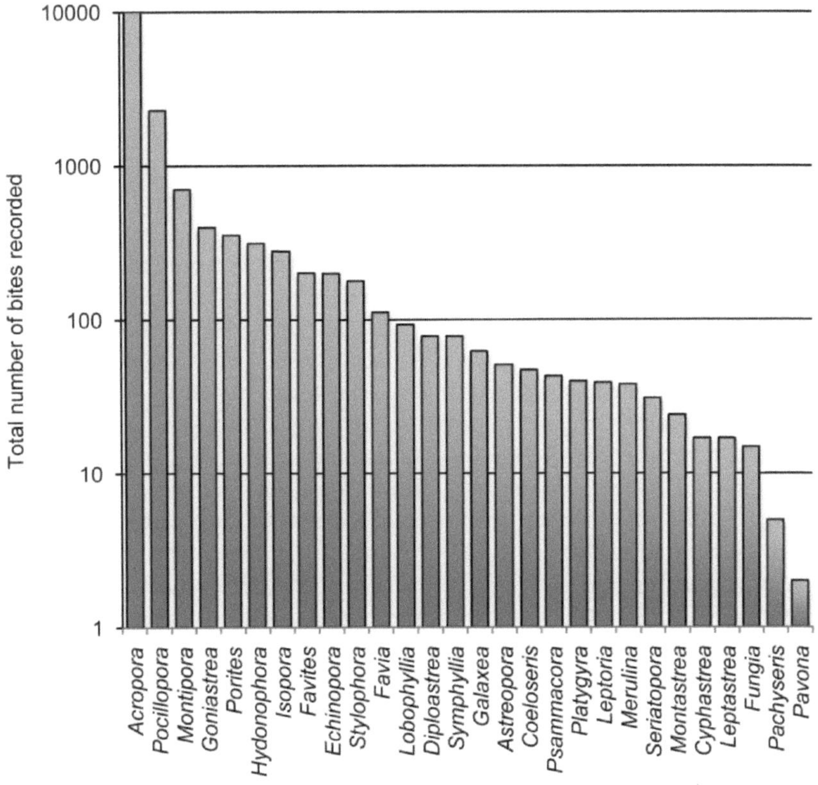

Fig. 6.2 Total number of bites recorded for different coral genera based on 1,272 feeding distinct feeding observations (3-minutes) across all *Chaetodon* butterflyfishes (20 species) at Lizard Island, northern GBR. The vertical axis is on a log scale, so there is a 4-fold difference in the number of bites taken from *Acropora*, compared to the next most frequently consumed coral, *Pocillopora*.

It might be the case, for example, that feeding is distributed among corals in approximate accordance with their local abundance. Certainly, the predominant corals used by most coral-feeding butterflyfishes (e.g., *Acropora*, *Montipora*, *Pocillopora* and *Porites*) are often dominant in shallow reef environments. However, Cole and Pratchett (2011) showed that the average rate of feeding on individual colonies of *A. hyacinthus* (44.7–50.1 bites per hour) was significantly higher than feeding rates on any other species, including *P. damicornis* (25.2–31.5 bites per hour).

Intraspecific Differences in Prey use

Resource selection and specialisation are often considered to be species-level attributes, but localised patterns of resource use can be constrained by a number of factors (e.g., prey availability and competition) that lead to marked intraspecific variation in dietary composition (e.g., Stachowicz and Hay, 2000). For highly specialised species, patterns of resource use are expected to be invariant, such that specialist species can only exist in habitats and locations that provide adequate access to necessary resources. Indeed, highly conserved patterns of prey use across a wide range of locations or along marked resource gradients is one indicator of highly specialised species (Devictor et al., 2010). Conversely, when species are shown to use only a very restricted range of resources at a single time and location, this may be due to inherent prey preferences, high abundance of a particular prey type, or strong constraints on prey use due to environmental factors or biological interactions (Fox and Morrow, 1981). In San Blas, for example, two species of toadfishes (*Amphichthy scryptocentrus* and *Sanopus barbatus*) had been reported to feed almost exclusively on the sea urchin *Diadema antillarum*, and expected therefore, to decline in abundance following mass mortality of *Diadema* in 1983–84 (Robertson, 1987). However, Robertson (1987) actually saw marked changes in the diets of these toadfishes following declines in availability of their previously dominant prey, and warned that one must be cautious when interpreting narrow diets. Still however, very few studies have explicitly tested for variation in resource use along gradients of resource availability (but see Lawton et al., 2012a).

Dietary preferences and specialisation can only really be established through manipulative experiments, testing an individual's use and performance across a range of different resources, independent of vagaries in resource access (Devictor et al., 2010). However, these experiments are very involved and rarely consider the full range of possible resources. For example, Berumen and Pratchett (2008) undertook experimental tests of individual performance (specifically, juvenile growth) for two species of *Chaetodon* butterflyfishes (discussed later), but only looked at three alternative coral species. Testing for intra-specific variation in dietary composition across broad geographic locations with known differences in resource availability may therefore, be the best way to assess the extent to which patterns of prey use are constrained by extrinsic (mostly prey availability and competition for resources) or intrinsic factors (specific adaptations or individualistic feeding preferences).

Direct comparisons of dietary composition across geographically distinct locations have been conducted for four species of *Chaetodon* butterflyfishes, including two species of obligate corallivore, *C. lunulatus* and *C. trifascialis* (Lawton et al., 2012a). Lawton et al. (2012a) quantified dietary composition for at least 60 individuals (using 3-minute *in situ* feeding observations) of each species at each of five geographically distinct locations throughout the Pacific; Papua New Guinea, northern GBR (Lizard Island), southern GBR (Heron Island), Noumea and French Polynesia. For

C. lunulatus, dietary composition varied greatly among locations, with low rates of feeding on *Acropora* at Papua New Guinea, and especially French Polynesia (Fig. 6.3). Conversely, *C. lunulatus* took more bites on *Porites* at

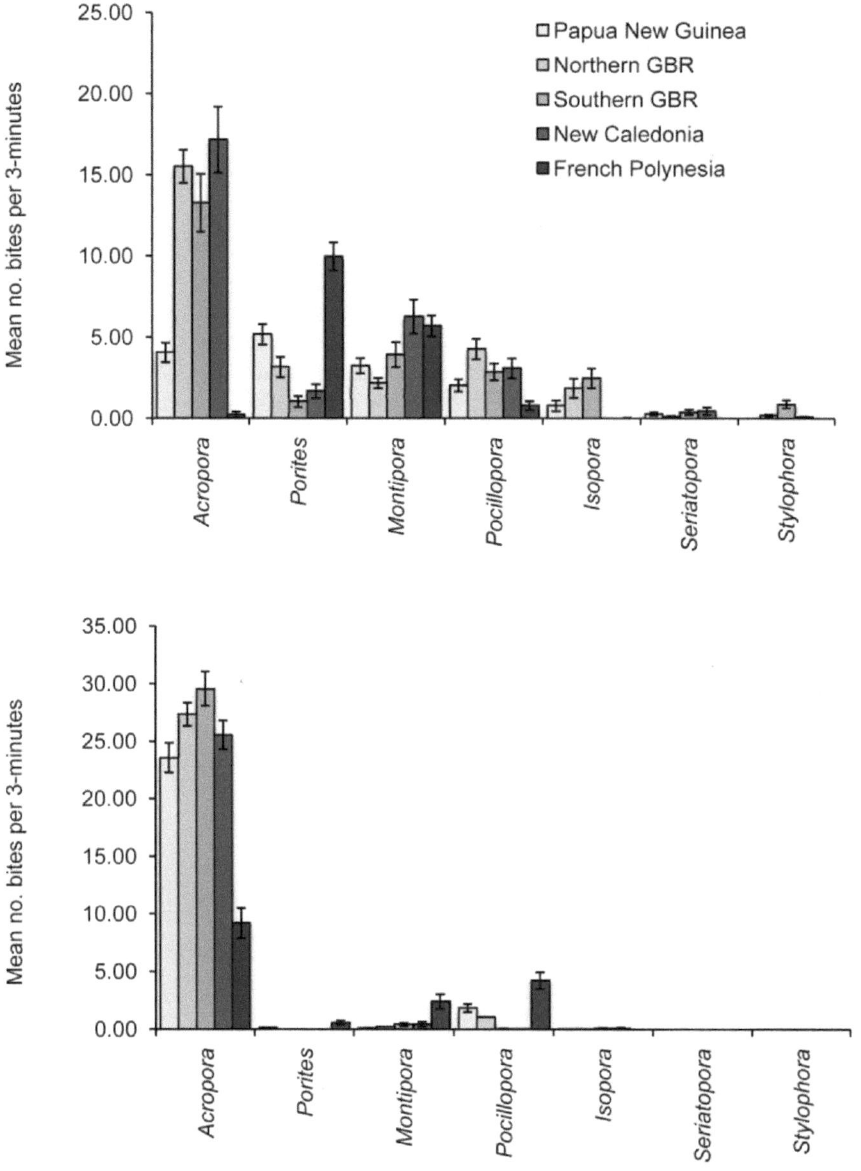

Fig. 6.3 Geographical variation in dietary composition of *Chaetodon lunulatus* (top), and *Chaetodon trifascialis* (bottom). Redrawn based on data presented in Lawton et al. (2011).

both these locations. For *C. trifascialis*, dietary composition varied very little among the five locations (Lawton et al., 2012a). With the exception of French Polynesia, over 92% of bites were taken from *Acropora* in each location. In French Polynesia, only 56.6% of bites were taken from *Acropora*, but this reflected the extremely low availability (<2%) of *Acropora* at this location. Interestingly, *C. trifascialis* still managed to maintain high levels of feeding on the few remaining *Acropora* colonies, whereas *C. lunulatus* did not consume any *Acopora* at French Polynesia. To compensate for low availability of *Acropora* in French Polynesia, *C. trifascialis* also fed on *Montipora* and *Pocillopora*, but overall bite rates (16.42 bites per 3-minutes) were 40% lower than at all other locations (Fig. 6.3).

Rigorous comparisons of prey use among geographic locations (e.g., Lawton et al., 2012a) need to be undertaken for a much greater range of species. By comparing published data (Table 6.1) it appears that (in addition to *C. lunulatus*) marked geographical differences in prey preferences may also exist for *C. baronessa*, *C. ornatissimus*, *C. trifasciatus*, *C. unimaculatus*, and *C. zanzibarensis*. *Chaetodon baronessa* does feed predominantly on *Acropora* in both locations studied, but very high levels of feeding on *Acropora* (and especially, *A. hyacinthus*) reported at Lizard Island (Pratchett, 2005) are not apparent in Borneo (Nagelkerken et al., 2009). Similarly, Berumen et al. (2005) showed that relative use of *A. hyacinthus* and other *Acropora* by *C. baronessa* varies even at local scales, among reef habitats with different coral assemblages. Diet composition for *C. trifasciatus* varies markedly with differences in the local abundance of corals, and is one of the few butterflyfishes (along with other closely related sister species—*C. austriacus*, *C. lunulatus*, *C. melapterus*) that will take a high proportion of bites from massive *Porites* when locally abundant (Montanari et al., 2012). *Chaetodon ornatissimus* also takes a high proportion (30–45%) of bites from *Porites*, but the other major dietary components vary regionally from *Acropora* on the GBR, *Pocillopora* in French Polynesia, to *Pavona varians* in Hawaii (Table 6.1).

The two species that do show consistent patterns of prey use (for coral genera) across different studies are *C. trifascialis* and *C. bennetti*. As shown by Lawton et al. (2012a) for the Pacific, prey preferences for *C. trifascialis* (at least at the level of coral genera) appear highly conserved across the entire Indo-Pacific (Table 6.1). Despite its broad geographic range throughout the entire Indo-west Pacific, *C. trifascialis* is always seen to feed mainly, if not exclusively on *Acropora* corals (Reese, 1981; Irons, 1989; Tricas, 1989; Graham, 2007; references in Table 6.1). Even where there has been significant coral loss (e.g., after catastrophic bleaching), *C. trifascialis* persists mainly in areas with remnant colonies of *Acropora* and continues to feed mainly on *Acropora* (Samways, 2005; Berumen and Pratchett, 2006a; Graham, 2007). When *Acropora* is extremely scarce, *C. trifascialis* does appear to increase its intake of *Pocillopora* (e.g., Graham, 2007), but *Acropora* remains the major component of the diet. The relative use of different corals also seems to be consistent among studies and locations for *C. bennetti* (Table 6.1), but only in so much that it tends to avoid *Acropora*, *Montipora*, *Pocillopora* and *Porites*; if we consider the specific coral types used by *C. bennetti* at each location, regional differences in prey preferences do emerge. In Moorea, for example, *C. bennetti* consumed *Leptastrea*, *Montastrea*, *Pavona* and *Pocillopora* (Berumen and Pratchett, unpublished data), whereas Nagelkerken et al. (2009) reported it to feed mainly on *Cyphastrea*, *Physogyra* and *Montipora*. Data used to assess inter- and intraspecific differences in prey preferences (Table 6.1) are very crude, based solely on the proportional use of just the four main genera. While more rigorous comparisons of the published data may be possible, any results might be confounded by inconsistencies in the sampling methodologies used between studies. Importantly, Lawton et al. (2012a) used a standardised sampling method at all locations (including highly resolved measures of prey use and prey availability), and also tested for variation within and among locations by sampling at three distinct sites per location.

Energetic Value of Preferred Corals

An animal's choice of prey, along with the rate of feeding, regulates the energy available to all other activities (e.g., maintenance, growth, and reproduction) thereby affecting their survival and overall fitness. It is expected therefore, that foraging strategies that yield the greatest energetic intake (and thereby maximise individual fitness) will be favoured (MacArthur and Pianka, 1966; Emlen, 1968; Schoener, 1971). Tests of optimal foraging theory on butterflyfishes have shown that prey selection may be correlated with several different measures of food quality, such as protein content or caloric content (Tricas, 1985, 1989, but see Graham, 2007). However, selection of profitable prey may also be moderated by several other factors, including chemical (mostly, secondary metabolites) or physical (e.g., nematocysts or tissue depth) defence mechanisms of the prey. Among herbivorous fishes, for example, the consumption of different macroalgae is moderated by the presence of toxic compounds (e.g., Lobel and Ogden, 1981).

While numerous studies have documented distinct patterns of prey use among coral-feeding and non-coral feeding butterflyfishes (e.g., Birkeland and Neudecker, 1981; Gore, 1984; Lasker, 1985; Alino et al., 1988; Irons, 1989; Tricas, 1989), few have provided satisfactory explanations for the observed patterns of prey preference (but see Tricas, 1989; Alino et al., 1988). As such, the underlying basis of feeding preferences remains largely unclear (Hourigan et al., 1988; Pratchett, 2007). Tricas (1989) related the energetic content (caloric yield) of three corals (*Pocillopora meandrina*, *Porites lobata* and *Porites compressa*) to prey selection by *C. multicinctus* and did reveal a positive relationship. However, feeding preferences were also inversely related to the density of nematocysts within the each of the three prey corals (Tricas, 1989). Subsequent studies across a broader range of butterflyfishes have focussed either on the nutritional content of prey corals (e.g, Carbon-Nitrogen ratios: Graham, 2007) or feeding deterrents (e.g., Alino et al., 1988; Gochfeld, 2004), but have not revealed consistent differences between preferred versus non-preferred corals.

Highly convergent feeding preferences are strongly suggestive of marked differences in the food value of corals (Ebeling and Laur, 1986). Based on findings to date, *Acropora* and *Pocillopora*, would be expected to provide butterflyfishes with the highest energetic return (e.g., Graham, 2007). However, Graham (2007) found that *Pocillopora* was the least nutritious of different corals tested from Chagos. Moreover, the nutritional value of *Acropora* was no different to that of *Lobophyllia* or *Porites*, which are among the least preferred by most coral feeding butterflyfishes. An alternative explanation to potentially explain differences in prey use by coral-feeding butterflyfishes is that the rate of feeding (specifically, the amount of tissue removed per bite) differs among coral types (Cole et al., 2011), based on the tissue depth or specific colony morphology. Physical constraints on feeding have been proposed previously to explain differences in feeding preferences of butterflyfishes (e.g., Hourigan et al., 1988; Tricas, 1989), but it has never been tested whether a given butterflyfish can feed more efficiently from one coral over another. Cole et al. (2011) showed that the amount of tissue consumed per bite varied with size and species of butterflyfish, but didn't explicitly test for differences in the amount of tissue removed from each coral. In this study, fishes were collected from the field and maintained in aquaria to quantify bite rates on a specific coral and then related the number of bites taken to the weight of stomach contents (after controlling for the weight of gastric juices), to estimate the average weight (mg) of coral tissue consumed per bite (Cole et al., 2011). These trials were conducted with a total of 20 fishes from each of four different species, *C. baronessa*, *C. lunulatus*, *C. plebeius*, and *C. aureofasciatus*. Half of the fishes (n = 10 per species) were allowed to feed on *A. hyacinthus*, while the other half were provided with *P. damicornis*. This data was re-analysed here to specifically test for differences in feeding efficiency between corals (Fig. 6.4), and revealed a difference for *C. plebeius*, but not any of the other butterflyfishes. These findings are consistent with observed patterns of prey preferences, whereby *C. plebeius* exhibits the strongest preference for *P. damicornis* (Pratchett, 2007). Moreover, *C. plebeius* is the only butterflyfish that prefers *P. damicornis* over *A. hyacinthus* (Pratchett, 2007), and differences in feeding efficiency may be even more apparent when looking at generally

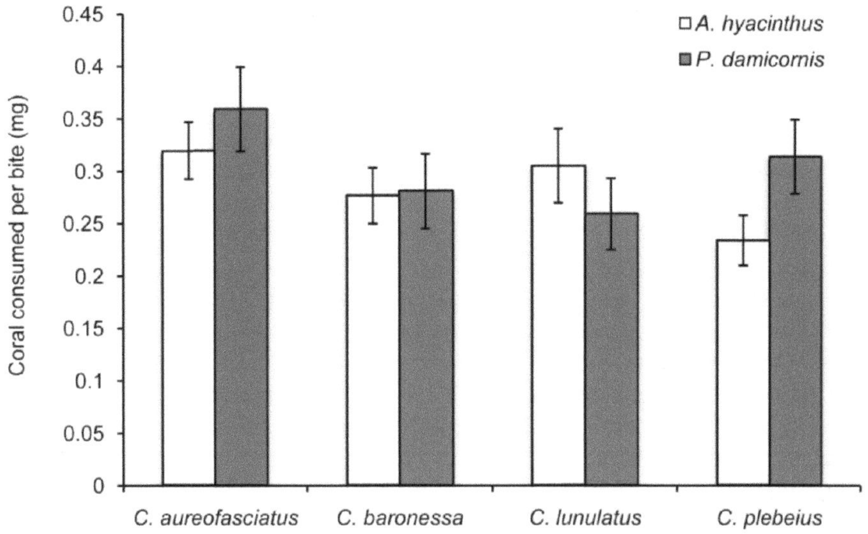

Fig. 6.4 Amount of coral tissue (mg) removed per bite by four species of butterflyfish when feeding on either *Acropora hyacinthus* or *Pocillopora damicornis*. Coral consumption was calculated based on the net weight of stomach contents after individual fish (n = 10) were allowed to take 400–600 bites on one or other corals in aquaria. All fishes were starved overnight to ensure the stomachs were empty prior to experiment (Cole et al., 2011).

non-preferred corals (e.g., *Porites*). Limitations to rates of ingestion on live corals may also explain why butterflyfishes (both obligate and facultative corallivores) feed voraciously on colonies that have been recently injured or damaged to expose coral tissue (e.g., Pratchett, 2005). Variation in feeding efficiencies among butterflyfishes for the same corals may also account for interspecific differences in prey preferences (e.g., Cox, 1994; Pratchett, 2005). Ideally, estimates of feeding efficiency or rates of ingestion should be combined with prey quality, to better reflect the nutritional value of different corals as experienced by the fishes themselves.

ECOLOGICAL SPECIALISATION

Ecological specialisation is a relative (and often highly subjective) term, used to describe generally narrower ranges in the extent of resources used by organisms (Futuyma and Moreno, 1988), which may be applied at the scale of individuals, populations or species (Devictor et al., 2010). There is considerable ongoing scientific interest surrounding specialisation, focusing on the role of specialisation in diversification and speciation (e.g., Salisbury et al., 2012), the evolutionary costs associated with generalist versus specialist habits (Colles et al., 2009), and macroecological consequences of ecological specialisation (e.g., Williams et al., 2009; Belmaker et al.,

2012). Accordingly, there have also been several recent reviews looking at refining and explicitly quantifying what is meant by specialisation (e.g., Devictor et al., 2010; Poisot et al., 2012). There are multiple definitions in common use, and also a number of different typologies used to account for differences in the evolution, underlying basis and/or rigidity of specialisation (Ferry-Graham et al., 2002). The most important distinction is whether specialisation is a mechanistic or otherwise evolved trait (Ferry-Graham et al., 2002), which would therefore, be fixed regardless of spatial or temporal variation in availability of alternative resources. These species are termed fundamental specialists (Devictor et al., 2010). The alternative is that seemingly specialist species (realised specialists) are actually capable of feeding on a much wider range of different species, but are ecologically constrained to feeding on a restricted set of prey (due to either intrinsic or extrinsic factors). Further, Devictor et al. (2010) point out that specialisation can be viewed both in terms of the impact that an organism has on the resources it exploits and on the broader environment (*Eltonian* specialisation, after Elton, 1927) or the benefit derived from using specific resources (*Grinnellian specialisation*, after Grinnell, 1917). There is little doubt that coral-feeding butterflyfishes do impact on the corals upon which they feed (Cox, Chapter 7). Aside from direct tissue removal for the coral, butterflyfishes may influence the outcome of competitive interactions among corals (Cox, 1986) and also influence the spread and/or progression of coral diseases (Cole et al., 2009; Raymundo et al., 2009), and these effects depend on the level of dietary specialisation. However, the primary focus of this chapter is on the butterflyfishes themselves, comparing the realised niches of sympatric species and exploring macroecological consequences (or correlates) of specialisation among obligate coral-feeding butterflyfishes (*Grinnellian* specialisation).

Ecological specialisation is determined first and foremost by the variety of resources that a population uses (e.g., resource width, Roughgarden, 1972), but it is also important to consider the proportional use of different resources. Furthermore, estimates of specialisation should really take account of the relative availability of different resources (Devictor et al., 2010). There are many metrics use to explicitly quantify or infer variation in ecological and realised specialisation, ranging from straight forward accounts of the variety of resources used by a population or species (e.g., Eeeley and Foley, 1999), to diversity indices applied to resource use (e.g., Munday, 2004), and complex measures of specialisation that account for proportional use and availability of different resources (e.g., the χ^2 log likelihood statistic χ^{L22}, Manly et al., 2002). There are also a number of proxies being used to infer (rather than directly measure) specialisation, such as variation in resource use among individuals from the same population or species (Araujo and Conzaga, 2007) or the co-occurrence of specific species

(Fridely et al., 2007). For most part, the choice of method(s) depends on data availability (Devictor et al., 2010), and depending on the specific research question, different metrics can often yield similar results, at least in terms of relative specialisation among populations and species (Lawton et al., 2012b). Strictly speaking, *Grinnellian* specialisation should be measured based on the fitness benefits arising from using different resources, rather than simply documenting relative use of prey (Devictor et al., 2010), but this is rarely done.

Obligate hard coral feeding butterflyfishes all exhibit high levels of dietary specialisation. In the first instance, there does not seem to be any obvious constraint (morphological or otherwise) that would prevent these species from feeding on other abundant sources of prey. Motta (1988) for example, showed that morphological similarities among *Chaetodon* butterflyfishes poorly reflected similarities in diet or feeding behaviour. There are extreme examples of morphological specialisation among butterflyfishes, but the morphology and feeding mode by *Chaetodon* species is not obviously different between corallivores and non-corallivores (Konow and Ferry-Graham, Chapter 2). Corallivorous species have been shown to have longer digestive tracts (Berumen et al., 2011) compared to non-coral feeders, presumably to maximise energetic gain from coral prey, but this should not preclude them from eating other sessile organisms (e.g., soft corals, clams, polychaetes) consumed by congenerics. Yet coral-feeding butterflyfishes feed almost exclusively on, or from the surface of, scleractinian corals. With the exception of *C. unimaculatus*, obligate coral feeding butterflyfishes consistently take close to 100% of bites from the surface of scleractinian corals, even after extensive local depletion of coral prey (Pratchett et al., 2006). It is possible that when these butterflyfishes bite from the surface of corals they sometimes consume non-coral prey (Nagelkerken et al., 2009). In Borneo for example, *C. punctatofasciatus* took 86.9% of bites from the surface of corals, but small crustaceans dominated stomach contents. Based on analyses of stable isotopes, *C. baronessa* also appears to ingest annelid worms during the course of coral feeding (Nagelkerken et al., 2009), but this requires further investigation.

Despite the wide range of scleractinian corals that are available, obligate coral-feeding butterflyfishes exhibit very high levels of dietary specialisation, either consuming a very restricted range of different coral prey (e.g., Irons, 1989; Tricas, 1989; Pratchett, 2005) or using different corals disproportionately to their availability (Cox, 1994; Pratchett, 2007; Lawton et al., 2012a). The more important and challenging question, however, is how does specialisation vary among obligate coral-feeding butterflyfishes? There have now been several different studies that quantify and compare dietary specialisation among distinct suites of sympatric butterflyfishes, often using recommended metrics that incorporate proportional use of different prey as well as the local abundance of these prey (Pratchett, 2007; Lawton et al., 2012a). However, these metrics are not directly comparable among studies because they are very sensitive to the categorisation of available resources, which varies among studies depending on the local diversity of corals and taxonomic resolution used in distinguishing prey. Three metrics (cumulative number of coral genera, mean number of genera per feeding observation, and Shannon-Wiener diversity) were selected therefore, to compare ecological specialisation across the full range of obligate coral-feeding butterflyfishes (22 species), for which there is information on dietary composition (Table 6.1). Firstly, the cumulative number of coral genera consumed by each species of butterflyfish was calculated based on a comprehensive search of all published records. This provides a simple measure of realised niche width, and genera (as opposed to species) were used simply to provide a more reliable and comparable measure of resource use across different studies. It is recognised however, that the number of genera recorded to be consumed by each butterflyfish is likely to increase with increases in sampling intensity, as well as the range of habitats and locations where feeding observations have been conducted. For example, the range of coral genera used by *C. octofasciatus* was documented during 5 replicate 3-minute feeding observations on a single inshore reef in Kimbe Bay, Papua New Guinea (Pratchett and Berumen, Unpublished Data). For *C. trifascialis*, however, >300 feed observations have been undertaken in a range of locations from the Red Sea to French Polynesia. Both these species have been recorded to feed on a relatively restricted range of different coral genera (9–10 genera), but it seems likely that *C. octofasciatus* is much less specialised, and would be seen to consume a wider range of coral genera following further detailed feeding observations in a wider range of habitats and locations.

To take account of marked differences in sampling intensity our second metric of specialisation was the mean number of genera consumed during replicate 3-minute feeding observations. While 3-minutes is a relatively short period with which to document individual patterns of prey use (initially selected to maximise replication within species rather than explicitly testing for individual variation in prey use; Berumen et al., 2005), individual butterflyfishes can travel >10 metres and visit multiple colonies during this period. It is expected therefore, that generalist species will consistently use a greater number of different coral types during each observation, compared to specialist species. While this does not account for geographic differences in coral diversity, it is expected that the number of genera used will be more constrained by the limited duration of feeding observations, allowing for more robust comparisons among species, regardless of location or sampling intensity. The final metric makes use of data on the proportional use of different coral genera, which is reported in a wide range of studies (Table 6.1), to actually calculate Shannon-Wiener diversity, following Lawton et al. (2012b).

Regardless of the metric used (cumulative number of coral genera, mean number of genera per feeding observation, and Shannon-Wiener diversity), marked interspecific differences in ecological specialisation were apparent among obligate coral feeding butterflyfishes (Table 6.2). The most specialised butterflyfish was *C. trifascialis* (see also Reese, 1981; Pratchett, 2007), which consumed the fewest different corals per individual feeding observation (Fig. 6.5), and also showed the lowest diversity (H′) in the coral genera it consumed (Fig. 6.6). As discussed previously, the cumulative number of genera consumed across all locations was higher for *C. trifascialis* (10 genera) than for some other butterflyfishes. However, the only butterflyfishes recorded to feed on ≤10 coral genera (*C. austriacus*, *C. larvatus*, *C. melapterus*, *C. octofasciatus*, *C. reticulatus*) were those that had only been studied at a single geographic location (Table 6.2). Other butterflyfishes revealed to be ecologically specialised by both measures were *C. unimaculatus*, *C. ornatissimus*, *C. pelewensis*, and *C. baronessa*. The least specialised (most generalist) species were *C. zanzibarensis* and *C. rainfordi*, though *C. octofasciatus* consumed the highest number of

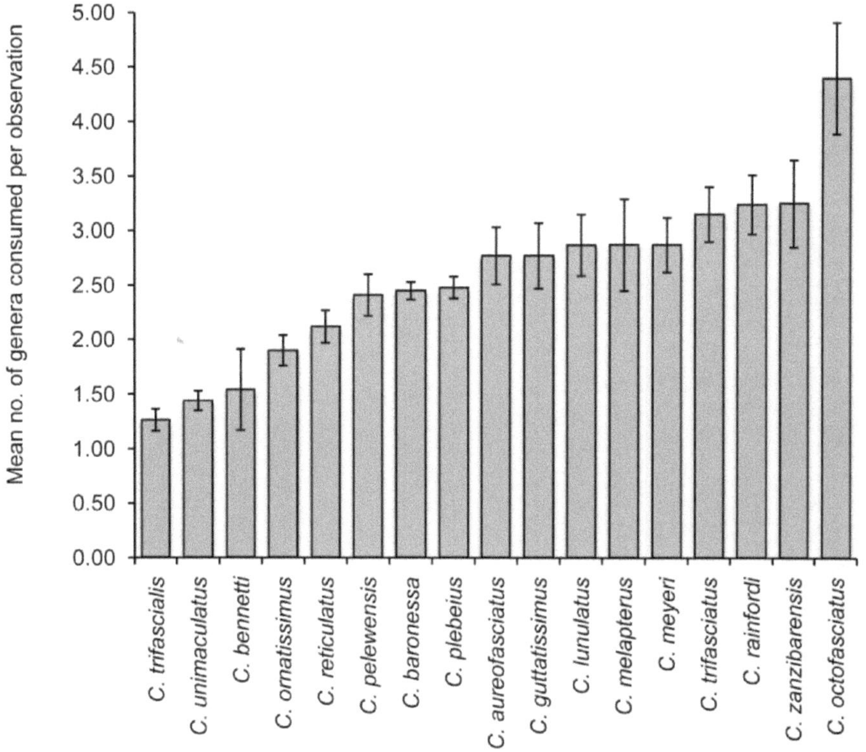

Fig. 6.5 Mean (± SE) number of genera consumed per 3-minute feeding observation by obligate coral-feeding butterflyfishes. Data was compiled across multiple locations (Table 6.1).

Table 6.2 Ecological specialization among obligate coral feeding butterflyfishes, showing on the i) the cumulative number of different coral genera consumed across geographic locations throughout the Indo-Pacific, ii) the mean number of genera consumed per 3-minute feeding observation (where possible), and iii) Shannon-Weiner diversity (H') prime calculated based on proportional use of different genera. "*" indicate species for which diet was comprised only partly of scleractinian corals, leading to greatly reduced number of genera consumed per 3-minute observation.

Species	Location	n	No. of genera	Cummulative no. of genera	No. genera per obs. (3-min)	H'	Reference
C. andamanensis	Sumatra	20	7	-	2.50 (0.21)	0.65	Hoey and Pratchett (Unpub.)
C. aureofasciatus	Lizard Island, GBR	35	20	-	2.77 (0.26)	1.01	Pratchett (2005)
C. austriacus	Aqaba, Egypt		9	-	-	0.69	Alwany et al. (2003)
C. baronessa	Lizard Island, GBR	152	19	-	2.45 (0.08)	0.31	Pratchett (2005)
	Derawan, Borneo	5	6	19	-	0.61	Nagelkerken et al. (2009)
C. bennetti	Chagos	9	6	-	1.67 (0.37)	0.36	Pratchett et al. (2013a)
	Derawan, Borneo	5	10	13	-	0.93	Nagelkerken et al. (2009)
	Moorea, French Polynesia	4	3	15	1.25 (0.25)*	0.47	Pratchett and Berumen (Unpub.)
C. guttatissimus	Chagos	26	13	-	2.77(0.30)	0.79	Pratchett et al. (2013a)
C. larvatus	Eritrea	15	6	-	-	0.65	Zekeria (2003)
C. lunulatus	Lizard Island, GBR	146	24	-	3.18 (0.08)	0.75	Pratchett (2005)
	Derawan, Borneo	6	8	25	-	0.50	Nagelkerken et al. (2009)
	Hawaii	4–19	5	25	-	-	Cox (1994)
	Moorea, French Polynesia	32	6	26	2.00 (0.20)	0.43	Berumen and Pratchett (Unpub.)
	Pacific—multiple locations	436	24	29	-	-	Lawton et al. (2011)
C. melapterus	Arabian Peninsula	13	6	-	2.87 (0.42)	0.67	Pratchett et al. (2013b)
C. meyeri	Chagos	23	13	-	2.87 (0.25)	0.72	Pratchett et al. (2013a)
C. multicinctus	Hawaii	4–19	5	-	-	-	Cox (1994)
C. octofasciatus	Kimbe Bay, PNG	5	9	-	4.40 (0.51)	0.80	Pratchett (Unpub.)
C. ornatissimus	Kimbe Bay, PNG	10	10	-	-	-	Pratchett and Berumen (Unpub.)

	Hawaii	4–19	5	10	-	-	Cox (1994)
	Moorea, French Polynesia	41	7	12	1.90 (0.14)	0.50	Berumen and Pratchett (Unpub.)
C. pelewensis	Moorea, French Polynesia	39	8	-	2.41 (0.19)	0.46	Berumen and Pratchett (Unpub.)
C. plebeius	Lizard Island, GBR	105	22	-	2.48 (0.10)	0.80	Pratchett (2005)
C. punctatofasciatus	Derawan, Borneo	5	10	-	-	0.87	Nagelkerken et al. (2009)
C. rainfordi	Lizard Island, GBR	43	21	-	3.24 (0.27)	1.02	Pratchett (2005)
C. reticulatus	Moorea, French Polynesia	41	10	-	2.12 (0.15)	0.42	Berumen and Pratchett (Unpub.)
C. trifascialis	Aqaba, Egypt		2	-	-	-	Alwany et al. (2003)
	Chagos	28	4	6	1.11 (0.06)	0.03	Pratchett et al. (2013a)
	Chagos	56	3	7	-	-	Graham (2007)
	Lizard Island, GBR	71	2	8	1.10 (0.04)	0.03	Pratchett (2005)
	Moorea, French Polynesia	26	3	8	1.88 (0.10)	0.31	Berumen and Pratchett (Unpub.)
	Pacific - multiple locations	396	6	10	-	-	Lawton et al. (2011)
C. trifasciatus	Chagos	26	13	-	3.15 (0.25)	0.61	Pratchett et al. (2013a)
C. unimaculatus	Lizard Island, GBR	31	7		0.61 (0.14)*	0.61	Pratchett (2005)
	Hawaii	4–19	5	9	-	-	Cox (1994)
	Moorea, French Polynesia	41	4	10	1.44 (0.09)	0.23	Berumen and Pratchett (Unpub.)
C. zanzibarensis	Chagos	20	15	-	3.25 (0.40)	1.09	Pratchett et al. (2013a)

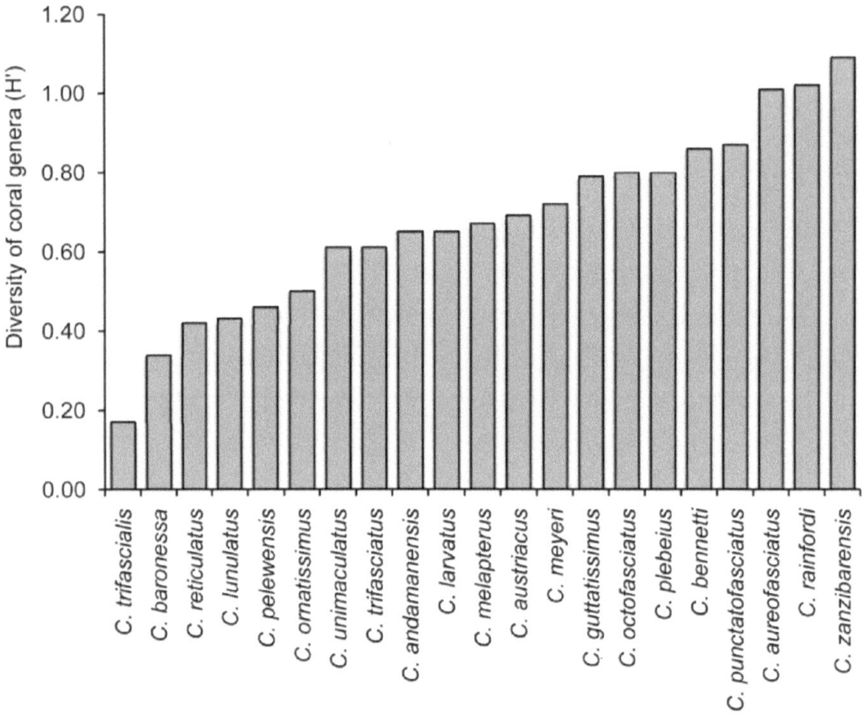

Fig. 6.6 Shannon-Weiner diversity (H') calculated based on proportional use of different genera by obligate coral-feeding butterflyfishes. Data was compiled across multiple locations (Table 6.1).

different genera (>4) per 3-minute feeding observation. The relative order of other butterflyfishes varied depending on the particular metric used (Table 6.2). Interestingly, however, the closely related sister species, *C. austriacus*, *C. lunulatus*, *C. melapterus*, and *C. trifasciatus* exhibited very similar levels of dietary specialisation regardless of the metric used (Figs. 6.5, 6.6).

Having shown marked interspecific differences in specialisation among coral-feeding butterflyfishes, the obvious question is why are some species (e.g, *C. trifascialis*) so specialised, while others (e.g., *C. zanzibarenesis*) are generalist in their use of coral prey? Specifically, what are the costs and benefits associated with increased levels of dietary specialisation? For species with specific adaptations that constrain resource-use (*adaptive specialisation*, Ferry-Graham et al., 2002), these adaptations are expected to increase efficiency in the capture or assimilation of prey resources (e.g., Futuyma and Moreno, 1988), or enable utilisation of resources that cannot be accessed by generalist counterparts. Generally, specialist species are expected to outperform generalist species whenever their primary resources are available and abundant (e.g., Caley and Munday, 2003). Berumen and

Pratchett (2008) explicitly tested this by comparing the growth rates of *C. trifascialis* versus *C. plebeius*, when each butterflyfish was provided with preferred versus non-preferred prey. Contrary to expectations, the more specialized species (*C. trifascialis*) did not outperform the more generalist species (*C. plebeius*) when both consumed their preferred prey (*A. hyacinthus* and *P. damicornis*, respectively). However, both species did selectively target the coral prey that maximised juvenile growth, as would be expected if resource specialisation is simply due to strong selectivity in resource use by species that are otherwise able to exploit a wide range of resources. Interestingly, these particular species (*C. trifascialis* and *C. plebeius*) do exhibit contrasting feeding preferences (Berumen and Pratchett, 2008), whereas many other coral-feeding butterflyfishes (*C. austriacus*, *C. guttatissimus*, and *C. meyeri*) all feed preferentially on *Acropora* corals. In these instances, high levels of specialisation would necessitate competitive superiority in order to monopolise access to preferred resources, thereby questioning whether specialisation is a cause or consequence of a species' dominance. Specifically, does *C. trifascialis* exhibit high levels of specialisation because it is competitively dominant over all other butterflyfishes and can thereby monopolize access to *A. hyacinthus*, or is *C. trifascialis* adapted to feed only on *Acropora* and *Pocillopora* corals and therefore, has most to loose if it does not successfully defend this resource? Both ecological and evolutionary specialisation should typically arise (or persist) only where preferred resources are abundant or widespread. However, the extent to which fishes are evolutionarily specialised will greatly affect their vulnerability to disturbances, and resource depletion.

Vulnerability to Disturbance

Specialisation is increasingly considered alongside population size and geographical range as one of the key predictors of a species' extinction risk (e.g., McKinney, 1997; Graham et al., 2011). In general, highly specialised species are expected to be constrained by the availability of resources and less able to cope with fluctuations in resource availability compared to generalist counterparts, making them more susceptible to population collapse or extinction during disturbances ("the specialization-disturbance hypothesis", Vazquez and Simberloff, 2002). In support of this hypothesis, butterflyfishes that consume a smaller range of different coral prey tend to exhibit much more pronounced declines in abundance during extensive coral loss compared to those species (e.g., *C. lunulatus*) that consume a wider range of different corals (Pratchett et al., 2008; Wilson et al., Chapter 9). The specialisation disturbances hypothesis presupposes however, that specialist species are actually constrained to feeding on a restricted set of prey types (e.g., Liem, 1980; Robinson and Wilson, 1998), otherwise seemingly specialist species would respond to disturbances simply by switching prey and/or increasing the range of prey items on which they feed (e.g., Pratchett et al., 2004; Berumen et al., 2005). If however, dietary specialisation results from specific morphological or physiological adaptations, then opportunities for expanding the range of prey consumed would be very limited (e.g., Wainwright, 1988; Clifton and Motta, 1998).

While there has been limited experimental testing of fundamental niches among coral-feeding butterflyfishes, one way to distinguish fundamental specialists is to test for consistent patterns of prey use across gradients of resource availability (Devictor et al., 2010). High levels of dietary specialisation, and selective use of the same resources across locations with marked differences in coral availability (Table 6.1), strongly suggests that *C. trifascialis* is a fundamental specialist (Lawton et al., 2011). This implies that *C. trifascialis* is morphologically or physiologically constrained to feed on *Acropora* and to a lesser extent *Pocillopora*, but is physically incapable of ingesting or assimilating other coral prey. This is supported by results from the above mentioned feeding experiments conducted in aquaria at Lizard Island, where juvenile *C. trifascialis* (<50 mm total length) were fed on exclusive diets of *A. hyacinthus*, *P. damicornis*, or *Porites cylindrica* (Berumen and Pratchett, 2008). Juvenile *C. trifascialis* grew fastest when fed on *A. hyacinthus*, and actually lost weight (and often died) when restricted to feeding on alternate diets of *P. damicornis* and *P. cylindrica*. Marked differences in growth occurred despite constant and consistent rates of feeding across all coral species (Berumen and Pratchett, 2008), suggesting that while *C. trifascialis* will ingest prey from non-preferred coral it simply cannot assimilate sufficient energy for juvenile growth and survival.

For some coral reef fishes, it is clear that specialisation has resulted from distinct morphological adaptations (e.g., Wainwright, 1988). Among wrasses for example, fundamental differences in the pharyngeal jaw morphology impose significant limits on the ability to utilise hard-shelled prey (Wainwright, 1988; Clifton and Motta, 1998). For many other families of fishes however, the role of ecomorphology in structuring prey use is equivocal (e.g., Motta, 1988). Among butterflyfishes, interspecific differences in dietary specialisation are only just beginning to become apparent, and so there have not been any comparative analyses of ecomorphology specifically comparing among specialist and generalist. Compared to other butterflyfishes, *C. trifascialis* does have a fairly unique body shape (Fig. 6.6), a small mouth, and restricted flexion of the lower jaw (Motta, 1985; Konow and Ferry-Graham, Chapter 2). It is unknown whether this limits effective use of non-*Acropora* prey, but this could be readily tested. *Chaetodon trifascialis* also has one of the shortest digestive tracts for a corallivorous species (Motta, 1988; Berumen et al., 2011), which might limit assimilation of poor quality prey. However, interspecific differences in gut morphology could equally be attributed to consistently high levels of *Acropora* consumption relative to other obligate hard coral feeders (Elliot and Bellwood, 2003), as alimentary tract morphology can change in response to changes in diet.

The vulnerability of ecologically specialised species to environmental fluctuations has been documented across a range of different organisms, including birds (Aitken and Martin, 2008), mammals (Harcourt et al., 2002; Fisher et al., 2003) and insects (Kotze and O'Hara, 2003). For butterflyfishes, both Pratchett et al. (2008) and Wilson et al. (Chapter 9) provide comprehensive syntheses of interspecific differences in vulnerability to disturbances, relating species-specific declines in abundance during coral depletion to i) coral dependence and ii) ecological specialisation. Not surprisingly, *C. trifascialis* emerges as the species of butterflyfish that is most vulnerable to sustained and ongoing coral loss (Pratchett et al., 2008; Wilson et al., Chapter 9), and may therefore, be particularly vulnerable to global extinction (Carpenter and Pratchett, 2010). Extinction risk, may however be offset by large geographical range and high levels of connectivity among widespread populations (Lawton et al., 2011), and so it is important to understand whether range size and abundance are positively (e.g., Charrette et al., 2006) or negatively correlated with ecological specialisation (e.g., Williams et al., 2006).

Range Size and Abundance

The degree to which species are specialised (e.g., niche width) is expected to be negatively related to their effective population size (comprising both geographic range and local abundance) for three reasons; i) specialist species are more constrained by resource availability (as discussed above), ii) restricted range species are locally adapted and therefore increasingly specialised, and iii) localised adaptations among wide-range species is expected to lead to broadly differentiated patterns of resource-use among locations, contributing to high niche width for the species (Gaston et al., 1997). Highly specialised species are expected, therefore, to have relatively narrow geographic ranges (e.g., Charrette et al., 2006), which would further increase their risk of extinction. Williams et al. (2006) however, revealed a strong negative relationship between dietary evenness and geographical range for frogs (*Cophixalus* spp.) in northern Australia, showing that the most specialised species actually had the largest ranges. Similarly for obligate coral feeding butterflyfishes, the most specialised species, *C. trifascialis* is distributed throughout the Indo-west Pacific and has one of the largest geographical ranges of all butterflyfishes. In contrast, *C. rainfordi* is endemic to north-east Australia, and has one of the lowest levels of dietary specialisation (Table 6.2). All three species (*C. aureofasciatus*, *C. rainfordi* and *C. zanzibarensis*) that had diet diversity (H')>1.00 were restricted to small distinct geographic locations within either the Pacific or Indian Ocean. In contrast, three out of four species that consumed an average of <2.00 coral genera per 3-minute feeding observation (*C. trifascialis*, *C. bennetti* and *C. ornatissimus*) are distributed throughout the Indo-Pacific. Jones et al. (2002) suggested that there was no clear relationship between geographic range size and specialisation among butterflyfishes, but it is clear that if a relationship does exist, it is negative rather than positive. Williams et al. (2006) attribute the inverse relationship between ecological specialisation and geographic range to extinction filtering, whereby restricted-range and therefore extinction-prone species are more likely to persist if they are diet generalists. Conversely, specialist species will only persist on evolutionary time-scales if they are sufficiently widespread to withstand environmental stochasticity and localized catastrophes.

In addition to suggesting that specialist species should have relatively narrow geographic ranges, Brown (1984) argued that specialist species should be less abundant compared to generalist counterparts. The relative abundance of generalist versus specialist species within a specific location or habitat will depend mainly upon the availability of preferred resources, and specialist species may be more abundant if preferred resources are abundant. However, given inherent variation in resource availability at larger scales, it is expected that generalists will be more abundant overall (Fox and Morrow, 1981). While rarely tested (especially at very large scales), specialist butterflyfishes may actually be more abundant than generalist species (e.g., Pratchett et al., 2013a). It is clear for example, that obligate coral-feeding butterflyfishes are generally more abundant than facultative corallivores (Cole and Pratchett, Chapter 5). In Chagos, Pratchett et al. (2013a) explicitly compared the overall abundance of different butterflyfishes, based on sampling at two different depth zones (2–4 metres, 7–10 meters), at both exposed and sheltered aspects, on three different atolls (the Chagos Bank, Peros Banhos, and Salamon Atoll). While there was evidence of greater spatial variation in the abundance of specialist species (e.g., *C. trifascialis*) compared to generalist species (see also Pratchett and Berumen, 2008; Lawton and Pratchett, 2012), abundance of individual species was strongly and positively correlated with dietary specialisation (Fig. 6.7; Pratchett et al., 2013a). Even in the least favourable habitats (e.g., shallow reef crests with limited coral cover and diversity), *C. trifascialis* was among the most abundant species.

Several different factors (e.g., narrow ranges, low abundance, ecological specialisation, and short lifespan) combine to determine a species' overall extinction risk. Therefore, species that are particularly vulnerable to localised disturbances (e.g., *C. trifascialis*) due to high levels of ecological specialisation, may exhibit ecological compensation, whereby

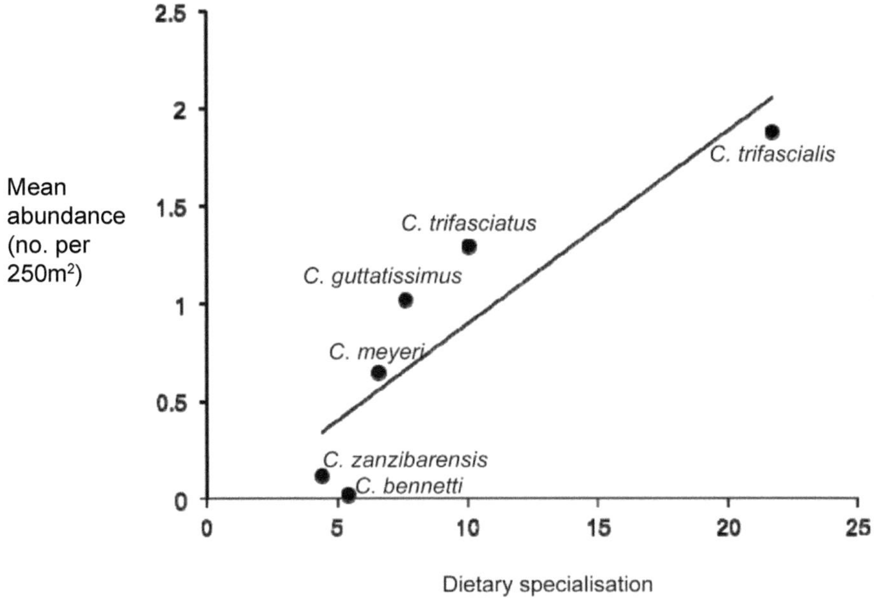

Fig. 6.7 Relationship between the overall abundance (averaged across all locations and zones) of each species of butterflyfish and their dietary specialisation (the mean number of different corals consumed per 100 bites). The relationship for all butterflyfishes (Pratchett et al., 2013a) was constrained at low levels of dietary specialisation for facultative corallivores, but even when restricting attention to just obligate coral feeding species there is a strong positive relationship.

there are other factors (e.g., large range size) that reduce extinction risk (e.g., Williams et al., 2006). Among butterflyfishes, there are very few species that have narrow geographical ranges, low abundance, and are also ecologically specialized (Pratchett et al., 2008), suggesting that there is strong ecological compensation within this group. The distribution and abundance of *C. trifascialis* does appear to be strongly constrained by availability of *Acropora* (Reese, 1981; Lawton et al., 2011), which is currently widespread and very abundant. Increases in the diversity, frequency, and severity of anthropogenic disturbances are expected to have a disproportionate impact on *Acropora*, which may in turn, threaten the persistence of *C. trifascialis* across an increasing range of locations (Carpenter and Pratchett, 2010). However, Lawton et al. (2011) found that *C. trifascialis* is not only widespread, but has maintained high levels of gene flow among geographical regions (i.e., Papua New Guniea, Australia, New Guinea and French Polynesia) over both recent and historical timeframes. High levels of gene flow at this scale provide increased opportunity for *C. trifascialis* populations to recover from local declines through the dispersal of individuals from other less affected populations.

CONCLUSIONS AND FUTURE DIRECTIONS

In conclusion, obligate coral-feeding butterflyfishes clearly distinguish among different coral species, with *Acropora* and *Pocillopora* being the most preferred coral genera. Strong and consistent feeding preferences among coral-feeding butterflyfishes lead to high levels of chronic predation on highly preferred prey, which may have significant demographic consequences for these corals (Cox, 1986; Cole et al., 2012). Moreover, highly convergent prey preferences are indicative of significant differences in prey quality, whereby tissues of *Acropora* and *Pocillopora* are expected to have higher nutritional content, though biochemical analyses have failed to show this. One possibility that warrants much further research is whether rates of ingestion (equivalent to feeding efficiency) differ among corals, and thereby account for observed discrepancies between nutritional content and feeding preferences. It is also possible that failure to adequately account for dietary preferences of coral-feeding butterflyfishes is due to failings in accounting for the full range of macronutrients that may influence preference and avoidance (*sensu* Raubenheimer et al., 2005). Some "beneficial" compounds can actually become harmful if taken in excess and some "detrimental" compounds are essential in low levels (Raubenheimer et al., 2005). Differences in specific nutritional requirement may therefore lead to marked differences in levels of specialisation, even if preferred prey are generally equivalent, as was observed at Lizard Island (Pratchett, 2005). Integrating information on the content of different macronutrients in available prey with contrasting physiological requirements of sympatric species may help to understand prey preferences, as well as predicting species or community responses to the declines in availability of specific prey types (Raubenheimer and Simpson, 1999).

Despite generally high levels of dietary specialisation and substantial opportunity for resource partitioning, there is little evidence of complementary patterns of prey use among coral-feeding butterflyfishes. Importantly, the restricted set of coral genera used by specialist species (*C. trifascialis*) are a subset of the coral used by more generalist species (*C. lunulatus*). While interspecific competition does not appear to have lead to niche diversification, competition may be important in moderating the extent to which subordinate species can feed on preferred corals, such as *Acropora*. The question remains however, are specialist butterflyfishes able to exploit a limited range of corals because they are competitively dominant, or are these species evolutionarily constrained to feeding on a restricted set of corals, and must therefore aggressively defend these resources from generalist counterparts?

Very high dietary specialization appears to be a questionable evolutionary strategy for *C. trifascialis* because there is no apparent fitness benefit arising from using predominantly just one coral genus. More specifically, Berumen and Pratchett (2008) could not find any difference in feeding and assimilation efficiencies between *C. trifascialis* and *C. plebeius* when each was given exclusive access to the most preferred prey (*A. hyacinthus*), whereas it is generally expected that specialist species should outperform generalist species when using preferred prey (e.g., Futuyma and Moreno, 1988). Benefits of specialisation may therefore, only be apparent during fluctuations in availability of certain coral prey. Specialised butterflyfishes (particularly, *C. trifascialis*) do appear to be both competitively and numerically dominant at locations with high coral cover (Pratchett et al., 2013a). As coral cover fluctuates through time, dominant competitors will have more constant access to available resources, whereas generalists and competitively sub-ordinate species may be excluded from feeding on preferred prey, when corals are scarce.

While very preliminary, interspecific differences in dietary specialisation among coral-feeding butterflyfishes appear positively correlated with both geographic range and overall abundance. Specialist species (especially, fundamental specialists) are expected to be extremely constrained by their reliance on specific prey (e.g., Brown, 1984), however, the most specialised of all butterflyfishes, *C. trifascialis,* is one of the few butterflyfishes distributed throughout the Indo-west Pacific. *Chaetodon trifascialis* is also among the most abundant species at many locations throughout its' range (Pratchett et al., 2013a). These findings challenge well-established paradigms of specialisation and biogeography (e.g., Brown, 1984), and warrant much more careful and detailed consideration. Importantly, the coral-feeding butterflyfishes that are at greatest risk of extinction are probably not those that have the most specialised diet, if other factors (large geographic range, large effective population size, and high levels of gene flow among disparate populations) consistently offset increased extinction risk (Lawton et al., 2011).

ACKNOWLEDGEMENTS

This chapter represents a culmination of many years of research, which was aided greatly by the efforts of three exceptional graduate students, Michael Berumen, Andrew Cole and Rebecca Lawton. There are many other people who have also assisted in compiling extensive diet and behavioural information presented here, especially Andrew Hoey and Shaun Wilson. I am also grateful to Andrew Hoey and Shane Blowes for suggested improvements to the structure and content of this chapter.

REFERENCES

Aitken, K.E.H. and K. Martin. 2008. Resource selection plasticity and community responses to experimental reduction of acritical resource. *Ecology* 89: 971–980.

Araujo, M.S. and M.O. Gonzaga. 2007. Individual specialization in the hunting wasp *Trypoxylon (Trypargilum) albonigrum* (Hymenoptera, Crabronidae). *Behavioral Ecology and Sociobiology* 61: 1855–1863.

Aliño, P.M., P.W. Sammarco and J.C. Coll. 1988. Studies of the feeding preferences of *Chaetodon melannotus* (Pisces) for soft corals (Coelenterata: Octocorallia) *Proceedings of the 6th International Coral Reef Symposium, Townsville* pp. 31–36.

Alwany, M., E. Thaler and M. Stachowitsch. 2003. Food selection in two corallivorous butterflyfishes, *Chaetodon austriacus* and *Chaetodontrifascialis,* in the northern Red Sea. *Marine Ecology* 24: 165–177.

Anderson, G., A. Ehrlich, P. Ehrlich, J. Roughgarden, B. Russell and F. Talbot. 1981. The community structure of coral reef fishes. *American Naturalist* 117: 476–495.

Berumen, M.L. and M.S. Pratchett. 2006a. Recovery without resilience: Persistent disturbance and long-term shifts in the structure of fish and coral communities at Tiahura Reef, Moorea. *Coral Reefs* 25: 647–653.

Berumen, M.L. and M.S. Pratchett. 2006b. Effects of resource availability on the competitive behaviour of butterflyfishes (Chaetodontidae). *Proceedings of the 10th International Coral Reef Symposium, Okinawa*, pp. 644–650.

Berumen, M.L. and M.S. Pratchett. 2008. Trade-offs associated with dietary specialisation in corallivorous butterflyfishes (Chaetodontidae: *Chaetodon*). *Behavioral Ecology and Sociobiology* 62: 989–994.

Berumen, M.L., M.S. Pratchett and M.I. McCormick. 2005. Within-reef differences in diet and body condition of coral-feeding butterflyfishes (Chaetodontidae). *Marine Ecology Progress Series* 287: 217–227.

Berumen, M.L., M.S. Pratchett and B.A. Goodman. 2011. Relative gut lengths of coral reef butterflyfishes (Pisces: Chaetodontidae). *Coral Reefs* 30: 1005–1010.

Birkeland, C. and S. Neudecker. 1981. Foraging behavior of two Caribbean chaetodontids: *Chaetodon capistratus* and *C. aculeatus*. *Copeia* 1: 169–178.

Bellwood, D.R., A.S. Hoey and J.H. Choat. 2003. Limited functional redundancy in high diversity systems: resilience and ecosystem function on coral reefs. *Ecology Letters* 6: 281–285.

Bellwood, D.R., A.S. Hoey, J.L. Ackerman and M. Depczynski. 2006a. Coral bleaching, reef fish community phase shifts and the resilience of coral reefs. *Global Change Biology* 12: 1587–1594.

Bellwood, D.R., T.P. Hughes, and A.S. Hoey. 2006b. Sleeping function group drives coral-reef recovery. *Current Biology* 16: 2434–2439.

Bellwood, D.R., P.C. Wainwright, C.J. Fulton and A.S. Hoey. 2006c. Functional versatility supports coral reef biodiversity. *Proceedings of the Royal Society B* 273: 101–107.

Bellwood, D.R., S. Klanten, P.F. Cowman, M.S. Pratchett, N. Konow and L. van Herwerden. 2009. Evolutionary history of the butterflyfishes (f: Chaetodontidae) and the rise of coral feeding fishes. *Journal of Evolutionary Biology* 23: 335–349.

Belmaker, J., C.H. Sekercioglu and W. Jetz. 2012. Global patterns of specialization and coexistence in bird assemblages. *Journal of Biogeography* 39: 193–203.

Bouchon-Navaro, Y. 1986. Partitioning of food and space resources by chaetodontid fishes on coral reefs. *Journal of Experimental Marine Biology and Ecology* 103: 21–40.

Brown, J.H. 1984. On the relationship between abundance and distribution of species. *American Naturalist* 124: 255–279.

Caley, M.J. and P.L. Munday. 2003. Growth trades off with habitat specialization. *Proceedings of the Royal Society of London. Series B: Biological Sciences* 270: S175–S177.

Carpenter, K. and M.S. Pratchett. 2010. *Chaetodon trifascialis*. *In:* IUCN 2012. IUCN Red List of Threatened Species. Version 2012.2.
Charrette, N.A., D.F.R. Cleary and A.Ø. Mooers. 2006. Range-restricted, specialist Bronean butterflies and less likely to recover form ENSO-induced disturbance. *Ecology* 87: 2330–2337.
Clifton, K.B. and P.J. Motta. 1998. Feeding morphology, diet, and ecomorphological relationships among five Caribbean labrids (Teleostei, Labridae). *Copeia* 1988: 953–966.
Cole, A.J., M.S. Pratchett and G.P. Jones. 2008. Diversity and functional importance of coral-feeding fishes on tropical coral reefs. *Fish and Fisheries* 9: 286–307.
Cole, A.J., K.M. Chong-Seng, M.S. Pratchett and G.P. Jones. 2009. Coral-feeding fishes slow progression of black-band disease. *Coral Reefs* 28: 264
Cole, A.J. and M.S. Pratchett. 2011. Inter-specific variation in suceptibility to grazing among common reef corals. *Marine Ecology Progress Series* 422: 155–164.
Cole, A.J., R.J. Lawton, M.S. Pratchett and S.K. Wilson. 2011. Chronic coral consumption by butterflyfishes. *Coral Reefs* 30: 85–93.
Cole, A.J., R.J. Lawton, S.K. Wilson, and M.S. Pratchett. 2012. Consumption of tabular acroporid corals by reef fishes: a comparison with plant-herbivore interactions. *Functional Ecology* doi: 10.1111/j.1365–2435.2011.01935.x
Colles, A., L.H. Liow and A. Prinzing. 2009. Are specialists at risk under environmental change? Neoecological, paleoecological and phylogenetic approaches. *Ecology Letters* 12: 849–863.
Cox, E.F. 1986. The effects of a selective corallivore on growth rates and competition for space between two species of Hawaiian corals. *Journal of Experimental Marine Biology and Ecology* 101: 161–174.
Cox, E.F. 1994. Resource use by corallivorous butterflyfishes (Family Chaetodontidae) in Hawaii. *Bulletin of Marine Science* 54: 535–545.
Croze, H.J. 1970. Searching image in carrion crows. Z. Tierpsychol. Beih. 5: 1–85.
Devictor, V., J. Clavel, R. Julliard, S. Lavergne, D. Mouillot, W. Thuiller, P. Venail, S. Villéger and N. Mouquet. 2010. Defining and measuring ecological specialization. *Journal of Applied Ecology* 47: 15–25.
Ebeling, A.W. and D.R. Laur. 1986. Foraging in serfperches: resource partitioning or individualistic responses? *Env. Biol. Fishes* 16: 123–133.
Eeley, H.A.C. and R.A. Foley. 1999. Species richness, species range size and ecological specialisation among African primates: geographical patterns and conservation implications. *Biodiversity and Conservation* 8: 1033–1056.
Elliott, J.P. and D.R. Bellwood. 2003. Alimentary tract morphology and diet in three coral reef fish families. *Journal of Fish Biology* 63: 1598–1609.
Elton, C. 1927. *Animal Ecology*. Sidgwick and Jackson, London.
Emlen, J.M. 1968. Optimal choice in animals. *American Naturalist* 385–389.
Fautin, D.G. 1985. Competition by anemonefishes for host actinians.*Proceedings of the Fifth International Coral Reef Congress* 5: 373–377.
Ferry-Graham, L.A., D.I. Bolnick and P.C. Wainwright. 2002. Using functional morphology to examine the ecology and evolution of specialization. *Integrative and Comparative Biology* 42: 265–277.
Fox, L.R. and P.A. Morrow. 1981. Specialization–species property or local phenomenon.*Science* 211: 887–893.
Fridley, J.D., D.B. Vandermast, D.M. Kuppinger, M.Manthey and R.K. Peet. 2007. Co-occurrence based assessment of habitat generalists and specialists: a new approach for the measurement of niche width. *Journal of Ecology* 95: 707–722.
Fisher, D.O., S.P. Blomberg and I.P. Owens. 2003. Extrinsic versus intrinsic factors in the decline and extinction of Australian marsupials. *Proceedings of the Royal Society of London. Series B: Biological Sciences* 270: 1801–1808.
Futuyma, D.J. and G. Moreno. 1988. The evolution of ecological specialization. *Annual Review of Ecology and Systematics* 1988: 207–233.

Gardiner, N.M. and G.P. Jones 2005. Habitat specialisation and overlap in a guild of coral reef cardinalfishes (Apogonidae). *Marine Ecology Progress Series* 305: 163–175.
Gaston, K.J., T.M. Blackburn and J.H. Lawton. 1997. Interspecific abundance-range size relationships: an appraisal of mechanisms. *Journal of Animal Ecology* 579–601.
Gochfeld, D.J. 2004. Predation-induced morphological and behavioral defences in a hard coral: implications for foraging behavior of coral-feeding butterflyfishes. *Marine Ecology Progress Series* 267: 145–158
Gore, M.A. 1984. Factors affecting the feeding behaviour of a coral reef fish, *Chaetodon capistratus.Bulletin of Marine Science* 35: 211–220.
Graham, N.A.J. 2007. Ecological versatility and the decline of coral feeding fishes following climate driven coral mortality. *Marine Biology* 153: 119–127.
Graham, N.A.J., S.K. Wilson, S. Jennings, N.V.C. Polunin, J.P. Bijoux and J. Robinson. 2006. Dynamic fragility of oceanic coral reef ecosystems. *Proceedings of the National Academy of Sciences of the United States of America* 103: 8425–8429.
Graham, N.A.J., P. Chabanet, R.D. Evans, S. Jennings, Y. Letourneur, A.M. MacNeil, T.R. McClanahan, M.C. Öhman, N.V.C. Polunin and S.K. Wilson. 2011. Extinction vulnerability of coral reef fishes. *Ecology Letters* 14: 341–348.
Grinnell, J. 1917. The nich relationship of the California Thresher. *The Auk* 34: 427–433.
Harcourt, A.H., S.A. Coppeto and S.A. Parks. 2002. Rarity, specialization and extinction in primates. *Journal of Biogeography* 29: 445–456.
Harmelin-Vivien, M.L. and Y. Bouchon-Navaro. 1981. Trophic relationships among chaetodontid fishes in the Gulf of Aqaba (Red Sea). *Proceedings of the 4th International Coral Reef Symposium* 2: 537–544.
Harmelin-Vivien M.L. and Y. Bouchon-Navaro. 1983. Feeding diets and significance of coral feeding among Chaetodontid fishes in Moorea (French Polynesia). *Coral Reefs* 2: 119–127.
Hutchinson, G.E. 1959. Homage to Santa Rosalia or why there are so many species? *American Naturalist* 93: 145–159.
Hourigan, T.F., T.C. Tricas and E.S. Reese. 1988. Coral reef fishes as indicators of environmental stress in coral reefs. In *Marine Organisms as Indicators* (Soule, D.F. and G.S. Kleppeleds), New York: Springer-Verlag pp. 107–135.
Irons, D.K. 1989. Temporal and areal feeding behaviour of the butterflyfish, *Chaetodon trifascialis,* at Johnston Atoll. *Environmental Biology of Fishes* 25: 187–193.
Jones, G.P., M.J. Caley and P.L. Munday. 2002. Rarity in coral reef fish communities. In *Coral Reef Fishes: Dynamics and Diversity in a Complex Ecosystem,* P.F. sale (ed.). San Diego, California: Academic Press 81–101.
Kotze, D.J. and R.B. O'Hara. 2003. Species decline—but why? Explanations of carabid beetle (Coleoptera, Carabidae) declinesin Europe. *Oecologia* 135: 138–148.
Lasker, H.R. 1985. Prey preferences and browsing pressure of the butterflyfish *Chaetodoncapistratus* on Caribbean gorgonians. *Marine Ecology Progress Series* 21: 213–220.
Lawton, R.J., V. Messmer, L.K. Bay and M.S. Pratchett. 2011. High gene flow across large geographic scales reduces extinction risk for highly specialised coral feeding butterflyfishes. *Molecular Ecology* 17: 3584–3598.
Lawton, R.J. and M.S. Pratchett. 2012. Influence of dietary specialisation and resource availability on geographical variation in abundance of butterflyfishes.*Ecology and Evolution* 7: 1347–1361.
Lawton, R.J., A.J. Cole, M.L. Berumen and M.S. Pratchett. 2012a. Geographical variation in resource use by specialist versus generalist butterflyfishes. *Ecography* 35: 566–576.
Lawton, R.J., M.S. Pratchett, and M.J. Berumen. 2012b. The use of specialisation indices to predict vulnerability of coral-feeding butterflyfishes to environmental change. *Oikos* 121: 191–200.
Liem, K.F. 1980. Adaptive significance of intra-and interspecific differences in the feeding repertoires of cichlid fishes.*American Zoologist* 20: 295–314.

Lobel, P.S. and J.C. Ogden. 1981. Foraging by the herbivorous parrotfish *Sparisoma radians*. *Marine Biology* 64: 173–183.

MacArthur, R.H. 1972. *Geographical Ecology. Patterns in the Distribution of Species*. Harper and Row, New York.

MacArthur, R.H and E.R. Pianka. 1966. On optimal use of a patchy environment. *American Naturalist* 603–609.

Manly, B.F.J., L.L. McDonald, D.L. Thomas, T.L. McDonald and W.P. Erickson. 2002. Resource Selection by Animals: Statistical Analysis and Design for Field Studies. Kluwer: Boston, MA.

McCormick, M.I. 2003. Consumption of coral propagules after mass spawning enhances larval quality of damselfish through maternal effects. *Oecologia* 136: 37–45.

McKinney, M.L. 1997. Extinction vulnerability and selectivity: combining ecological and paleontological views. *Annual Review of Ecology Evolution and Systematics* 495–516.

Montanari, S.R., L. van Herwerden, M.S. Pratchett, J.A. Hobbs and A. Fugedi. 2012. Reef fish hybridisation: lessons learnt from butterflyfishes (genus *Chaetodon*). *Ecology and Evolution* 2: 310–328.

Motta, P.J. 1985. Functional morphology of the head of Hawaiian and Mid-Pacific butterflyfishes (Perciformes, Chaetodontidae). *Environmental Biology of Fishes* 13: 253–276.

Motta, P.J. 1988. Functional morphology of the feeding apparatus of ten species of Pacific butterflyfishes (Perciformes, Chaetodontidae): an ecomorphological approach. *Environmental Biology of Fishes* 22: 39–67.

Munday, P.L., G.P. Jones and M.J. Caley. 1997. Habitat specialisation and the distribution and abundance of coral-dwelling gobies. *Mar. Ecol. Prog. Ser.* 152: 227–239.

Munday, P.L. 2001. Fitness consequences of habitat use and competition among coral-dwelling fishes. *Oecologia* 128: 585–593.

Munday, P.L. 2004. Habitat loss, resource specialization, and extinction on coral reefs. *Global Change Biology* 10: 1642–1647.

Nagelkerken, I., G. van der Velde, S.L.J. Wartenbergh, M.M. Nugues and M.S. Pratchett. 2009. Cryptic dietary components reduce dietary overlap among sympatric butterflyfishes (Chaetodontidae). *Journal of Fish Biology* 75: 1123–1143.

Neudecker, S. 1979. Effects of grazing and browsing fishes on the zonation of corals in Guam. *Ecology* 60: 666–672.

Pitts, P.A. 1991. Comparative use of food and space by three Bahamian butterflyfishes. *Bulletin of Marine Science* 48:749–756.

Poisot, T., E. Canard, N. Mouquet and M.E. Hochberg. 2012. comparative study of ecological specialization estimators. *Methods in Ecology and Evolution* 3: 537–544.

Pratchett, M.S. 2005. Dietary overlap among coral-feeding butterflyfishes (Chaetodontidae) at Lizard Island, northern Great Barrier Reef. *Marine Biology* 148: 373–382.

Pratchett, M.S. 2007. Dietary selection by coral-feeding butterflyfishes (Chaetodontidae) on the Great Barrier Reef, Australia. *The Raffles Bulletin of Zoology* Supplement No. 14: 171–176.

Pratchett, M.S. and M.L. Berumen. 2008. Inter-specific variation in distributions and diets of coral reef butterflyfishes (Teleostei: Chaetodontidae). *Journal of Fish Biology* 73: 1730–1747.

Pratchett, M.S., N. Gust, G. Goby and S.O. Klanten. 2001. Consumption of coral propagulesre present a significant trophic link between corals and reef fish. *Coral Reefs* 20: 13–17.

Pratchett, M.S., S.K. Wilson, M.L. Berumen and M.I. McCormick. 2004. Sublethal effects of coral bleaching on an obligate coral feeding butterflyfish. *Coral Reefs* 23: 352–356.

Pratchett, M.S., S.K. Wilson and A.H. Baird. 2006. Declines in the abundance of *Chaetodon* butterflyfishes (Chaetodontidae) following extensive coral depletion. *Journal of Fish Biology* 69: 1269–1280.

Pratchett, M.S., P.L. Munday, S.K. Wilson, N.A.J. Graham, J.E. Cinner, D.R. Bellwood, G.P. Jones, N.V.C. Polunin and T.R. Mcclanahan. 2008. Effects of climate-induced coral bleaching on coral-reef fishes—Ecological and economic consequences. *Oceanography and Marine Biology: An Annual Review* 46: 251–296.

Pratchett, M.S., D.J. Coker, G.P. Jones and P.L. Munday. 2012. Specialisation in habitat use by coral reef damselfishes and their susceptibility to habitat loss. *Ecology and Evolution* 2: 2168–2180.

Pratchett, M.S., N.A.J. Graham and A.J. Cole. 2013a. Specialist corallivores dominate butterflyfish assemblages in coral-dominated reef habitats. *Journal of Fish Biology* 82: 1177–1191.

Pratchett, M.S., A.S. Hoey, D.A Feary, A.G. Bauman, J.A. Burt and B. Riegl. 2013b. Functional composition of *Chaetodon* butterflyfishes at a peripheral and extreme coral reef location, the Persian Gulf. *Marine Pollution Bulletin* DOI: 10.1016/j.marpolbul.2012.10.013.

Randall, J.E. and W.D. Hartman. 1968. Sponge-feeding fishes of the West Indies. *Marine Biology* 1: 216–225.

Raubenheimer, D. and S.J. Simpson. 1999. Integrating nutrition: a geometrical approach. *Entomologia Experimentaliset Applicata* 91: 67–82.

Raubenheimer, D., W.L. Zemke-White, R.J. Phillips and K.D. Clements. 2005. Algal macronutrients and food selection by the omnivorous marine fish *Girellatricuspidata*. *Ecology* 86: 2601–2610.

Raymundo, L.J., A.R., Halford, A.P. Maypa and A.M. Kerr. 2009. Functionally diverse reef-fish communities ameliorate coral disease. *Proceedings of the National Academy of Sciences of the United States of America* 106: 17067–17070.

Reese, E.S. 1977. Coevolution of corals and coral feeding fishes of the family Chaetodontidae. *Proceedings of the 3rd International Coral Reef Symposium* 1: 267–274.

Reese, E.S. 1981. Predation on corals by fishes of the family Chaetodontidae: implications for conservation and management of coral reef ecosystems. *Bulletin of Marine Science* 31: 594–604.

Robertson, D.R. 1987. Responses of two coral reef toadfishes (Batrachoididae) to the demise of their primary prey, the sea urchin, *Diademaantillarum*. *Copeia* 3: 637–642.

Robertson, R. 1970. Review of the predators and parasites of stony corals, with special reference to symbiotic prosobranch gastropods. *Pacific Science* 24: 43–54.

Robinson, B.W. and D.S. Wilson. 1998. Optimal foraging, specialization, and a solution to Liem's paradox. *American Naturalist* 151: 223–235.

Ross, S.T. 1986. Resource partitioning in fish assemblages: A review of field studies. *Copeia* 1986: 352–388.

Roughgarden, J. 1972. Evolution of niche width. *American Naturalist* 683–718.

Salisbury, C.L., N. Seddon, C.R. Cooney and J.A. Tobias. 2012. The latitudinal gradient in dispersal constraints: ecological specialisation drives diversification in tropical birds. *Ecology Letters* 15: 847–855.

Samways, M.J. 2005. Breakdown of butterflyfish (Chaetodontidae) territories associated with the onset of a mass coral bleaching event. *Aquatic Conservation: Marine and Freshwater Ecosystems* 15: S101–S107.

Sano, M. 1989. Feeding habits of Japanese butterflyfishes (Chaetodontidae). *Environmental Biology of Fishes* 25:195–203.

Schoener, T.W. 1971. Theory of feeding strategies. *Annual Review of Ecology and Systematics* 1971: 369–404.

Schoener, T.W. 1974. Resource partitioning in ecological communities.*Science* 185: 27–39.

Shokri, M.R., S.M.R. Fatemi and M.P. Crosby. 2005. The status of butterflyfishes (Chaetodontidae) in the northern Persian Gulf, I.R. Iran. *Aquatic Conservation: Marine and Freshwater Ecosystems* 15: S91–S99.

Schpigel, M. and L. Fishelson. 1989. Food habits and prey selection of three species of groupers from the genus *Cephalopholis* (Serranidae: Teleostei). *Environmental Biology of Fishes* 24: 67–73.

Stachowicz, J.J. and M.E. Hay. 2000. Geographic variation in camouflage specialization by a decorator crab. *American Naturalist* 156: 59–71.

Tricas, T.C. 1985. The economics of foraging in coral-feeding butterflyfishes of Hawaii. *Proceedings of the 5th International Coral Reef Symposium, Tahiti* pp. 409–414.

Tricas, T.C. 1989. Prey selection by coral-feeding butterflyfishes: strategies to maximize the profit. *Environmental Biology of Fishes* 25: 171–185.

Vázquez, D.P. and D. Simberloff. 2002. Ecological specialization and susceptibility to disturbance: conjectures and refutations. *American Naturalist* 159: 606–623.

Wainwright, P.C. 1988. Morphology and ecology: Functional basis of feeding constraints in Caribbean labridfishes. *Ecology* 69: 635–645.

Williams, D.M. 1986. Temporal variation in the structure of reef slope fish communities (central GBR): short term effects of *Acanthaster planci* infestation. *Marine Ecology Progress Series* 28: 157–164.

Williams, Y., S.E. Williams, R.A. Alford, M. Waycott and C.N. Johnson. 2006. Niche breadth and geographical range: ecological compensation for geological rarity in rainforest frogs. *Biology Letters* 2: 532–535.

Williams, S.E., Y.M. Williams, J. VanDerWal, J.L. Isaac, L.P. Shoo and C.N. Johnson. 2009. Ecological specialization and population size in a biodiversity hotspot: How rare species avoid extinction. *Proceedings of the National Academy of Sciences* 106: 19737–19741.

Zekeria, Z.A., Y. Dawit, S. Ghebremedhin, M. Naser and J.J. Videler. 2002. Resource partitioning among four butterflyfish species in the Red Sea. *Marine and Freshwater Research* 53: 163–168.

Zekeria, Z.A. 2003. *Butterflyfishes of the southern Red Sea: ecology and population dynamics*. PhD Thesis, Rijksuniversiteit Groningen: Groningen.

CHAPTER 7

Corallivory: The Coral's Point of View

Evelyn F. Cox

INTRODUCTION

At least since the time of Darwin, there has been controversy about the impact of fishes on corals. Although Darwin (1845) reported that two species of parrotfishes feed exclusively on coral and that their feeding behaviour produced copious amounts of sediment, Dana (1849) countered that he had never seen grazing by fishes on corals and that fish feeding could not produce the volume of sediment present on coral reefs. Since that time many species have been described to feed on corals, although the type of feeding and potential impacts on the corals themselves vary tremendously among corallivores. Even within the butterflyfishes (which have the greatest number of corallivorous species), the mode of feeding (ranging from extensive scraping to nipping of coral tentacles) and the extent of reliance on corals for food varies greatly (Cole and Pratchett, Chapter 5). There is no doubt that coral feeding can have a dramatic impact on coral populations, for example when there is an outbreak of corallivores, as has been seen repeatedly with population explosions of crown-of-thorn seastars (Branham et al., 1971; Endean, 1973; Glynn, 1974) or coral-feeding snails (Moyer et al., 1982). In these cases, total mortality of colonies can occur, and the recovery of the coral community may require decades (Moran, 1986; Endean and Cameron, 1990; Colgan, 1987; Sano, 2000). However, many forms of

University of Hawai'i West O'ahu, Pearl City, and Hawai'i Institute of Marine Biology, Kāne'ohe, Hawai'i.
E-mail: fcox@hawaii.edu

corallivory have less visible impacts on coral populations. Harmelin-Vivien and Bouchon-Navaro (1981, 1983) estimated that consumption by the butterflyfish populations at Aqaba was 10 g per 1000 m^2 each day and at Moorea 28 g per 1000 m^2 each day. They suggested that this low rate of coral feeding and the territorial behaviour of these butterflyfish species prevented significant impacts on the coral food resource. Reese (1991) suggested that although corals are a relatively poor food resource, corals are spatially and temporally stable in abundance and their aggregated distributions and ability to regenerate rapidly allow butterflyfishes to sustainably harvest coral resources within a defendable territory.

A similar debate regarding the impacts of herbivores on the fitness of plants has been waged (Wise and Abrahamson, 2007). For example, grazing was viewed as a positive impact on grasses due to the stimulation of regrowth from basal meristems and fertilisation of grasses with herbivore excretions. Alternatively, the regrowth was seen simply as a reaction to damage, not damage specific to grazing. The ability of a plant to tolerate herbivory must be evaluated within the context of plant resource availability, with a focus on what resources are limiting to the plants and how herbivory affects the ability of the plants to obtain those resources (Wise and Abrahamson, 2007). In addition, plant responses to herbivory can include both tolerance and resistance, and selection pressures for these two responses can vary with life stage, environmental factors, or mating system (Núñez-Farfán et al., 2007).

Corals share many physiological and ecological features with plants. Many corals rely on photosynthetically derived carbon from the zooxanthellae within their tissues. As with modular plants, colonial corals are composed of many polyps that are interconnected and can share resources. Similar to plants, corals have a wide array of growth forms and modes of reproduction, including asexual reproduction via fragmentation.

Comparable to plant herbivory, corallivory generally results in partial predation on the coral colony, and colonies can regenerate lost polyps and tissues. Gochfeld (1997) demonstrated that artificial polyp damage simulating a butterflyfish bite was regenerated in 7–10 days. Corallivores that produce larger scars (such as *Exallias brevis*) or break off branches (such as filefishes or puffers) may create coral damage that takes longer to recover. Although polyps were regenerated in 1 week, the scar formed by an *E. brevis* bite took 50 days to heal completely (Carlson, 1992), and broken *Porites* fingers can take over 40 days to heal (Jayewardene and Birkeland, 2006). This type of foraging resulting in partial mortality may be analogous to the grazing of grasses.

Beksly et al. (1993) proposed that grasses have the ability to tolerate herbivory because they possess 1) storage organs with carbohydrate and protein reserves, 2) dormant meristems or buds that can replace destroyed tissues, 3) the ability to reallocate energy and reserves from undamaged to damaged tissues, 4) the ability to increase photosynthesis in residual tissues, and 5) the ability to use photosynthate from newly recovered tissues. Colonial corals share these traits: 1) Most shallow water corals have excess energy/carbon production associated with photosynthesis by symbiotic zooxanthellae (Muscatine et al., 1981). Estimates of allocation of energy resources to organismal functions suggested that little of the daily energy budget is allocated to growth (6.3%) or reproduction (0.4%) in the Caribbean coral *Porites porites* (Edmunds and Davies, 1986); 2) Corals have populations of amoebocytes or stem cells that may serve an analogous role to dormant meristems (Rinkevich, 1996); 3) Corals have the ability to reallocate materials from one part of a colony to other areas (Rinkevich and Loya, 1983); 4) Corals have the ability to increase the density of zooxanthellae within tissues to respond to changes in light regimes (Chalker et al., 1983), and one species, *Montipora capitata*, has been demonstrated to increase the rate of heterotrophic feeding to compensate for the loss of zooxanthellae following bleaching (Grotolli et al., 2006); 5) Corals can use photosynthate from nearby tissues (Oren et al., 1997).

The biggest black box in understanding the impacts of corallivory revolves around the presumed trade-off in use of resources by corals for regeneration from damage and other metabolic activities. However, Rinkevich (1996) has proposed that three factors challenge the paradigm that reproduction and regeneration in corals must require a trade-off in resources: 1) sexual reproduction may be a hierarchical and programmed function that takes precedence over other metabolic functions; 2) shallow water corals may not be limited by energy resources as many coral species acquire excess carbon from their zooxanthellae and maintain high lipid levels in their tissue; and 3) regeneration is also a regulated process and may not necessarily drain energy available for other metabolic functions. He proposed instead that the allocation of stem cells between reproduction and regeneration of tissues would only require a trade-off when the number of amoebocytes was limited by chronic stress.

However, even partial mortality could still have significant consequences for the corals, especially if corallivory is chronic or coincides with other environmental stresses. Colonies usually regenerate tissues following corallivory, however this process of regeneration, especially when other stresses are acting on the colony, may divert resources from other important metabolic functions, such as tissue growth, competition with other organisms, or reproduction. These changes can impact the abundance and distribution of corals within the reef community.

EFFECTS OF CORALLIVORY ON CORAL PHYSIOLOGY

Energy Resources in Coral Tissues

Regeneration has been assumed to require diversion of metabolic resources from other activities for tissue repair. Regeneration in coral colonies from natural and artificial lesions is highly variable, depending on the size of the colony, the size of the lesion and its perimeter, the placement of the lesion, the shape of the lesion, as well as other environmental factors (Henry and Hart, 2005). Because colonies have modular growth, energetic resources can be translocated throughout the colony to assist areas being regenerated (Oren et al., 1997, 2001). Oren et al. (1998) also suggested that feeding by *Coralliophilla violacea* on poritid colonies was a form of prudent feeding by this snail corallivore. Poritid colonies normally translocate energy resources from central portions to the margins where production is usually lower but demand may be higher for growth. Therefore, feeding by the snail on the energy sinks at the margins would mininise tissue damage to other parts of the colony while allowing maximisation of energy resources available to the snail.

Two studies have shown no change in photosynthesis or respiration rates for entire colonies under conditions of corallivory (Cox, 1991; Gochfeld, 1991). Cox (1991) looked at medium-term (1 year) impacts of corallivory by *Chaetodon unimaculatus* on *Montipora capitata* (formerly *verrucosa*) and found no differences in rates of photosynthesis or respiration per unit of surface area, although corals under grazing showed changes in morphology and a decrease in surface area, with the potential for decreased total productivity due to reduced tissue area. Gochfeld (1991) looked at short-term (6 days) changes in photosynthesis and respiration rates in *Pocillopora meandrina* and found no differences in colonies fed on by the corallivorous damselfish *Plectroglyphidodon johnstonianus*. The rate of photosynthesis is usually controlled by the density of zooxanthellae, therefore, the lack of change in photosynthesis suggests that the removal of small amounts of tissue by corallivory in the short-term does not significantly change the productivity of the zooxanthellae throughout the colony's tissues. Increases in respiration would be expected if the colony needed to expend extra energy to repair tissues, although that increase would be greatest in the damaged area and adjacent areas, and hence may have an insignificant effect on measurements of the respiration of the entire colony.

Growth Rates of Corals

As a consequence of potential trade-offs between growth and other metabolic activities such as regeneration of damaged tissues, it has been assumed that colony growth will be depressed by the activities of corallivores. This has been demonstrated for a variety of corallivores feeding on various coral species. Neudecker (1977) measured a loss of 24% of the skeletal mass of *Pocillopora damicornis* colonies transplanted to the deeper reef areas off Guam, and used Alizarin staining and caging to demonstrate that growth rates in deeper areas were comparable to or greater than growth rates on the reef flat if colonies were protected from fish feeding. Glynn et al. (1979) estimated that the corallivorous sea urchin *Eucidaris thouarsii* removed 11–20% of the net $CaCO_3$ production of pocilloporid corals in high coral coverage areas (60–80%) in the Galapagos Islands, and up to 100% in areas where coral coverage was < 30%. Cox (1986) demonstrated 53%–66% higher vertical branch growth in *Montipora verrucosa* when protected from the butterflyfish *Chaetodon unimaculatus*, a species that shows preferential feeding on this coral species in this area. Witman (1988) showed that growth of *Millepora complanata* branch tips fed on by the corallivorous fireworm *Hermodice carunculata* had less than 5% of the vertical growth of non-consumed branch tips, however grazed branch tips were subsequently colonised by filamentous and coralline algae and that may have also prevented tissue regeneration and continued growth. Littler et al. (1989) found a 9% tissue area loss in 24 hours of exposure of *Porites porites* to grazing fishes on the reef crest in Belize. Kosaki (1999) found that unprotected *Pocillopora damicornis* had a 3% decrease in mass relative to control colonies protected from grazing by *Chaetodon lunulatus* (formerly *trifasciatus)* in Hawai'i. Burkepile and Hay (2007) found an increase of 36% in tissue removal on gorgonians when the gorgonian specialist snail *Cyphoma gibbosum* was protected from its predators.

Changes in growth rates for colonies have implications for population dynamics, as larger colonies typically have lower rates of mortality than small colonies, therefore slower growth may increase the period of time that colonies are exposed to other sources of mortality (Hughes, 1984; Jackson and Hughes, 1985; Hughes and Connell, 1987). Reproductive output is also linked with size, as the increase in number of polyps increases the number of gametes or planulae produced per colony. Models of population dynamics of corals are very sensitive to growth and increased size (Hughes, 1984; Hughes and Connell, 1987), although frequent disturbances and larval mortality also play major roles in population dynamics.

Transmission of Coral Diseases

Coral diseases were first described from the Atlantic in 1973 (Antonius, 1973) and the Indo-Pacific in 1985 (Antonius, 1985). Early reports correlated outbreaks of white syndrome in the Red Sea with an increase in the populations of a corallivorous snail *Drupella cornus* (Antonius and Reigl,

1997, 1998), although it was not clear to the investigators whether the disease attracted the snail or the feeding of the snail had aggravated the health of the corals. While there are many factors that affect the transmission of diseases, it is becoming increasingly apparent that feeding by corallivores may play a role in this process.

In Hawai'i, poritid corals are infected with a trematode parasite that causes the polyps to remain extended and discolored (Aeby, 1992, 2002), and are therefore, more vulnerable to predation by butterflyfishes. *Chaetodon multicinctus* feeds preferentially on the distended polyps (Aeby, 1992), serving as the next host in the life cycle of this parasite. Aeby (2002) estimated that about 8% of the ingested trematode cysts were able to metamorphose into juvenile worms in the butterflyfish gut, and that infection of the fish did not negatively impact body condition of the butterflyfishes. Fishes feeding on infected areas of the coral were able to obtain about 75% more coral tissue per bite than bites from uninfected polyps, although the total amount of food obtained was the same, as fishes feeding on uninfected polyps took more bites per unit time. Feeding by the butterflyfishes provides the infective stages for the next host, and these stages can be spread throughout the feeding territory of the butterflyfish (Aeby, 2007).

Likewise the polychaete corallivore, *Hermodice carunculata*, has been found to be the winter reservoir for the pathogenic bacteria *Vibrio shiloi*, which causes bleaching in *Oculina patagonica* during the warmer summer months in the Mediterranean (Sussman et al., 2003). The bacteria was not detected in the tissues of the coral during the winter months but was found in high numbers within the tissues of the corallivore. When laboratory infected specimens of the worm were placed in aquaria with healthy corals, the corals were totally bleached within 6 weeks, while corals exposed to worms infected with a related species (*V. mediterranei*) did not bleach. While the worms are not responsible for infection of the entire population (80–90% of *O. patagonica* bleach every summer), the worms may transmit the pathogen to the corals on which they feed, and population growth of *V. shiloi* within the coral tissues provides additional sources of infection for the *O. patagonica* population (Sussman et al., 2003).

A molluscan corallivore, *Coralliophila abbreviata*, has also been shown to transmit a coral white syndrome in the field from infected *Acropora cervicornis* colonies to healthy colonies in Florida (Williams and Miller, 2005). Williams and Miller (2005) transplanted snails found feeding on diseased or healthy corals to isolated *A. cervicornis* fragments and found rapid tissue loss in corals exposed to the snails that had previously fed on diseased corals. Although this corallivore has limited mobility, Williams and Miller (2005) suggest that the aggregation of snails on remnant populations of *Acropora* spp. can play a role in hastening the demise of the corals, both as a result of the concentrated feeding impacts and though the transmission of the pathogens for the coral disease.

Aeby and Santavy (2006) found that *Chaetodon capistratus* increased the rate of transmission of black-band disease between colonies of *Montastrea faveolata* in Florida, although any injury to coral tissues was also a mechanism of acquisition of disease if the pathogens were present. They suggest that dispersal of this disease could result from ingestion and then defecation of pathogens by butterflyfishes. The territoriality of butterflyfishes and the close proximity of colonies within the territories might increase their effectiveness as vectors for black-band disease (Aeby and Santavy, 2006). Grazing by other fishes, such as parrotfishes and damselfishes, may also cause damage to the corals, and the presence of these injuries may create sites for the infection of the coral by the pathogen for black-band disease.

Recovery from Bleaching

Corallivory may interfere with recovery from bleaching. McClanahan et al. (2005) transplanted colonies that had survived a bleaching event onto fished and unfished reefs. They found that colonies transplanted to the unfished reefs had low survivorship due to predation and suggested that survivorship following bleaching events would be partially determined by the susceptibility of the coral species to corallivores. The major corallivores associated with mortality or major damage in this study were *Acanthaster planci* and the puffer *Arothron nigropunctatus*. Rotjan et al. (2006) demonstrated decreased zooxanthellae density in *Montastrea* sp. tissue from colonies grazed by parrotfishes in Belize 5 months following a bleaching event. Interestingly, they also found an increased diversity in zooxanthellae type following bleaching in grazed colonies. Zooxanthellae diversity can play a role in resistance to bleaching or resilience following bleaching events.

EFFECTS OF CORALLIVORY ON CORAL POPULATION DYNAMICS

Survivorship of Coral Colonies and Fragments

As corals are modular organisms, colony survivorship is generally not impacted by feeding by corallivores, unless the colonies are small, are attacked by an outbreak of a corallivore population (Branham et al., 1971; Endean, 1973; Moyer et al., 1982), stressed by other factors such as bleaching (McIlwain and Jones, 1997), or placed in an environment where they normally do not occur and subsequently attract significant feeding activities (Neudecker, 1977, 1979; Wellington, 1982; Chasqui-Velasco et al., 2007). Even under "outbreak" conditions, individual corals are often not completely killed, as was noted by Branham et al. (1971) for *A. planci* feeding on their preferred prey species *Montipora capitata*. In general, smaller colonies have a higher chance to be killed by predation by some of the larger corallivorous fishes, such as parrotfishes (Miller and Hay, 1998; Rotjan and Lewis, 2005) or filefishes (Jayewardene and Birkeland, 2006).

Fragments of corals may be more susceptible to mortality due to their small size (Highsmith et al., 1980) and potential attraction of corallivores by chemical cues released from damaged coral tissues (Ormond et al., 1973; Sonoda and Paul, 1993; Teruya et al., 2001; Suenaga, 2004; Kita et al., 2005). McClanahan et al. (2005) noted that grazing by parrotfishes on transplanted fragments of *Porites* sp. had major impacts on coral transplant survivorship, while smaller butterflyfishes had few effects on the transplants. Fragments can also be created by fishes searching for other prey items within the branches of corals, and thus fragmentation could also be viewed as potential asexual reproduction.

Reproduction of Corals

Reproduction is another metabolic activity of corals that requires energy resources. Trade-offs between tissue repair in general and reproduction in corals have been demonstrated in *Montastrea* sp. (Szmant-Froehlich, 1984; Van Veghel and Bak, 1994), *Stylophora pistillata* (Rinkevich and Loya, 1989), *Pocillopora damicornis* (Zakai et al., 2000), *Acropora hyacinthus*, *A. gemmifera* and *Goniastrea retiformis* (Hall, 1997), *Acropora* sp. (Smith and Hughes, 1999), *Acropora palmata* (Lirman, 2000), and *Acropora formosa* (Okubo et al., 2007), suggesting that long-term or extensive corallivore damage could have a negative impact on reproduction. However, these studies include a wide range of injuries, from small to large lesions and fragmentation or transplantation. Lesions may decrease egg production adjacent to the lesion but have no effect on polyps further away from the lesion (Van Veghel and Bak, 1994). In fragmented corals, often the larger fragments would continue to produce eggs, however, eggs were sometimes smaller or less numerous than those produced by undamaged colonies. When comparing fragmented colonies of *Pocillopora damicornis* to non-fragmented colonies, Ward (1995) found no differences in larval production, although entire dislodged colonies, subjected to additional sediment stress as compared to intact colonies, did have reduced larval production. Okubo et al. (2007) found that fragmentation that occurred during the vitellogenic stages resulted in continued development of the oocytes, suggesting that reproduction in some species is programmed and will continue regardless of other demands on colony physiology.

In contrast to the previous studies, Cox (1991) found a 27% increase in the mass of gametes per unit of surface area produced by *Montipora capitata* following corallivory by *Chaetodon unimaculatus*, although these grazed corals showed a 48% decrease in linear growth. Williams (1975) proposed that clonal organisms should allocate resources to asexual reproduction and growth as long as space was available and switch to sexual reproduction when other factors (competition or predation) reduced the ability to capture new space (the strawberry-coral model). The response of *Montipora capitata* is consistent with the prediction that sexual reproduction should be increased in response to heavy predation to produce mobile propagules for dispersal to other habitats where predation levels would be different. Harvell and Helling (1993) also found that damage simulating predation on the growing edge of the bryozoan *Membranipora membranacea* lead to an increase in reproduction in zooids adjacent to the damage. They suggested that the normal source-sink relationship between centre and edge areas of the brozyoan was disrupted by the damage, allowing the zooids adjacent to the damage to utilise the resources that would have been translocated to the edge to initiate reproduction. Sánchez et al. (2004) studied transplanted *Montastrea annularis* that had been "marked" by parrotfish (a behaviour that removes coral tissue) and found spawning in these corals, suggesting that this species has the ability to maintain reproduction while recovering from partial mortality.

Studies of the impacts of damage to corals on their reproductive ability have shown negative, neutral, and positive effects. Some of the differences in these studies are related to the mode of sexual reproduction in the studied coral, the importance of asexual reproduction in the population dynamics of the studied species, the type of damage inflicted on the colony (small bites or large removals of tissue), and the alternate pathways for provision of resources for reproduction to the colony. Any evaluation of the effects of corallivory on reproduction must address all of these factors.

Recruitment of Corals

Predation can play a role in the recruitment of new individuals into a population and hence population dynamics of corals. Many coral species participate in mass spawning, providing an opportunity for feeding of many organisms on gametes that typically contain large amounts of lipid (Richmond, 1987; Arai et al., 1993). Fishes are predators of coral spawn (Westneat and Resing, 1988; Alino and Coll, 1989; Pratchett et al., 2001), which may lead to reductions in the subsequent supply of coral larvae. Westneat and Resing (1988) observed that damselfish planktivores switched almost entirely to feeding on coral spawn following a spawning even on the Great Barrier Reef. Alino and Coll (1989) described feeding of two butterflyfish species on octocoral spawn. Pratchett et al. (2001) observed 36 species of fish, including blennies, butterflyfishes, wrasses and damselfishes feeding on planktonic particles following coral spawning, and more than 90% of the plankton particles greater than 73 µm were coral eggs. Some of the species observed feeding in the plankton are not known as plankton feeders but abandoned benthic feeding to opportunistically benefit from the high-energy content of coral eggs. Pratchett et al. (2001) estimated that this consumption represents 20–36% of the coral propagules released. Clearly the consumption of coral eggs and larvae plays a major role in the cycling of materials in coral reef environments (Wild et al., 2008) with potential consequences for future larval recruitment. Pratchett et al. (2000) suggest that damselfishes feed preferentially on the eggs of acroporid species, finding that pellets made with eggs from *Pacyseris speciosa* were rejected in feeding trials. As acroporids are often the dominant species on Great Barrier Reefs, feeding selectivity may have an indirect effect on community composition by reducing the consumption of less abundant species and facilitating their survival.

Fishes and grazing urchins may also create space for recruitment of coral larvae. Appropriate benthic space is often a limiting resource for settlement in a coral reef environment (Jackson and Hughes, 1985). Grazing herbivores have been shown to create open benthic substrate for coral settlement (Brock, 1979; Birkeland and Randall, 1981; Sammarco, 1980, 1982), although Brock (1979) and Sammarco (1980) also demonstrated that these fish and echinoid grazers could kill newly settled spat. Carpenter and Edmunds (2006) demonstrated that the recovery of *Diadema* in the Caribbean has increased the amount of substratum suitable for coral settlement and thus is facilitating the recovery of scleractinian corals.

EFFECTS OF CORALLIVORY ON CORAL COMMUNITIES

Community Structure and Coral Diversity

Selective predation can play a role in the composition of coral communities. For example, Neudecker (1977, 1979) proposed that the high mortality and lower growth rates of *Pocillopora damicornis* associated with selective corallivory were the cause of the lack of *P. damicornis* on the deeper reef slopes. Cox (1986) suggested that selective predation on *Montipora capitata* altered the distributional pattern of this coral on a reef in Hawai'i where the selective corallivore *Chaetodon unimaculatus* was very abundant. Littler et al. (1989) suggested that the distribution of *Porites porites* on Belize reefs could be controlled by the grazing of parrotfishes on this species. Hoey and Bellwood (2008) indicated that the non-selective feeding of the large parrotfish *Bolbometopon muricatum* on the outer reefs of the Great Barrier Reef promotes coral diversity by reducing abundances of competitive dominant or fast growing coral species.

Effects of Corallivory on the Outcome of Competition

The effects of corallivory may be magnified if the coral is involved in competition with other benthic organisms. Corals may use mesenterial filaments and/or sweeper tentacles to kill tissues of other species that come in contact or near to them. Other species will grow in alternate directions to avoid interactions with species that may use mesenterial filaments against them. For example, Romano (1990) found that the slow growing *Cyphastrea ocellina* used digestion to prevent tissue loss to *Pocillopora damicornis* in field experiments in Hawai'i, with a subsequent loss in resources available for colony growth. The faster growing *P. damicornis* re-directed its growth and showed no difference in overall growth rates when in contact with *C. ocellina*. Therefore, any impacts of corallivory on coral species involved in competition could be magnified by the allocation of resources for competitive interactions. There have been no direct tests of the effects of corallivory on competition between corals, although Cox (1986) noted some reversals of the expected competitive outcome between *Montipora capitata (*formerly *verrucosa)* and *Porites compressa* when a specialist butterflyfish was allowed to feed on experimental colonies being monitored for changes in growth.

Competition may also impact reproduction in competing species. Rinkevich and Loya (1985) found a reduction in the number of female gonads per polyp in *Stylophora pistillata* colonies that were involved in intraspecific competition, demonstrating that the energy and resources required for successful competition decreased the ability of this species to continue planulation. In contrast, Tanner (1997) found no changes in reproductive parameters in *Acropora hyacinthus* involved in intraspecific competition, and the zone of non-fertile polyps actually decreased adjacent to the site of competition. Overall colony fecundity, however, was reduced as a result of the decrease in the total number of polyps due to a reduction in the rate of colony growth. These results suggest that corallivory, within the context of the additional stress of coral competition, could also affect the availability of resources for reproduction.

INDIRECT EFFECTS

In addition to direct effects of corallivory, there are other indirect impacts on corals that are associated with corallivory. Although crustacean guards may consume coral tissue, including fat bodies produced in the tentacles of pocilloporid corals (Stimson, 1990), they can also be an important defence of corals against additional feeding by *Acanthaster planci* (Glynn, 1987). Mithrax crabs, opportunistically feeding on coral tissues, assist *Osculina arbuscula* by removing seaweeds that could cover the coral colonies (Stachowicz and Hay, 1999). Symbiotic crabs may also assist corals in the removal of sediments (Stewart et al., 2006) that could result in bleaching of the tissues. In addition to defence, anemonefishes provide ammonia that can be absorbed by their host anemones (Roopin et al., 2008), providing a substantial nutritional benefit to the host.

Damselfishes are known algal gardeners, and their activities may kill coral tissues to facilitate algal growth, but their territorial behaviour may also protect corals growing within their territories (Vine, 1974; Kaufman, 1977; Potts, 1977; Sammarco and Williams, 1982; Glynn and Colgan, 1988; Jones et al., 2006). Wellington (1982) found complex changes in coral composition associated with damselfish killing of coral tissue to maintain their territories. In shallow water, damselfishes were more successful in removing *Pavona* polyps and thereby increased *Pocillopora* recruitment and successful competition against *Pavona*, while in deeper waters, shelters for damselfishes were reduced, and *Pavona* escaped the effects of damselfish polyp removal to become the dominant coral.

Feeding on corals may also provide benefit to corals in terms of a mechanism for the transmission of zooxanthellae between coral colonies. Consumption of coral tissue by some butterflyfishes does not kill symbiotic zooxanthellae, as they are viable when released in faeces (Parker, 1984).

HOW DO CORALS DEFEND THEMSELVES?

The coral skeleton has long been viewed as a defence mechanism against predation. Corals can withdraw their polyps and tentacles into the calyx, providing protection from corallivores such as small-mouthed butterflyfishes that feed by nipping at the polyps. Withdrawal of polyps may also limit the spread of feeding activities, as a wave of contraction of polyps spreads out from the area initially attacked. Reproductive materials, with high lipid content, can be protected from predation by their position along the mesenteries, sheltered deep within the calyx. In the soft corals, sclerites are thought to provide potential protection from predation (West, 1997), however Van Alstyne et al. (1994) suggested that sclerites in *Sinularia* were primarily used for support, as secondary metabolites provided antipredator defences. Likewise Kelman et al. (1999) found that the sclerites of *Parerythropodium fulvum* were palatable to generalised predators, although chemical defence did provide protection for this soft coral.

Small polyps are also viewed as a defence against feeding. Smaller polyps contain fewer energy resources for the predator than do larger, fleshier polyps. In addition, the low caloric content of coral tissues is also viewed as a deterrent to feeding on these organisms.

Many soft corals are known to produce toxic chemicals to deter feeding (Coll et al., 1982). However, these chemicals do not deter feeding by all predators (see review Ianora et al., 2006). In fact, coral feeding butterflyfishes were observed feeding on zoanthids containing palytoxin (Gleibs and Mebs, 1998), one of the most potent toxins in nature, and contained high levels of palytoxin in their organs without showing any deleterious effects. Soft corals can also produce water-borne cues that serve as anti-predatory signals (Goddard, 2006). Goddard (2006) found that attacks by a nudibranch predator on *Discophyton rudyi* caused the release of an alarm signal that lead to contraction of the polyps in adjacent colonies. However, release of water-borne substances has also been implicated in predator attraction to injured corals (Ormond et al., 1973; Teruya et al., 2001; Suenaga, 2004; Kita et al., 2005).

Sweeper tentacle and cnidarian nematocysts are also a potential mechanism for defence. Corals are known to produce sweeper tentacles that have large numbers of nematocysts and are used in defence of the colony from intra or interspecific competitors for space (Richardson et al., 1979; Wellington, 1980; Chornesky, 1983). Sweeper tentacles production may be induced by competition. Wellington (1980) suggested that the sweeper tentacles of pavonids were induced by competition, to reverse the initial dominance of pocilloporids over slower growing pavonids. Induction of sweeper tentacles has also been demonstrated in *Agaricia agaricities* (Chornesky, 1983), *Galaxea fascicularis* (Hidaka and Yamazato, 1984), and the corallimorpharian *Rhodactis rhodostoma* (Langmead and Chadwick-Furman, 1999). However, Peach and Hoegh-Guldberg (1999) did not find induction of sweeper tentacles in *Goniopora tenuidens*, however sweeper tentacles in this species are suspected of being instrumental in defense and not competitive interactions. Lapid and Chadwick (2006) found that sweeper tentacles in *Platygyra daedalea* were efficient mechanisms to prevent competition from *Favites complanata* and did not involve extra metabolic costs to *P. daedalea*.

Nematocyst densities may also be increased by predation. Gochfeld (2004) found that *Porites compressa* used polyp withdrawal for protection when intensively grazed by butterflyfishes in a laboratory situation, and that they also increased the nematocyst density in grazed parts of the colony over the longer term. Gochfeld (2004) suggested that in field situations the short-term withdrawal of polyps may be the signal for a butterflyfish to move on to adjacent coral colonies, allowing the grazed colony to then recover from the feeding bout and re-extend its polyps. The increase in nematocysts within intensively grazed portions of colonies in the lab study was followed by a greater number of head shakes following bites on grazed colonies, that was interpreted to mean that the butterflyfish was reacting to the nematocysts (Gochfeld, 2004).

Corals may also harbour crustacean symbionts that serve as coral guards (Glynn, 1974, 1976, 1987), protecting pocilloporids from attack by feeding *Acanthaster planci*. The crustaceans nip at the tube feet of the crown of thorns seastar, deterring feeding. Pratchett (2001) demonstrated that removal of the symbionts lead to increased feeding on the protected coral species. Although the coral hosts can be protected from feeding, the host may eventually be fed upon, especially when other prey items are limited (Pratchett, 2001). Anemones have anemonefishes that may also provide defences against coral-feeding fishes, including butterflyfishes (Godwin and Fautin, 1992). In the Red Sea, anemones with anemonefish expanded their tentacles 60% more time than those lacking anemonefish, and experimental removals of the anemonefish led to attacks by *Chaetodon fasciatus* (Porat and Chadwick-Furman, 2004). Anemones in French Polynesia with anemonefishes had 3 times faster growth rates than anemones without anemonefish (Holbrook and Schmitt, 2005), and anemones that were not defended by anemonefish suffered higher mortality.

CONCLUSIONS

Clearly there is a wide range of potential impacts associated with corallivory. When an outbreak of corallivores occurs or there are multiple stressors affecting corals simultaneously, these impacts can be severe. The type of feeding behaviour of the corallivore will also determine the severity of the injury to the coral species. Corals can resist some forms of corallivory by using polyp withdrawal into a hard skeleton or production of toxic chemicals in soft corals. Because corals are modular organisms, they also have the ability to tolerate, to some extent, partial mortality and regenerate lost tissues.

Corallivorous butterflyfishes in particular have additional behaviours that reduce their impact on corals. They typically feed by plucking a few tentacles from polyps or removing only a few polyps, and the coral response of withdrawal of adjacent polyps will reduce feeding activities on an individual colony. Butterflyfishes often defend feeding territories that may be stable for many years, reducing the abundance of other corallivores using the coral food resource.

There is a great need for more studies on the impacts of corallivores, and butterflyfishes in particular, on the physiology, population dynamics and community composition of corals. These studies need to also focus on what resources are limiting for coral survival, growth and reproduction and how corallivory affects the ability of corals to obtain those resources.

REFERENCES

Aeby, G.S. 1992. The potential effect the ability of a coral intermediate host to regenerate has had on the evolution of its association with a marine parasite. *Proceedings of the 7th International Coral Reef Symposium, Guam* pp. 809–815.

Aeby, G.S. 2002. Trade-offs for the butterflyfish, *Chaetodon multicinctus,* when feeding on coral prey infected with *trematode metacercariae. Behavioral Ecology and Sociobiology* 52: 158–165.

Aeby, G.S. 2007. Spatial and temporal patterns of *Porites* trematodiasis on the reefs of Kaneohe Bay, Oahu, Hawai'i. *Bulletin of Marine Science* 80: 209–218.

Aeby, G.S. and D.L. Santavy. 2006. Factors affecting susceptibility of the coral *Montastraea faveolata* to black-band disease. *Marine Ecology Progress Series* 318: 103–110.

Aliño, P.M. and J.C. Coll. 1989. Observations of the Synchronized Mass Spawning and Post Settlement Activity of Octocorals on the Great Barrier Reef, Australia: Biological Aspects. *Bulletin of Marine Science* 45: 697–707.

Antonius, A. 1973. New observations on coral destruction in reefs. *Proceedings of the 10th Meeting of the Association of Island Marine Laboratories* pp. 3.

Antonius, A. 1985. Coral Diseases in the Indo-Pacific: A First Record. *Marine Ecology* 6: 197–218.

Antonius, A. and B. Riegl. 1997. A possible link between coral diseases and a corallivorous snail (*Drupellacornus*) outbreak in the Red Sea. *Atoll Research Bulletin* 47: 1–9.

Antonius, A. and B. Riegl. 1998. Coral diseases and *Drupella* cornus invasion in the Red Sea. *Coral Reefs* 17: 48–48.

Arai, I., M. Kato, A. Heyward, Y. Ikeda, T. Iizuka and T. Maruyama. 1993. Lipid composition of positively buoyant eggs of reef building corals. *Coral Reefs* 12: 71–75.

Belsky, A.J., W.P. Carson, C.L. Jensen and G.A. Fox. 1993. Over compensation by plants: herbivore optimization or red herring? *Evolutionary Ecology* 7: 109–121.

Birkeland, C. and R. Randall. 1981. Facilitation of coral recruitment by echinoid excavations. *Proceedings of the 4th International Coral Reef Symposium, Manila* pp. 695–698.

Branham, J., S. Reed, J.H. Bailey and J. Caperon. 1971. Coral-eating sea stars *Acanthaster planci* in Hawai'i. *Science* 172: 1155.

Brock, R.E. 1979. An experimental study on the effects of grazing by parrotfishes and role of refuges in benthic community structure. *Marine Biology* 51: 381–388.

Burkepile, D.E. and M.E. Hay. 2007. Predator release of the gastropod *Cyphoma gibbosum* increases predation on gorgonian corals. *Oecologia* 154: 167–173.

Carlson, B. 1992. The life history and reproductive success of the coral blenny, *Exallias brevis* Kner, 1868. PhD dissertation, The University of Hawai'i, Honolulu, USA.

Carpenter, R.C. and P.J. Edmunds. 2006. Local and regional scale recovery of *Diadema* promotes recruitment of scleractinian corals. *Ecology Letters* 9: 271–280.

Chalker, B., W. Dunlap and J. Oliver. 1983. Bathymetric adaptations of reef-building corals at Davies Reef, Great Barrier Reef, Australia. II. Light saturation curves for photosynthesis and respiration. *Journal of Experimental Marine Biology and Ecology* 73: 37–56.

Chasqui-Velasco, L., E. Alvarado, A. Acero and F.A. Zapata. 2007. Effect of herbivorous and corallivorous fishes on the survival of transplanted corals in the Colombian Caribbean. *Revista de Biologia Tropical* 55: 825–837.

Chornesky, E.A. 1983. Induced development of sweeper tentacles on the reef coral *Agaricia agaricites*: a response to direct competition. *Biological Bulletin* 165: 569–581.

Colgan, M.W. 1987. Coral reef recovery on Guam (Micronesia) after catastrophic predation by *Acanthaster planci. Ecology*: 1592–1605.

Coll, J.C., S. La Barre, P.W. Sammarco, W.T. Williams and G.J. Bakus. 1982. Chemical Defences in Soft Corals (Coelenterata: Octocorallia) of the Great Barrier Reef: A Study of Comparative Toxicities. *Marine Ecology Progress Series* 8: 271–278.

Cox, E.F. 1986. The effects of a selective corallivore on growth rates and competition for space between two species of Hawaiian corals. *Journal of Experimental Marine Biology and Ecology* 101: 161–174.

Cox, E.F. 1991. Interactions between trophic levels on coral reefs: Scleractinian corals and corallivorous butterflyfishes in Hawai'i. PhD dissertation, The University of New Mexico, Albuquerque, USA.

Dana, J. 1849. United States exploration expedition 1838–1842. *Geology* 10: 109.

Darwin, C. 1845. *Journal of researches into the natural history and geology of the countries visited during the voyage of H.M.S. Beagle round the world: under command of Captain Fitz Roy*, John Murray, London, UK.

Edmunds, P.J. and P.S. Davies. 1986. An energy budget for *Porites porites* (Scleractinia). *Marine Biology* 92: 339–347.

Endean, R. 1973. Population explosions of *Acanthaster planci* and associated destruction of hermatypic corals in the Indo-West Pacific region. *Biology and Geology of Coral Reefs* 2: 389–438.

Endean, R. and A. Cameron. 1990. *Acanthaster planci* population outbreaks. *Ecosystems of the World* 25: 419–437.

Gleibs, S. and D. Mebs. 1998. Sequestration of a marine toxin. *Coral Reefs* 17: 338–338.

Glynn, P.W. 1974. The impact of *Acanthaster* on corals and coral reefs in the eastern Pacific. *Environmental Conservation* 1: 295–304.

Glynn, P.W. 1976. Some physical and biological determinants of coral community structure in the eastern Pacific. *Ecological Monographs*: 431–456.

Glynn, P.W. 1987. Some ecological consequences of coral-crustacean guard mutualisms in the Indian and Pacific Oceans. *Symbiosis* 4: 301–323.

Glynn, P.W. and M.W. Colgan. 1988. Defense of corals and enhancement of coral diversity by territorial damselfishes. *Proceedings of the 6th International Coral Reef Symposium, Townsville* pp. 157–63.

Glynn, P.W., G.M. Wellington and C. Birkeland. 1979. Coral reef growth in the Galapagos: limitation by sea urchins. *Science* 203: 47.

Gochfeld, D.J. 1991. Energetics of a predator-prey interaction: Corals and coral-feeding fishes. *Pacific Science* 45: 246–256.

Gochfeld, D.J. 2004. Predation-induced morphological and behavioural defenses in a hard coral: implications for foraging behaviour of coral-feeding butterflyfishes. *Marine Ecology Progress Series* 267: 145.

Gochfeld, D.J. and G.S. Aeby. 1997. Control of populations of the coral-feeding nudibranch *Phestilla sibogae* by fish and crustacean predators. *Marine Biology* 130: 63–69.

Goddard, J.H.R. 2006. Stealthy slugs and communicating corals: polyp withdrawal by an aggregating soft coral in response to injured neighbors. *Canadian Journal of Zoology* 84: 66–71.

Godwin, J. and D.G. Fautin. 1992. Defense of host actinians by anemonefishes. *Copeia* 1992: 902–908.

Grottoli, A.G., L.J. Rodrigues and J.E. Palardy. 2006. Heterotrophic plasticity and resilience in bleached corals. *Nature* 440: 1186–1189.

Hall, V.R. 1997. Effects of injury on growth, reproduction and survivorship for common reef-crest corals. *Proceedings of the 8th International Coral Reef Symposium, Panama*, pp. 571–574.

Harmelin-Vivien, M. and Y. Bouchon-Navaro. 1981. Trophic relationships among chaetodontid fishes in the Gulf of Aqaba (Red Sea). *Proceedings of the 4th International Coral Reef Symposium, Manila* pp. 537–544.

Harmelin-Vivien, M. and Y. Bouchon-Navaro. 1983. Feeding diets and significance of coral feeding among chaetodontid fishes in Moorea (French Polynesia). *Coral Reefs* 2: 119–127.

Harvell, C.D. and R. Helling. 1993. Experimental induction of localized reproduction in a marine bryozoan. *Biological Bulletin* 184: 286–295.

Henry, L.-A. and M. Hart. 2005. Regeneration from Injury and Resource Allocation in Sponges and Corals—a Review. *International Review of Hydrobiology* 90: 125–158.

Hidaka, M. and K. Yamazato. 1984. Intraspecific interactions in a scleractinian coral, *Galaxea fascicularis*: induced formation of sweeper tentacles. *Coral Reefs* 3: 77–85.

Highsmith, R.C., A.C. Riggs and C.M. D'Antonio. 1980. Survival of hurricane-generated coral fragments and a disturbance model of reef calcification/growth rates. *Oecologia* 46: 322–329.

Hoey, A.S. and D.R. Bellwood. 2008. Cross-shelf variation in the role of parrotfishes on the Great Barrier Reef. *Coral Reefs* 27: 37–47.

Holbrook, S.J. and R.J. Schmitt. 2005. Growth, reproduction and survival of a tropical sea anemone (Actiniaria): benefits of hosting anemonefish. *Coral Reefs* 24: 67–73.

Hughes, T.P. 1984. Population dynamics based on individual size rather than age: a general model with a reef coral example. *American Naturalist* 123: 778–795.

Hughes, T.P. and J.H. Connell. 1987. Population dynamics based on size or age? A reef-coral analysis. *American Naturalist* 129: 818–829.

Ianora, A., M. Boersma, R. Casotti, A. Fontana, J. Harder, F. Hoffmann, H. Pavia, P. Potin, S. Poulet and G. Toth. 2006. New trends in marine chemical ecology. *Estuaries and Coasts* 29: 531–551.

Jackson, J.B.C. and T.P. Hughes. 1985. Adaptive Strategies of Coral-Reef Invertebrates. *American Scientist* 73: 265–273.

Jayewardene, D. and C. Birkeland. 2006. Fish predation on Hawaiian corals. *Coral Reefs* 25: 328.

Jones, G.P., L. Santana, L.J. McCook and M.I. McCormick. 2006. Resource use and impact of three herbivorous damselfishes on coral reef communities. *Marine Ecology Progress Series* 328: 215–224.

Kaufman, L. 1977. The three spot damselfish: effects on benthic biota of Caribbean coral reefs. *Proceedings of the 3rd International Coral Reef Symposium, Miami* pp. 559–564.

Kelman, D., Y. Benayahu and Y. Kashman. 1999. Chemical defence of the soft coral *Parerythropodium fulvum fulvum* (Forskål) in the Red Sea against generalist reef fish. *Journal of Experimental Marine Biology and Ecology* 238: 127–137.

Kita, M., M. Kitamura, T. Koyama, T. Teruya, H. Matsumoto, Y. Nakano and D. Uemura. 2005. Feeding attractants for the muricid gastropod *Drupella cornus*, a coral predator. *Tetrahedron letters* 46: 8583–8585.

Kosaki, R.K. 1999. Behavioral mechanisms of coexistence among coral-feeding butterflyfishes. PhD dissertation, University of Hawai'i, Honolulu, USA.

Langmead, O. and N. Chadwick-Furman. 1999. Marginal tentacles of the corallimorpharian *Rhodactis rhodostoma*. 2. Induced development and long-term effects on coral competitors. *Marine Biology* 134: 491–500.

Lapid, E. and N. Chadwick. 2006. Long-term effects of competition on coral growth and sweeper tentacle development. *Marine Ecology Progress Series* 313: 115–123.

Lirman, D. 2000. Fragmentation in the branching coral *Acropora palmata* (Lamarck): growth, survivorship, and reproduction of colonies and fragments. *Journal of Experimental Marine Biology and Ecology* 251: 41–57.

Littler, M.M., P.R. Taylor and D.S. Littler. 1989. Complex interactions in the control of coral zonation on a Caribbean reef flat. *Oecologia* 80: 331–340.

McClanahan, T.R., J. Maina, C.J. Starger, P. Herron-Perez and E. Dusek. 2005. Detriments to post-bleaching recovery of corals. *Coral Reefs* 24: 230–246.

McIlwain, J.L. and G.P. Jones. 1997. Prey selection by an obligate coral-feeding wrasse and its response to small-scale disturbance. *Marine Ecology Progress Series* 155: 189–198.

Miller, M.W. and M.E. Hay. 1998. Effects of fish predation and seaweed competition on the survival and growth of corals. *Oecologia* 113: 231–238.

Moran, P. 1986. The Acanthaster phenomenon. *Oceanography and Marine Biology: An Annual Review* 24: 379–480.

Moyer, J.T., W.K. Emerson and M. Ross. 1982. Massive destruction of scleractinian corals by the muricid gastropod *Drupella* in Japan and the Philippines. *Nautilus* 96: 69–82.
Muscatine, L., L. McCloskey and R. Marian. 1981. Estimating the daily contribution of carbon from zooxanthellae to coral animal respiration. *Limnology and Oceanography* 26: 601–611.
Neudecker, S. 1977. Transplant experiments to test the effect of fish grazing on coral distribution. *Proceedings of the 3rd International Coral Reef Symposium, Miami* pp. 317–323.
Neudecker, S. 1979. Effects of grazing and browsing fishes on the zonation of corals in Guam. *Ecology* 60: 666–672.
Núñez-Farfán, J., J. Fornoni and P.L. Valverde. 2007. The Evolution of Resistance and Tolerance to Herbivores. *Annual Review of Ecology Evolution and Systematics* 38: 541–566.
Okubo, N., T. Motokawa and M. Omori. 2007. When fragmented coral spawn? Effect of size and timing on survivorship and fecundity of fragmentation in *Acropora formosa*. *Marine Biology* 151: 353–363.
Oren, U., B. Rinkevich and Y. Loya. 1997. Oriented intra-colonial transport of super (14) C labeled materials during coral regeneration. *Marine Ecology Progress Series* 161: 117–122.
Oren, U., I. Brickner and Y. Loya. 1998. Prudent sessile feeding by the corallivore snail, *Coralliophila violacea* on coral energy sinks. *Proceedings of the Royal Society of London. Series B: Biological Sciences* 265: 2043–2050.
Oren, U., Y. Benayahu, H. Lubinevsky and Y. Loya. 2001. Colony integration during regeneration in the stony coral *Favia favus*. *Ecology* 82: 802–813.
Ormond, R.F.G., A.C. Campbell, S.H. Head, R.J. Moore, P.R. Rainbow and A.P. Saunders. 1973. Formation and Breakdown of Aggregations of the Crown-of-Thorns Starfish, *Acanthaster planci* (L.). *Nature* 246: 167–169.
Parker, G.M. 1984. Dispersal of zooxanthellae on coral reefs by predators on cnidarians. *Biological Bulletin* 167: 159–167.
Peach, M.B. and O. Hoegh-Guldberg. 1999. Sweeper polyps of the coral *Gonioporatenuidens* (Scleractinia: Poritidae). *Invertebrate Biology* 118: 1–7.
Porat, D. and N. Chadwick-Furman. 2004. Effects of anemonefish on giant sea anemones: expansion behavior, growth, and survival. *Hydrobiologia* 530: 513–520.
Potts, D.C. 1977. Suppression of Coral Populations by Filamentous Algae within Damselfish Territories. *Journal of Experimental Marine Biology and Ecology* 28: 207–216.
Pratchett, M.S. 2001. Influence of coral symbionts on feeding preferences of crown-of-thorns starfish *Acanthaster planci* in the western Pacific. *Marine Ecology Progress Series* 214: 111–119.
Pratchett, M.S., A.H. Baird and C.P. Marquis. 2000. Comparative palatability among eggs of mass-spawning corals. *Proceedings of the 9th International Coral Reef Symposium, Bali* pp. 391–394.
Pratchett, M.S., N. Gust, G. Goby and S.O. Klanten. 2001. Consumption of coral propagules represents a significant trophic link between corals and reef fish. *Coral Reefs* 20: 13–17.
Reese, E. 1991. How behavior influences community structure of butterflyfishes (family Chaetodontidae) on Pacific coral reefs. *Ecology International Bulletin* 19: 29–41.
Richardson, C., P. Dustan and J. Lang. 1979. Maintenance of living space by sweeper tentacles of *Montastrea cavernosa*, a Caribbean reef coral. *Marine Biology* 55: 181–186.
Richmond, R. 1987. Energetics, competency, and long-distance dispersal of planula larvae of the coral *Pocillopora damicornis*. *Marine Biology* 93: 527–533.
Rinkevich, B. 1996. Do reproduction and regeneration in damaged corals compete for energy allocation? *Marine Ecology Progress Series* 143: 297–302.
Rinkevich, B. and Y. Loya. 1983. Oriented translocation of energy in grafted reef corals. *Coral Reefs* 1: 243–247.
Rinkevich, B. and Y. Loya. 1985. Intraspecific competition in a reef coral: effects on growth and reproduction. *Oecologia* 66: 100–105.

Rinkevich, B. and Y. Loya. 1989. Reproduction in regenerating colonies of the coral *Stylophora pistillata*. *Environmental Quality and Ecosystem Stability: proceedings of the 4th International conference of the Israel Society for Ecology & Environmental Quality Sciences, Jerusalem* pp. 257–265.

Romano, S.L. 1990. Long-term effects of interspecific aggression on growth of the reef-building corals *Cyphastrea ocellina* (Dana) and *Pocillopora damicomis* (Linnaeus). *Journal of Experimental Marine Biology and Ecology* 140: 135–146.

Roopin, M., R.P. Henry and N.E. Chadwick. 2008. Nutrient transfer in a marine mutualism: patterns of ammonia excretion by anemonefish and uptake by giant sea anemones. *Marine Biology* 154: 547–556.

Rotjan, R.D. and S.M. Lewis. 2005. Selective predation by parrotfishes on the reef coral *Porites astreoides*. *Marine Ecology Progress Series* 305: 193–201.

Rotjan, R.D., J. Dimond, D. Thornhill, J. Leichter, B. Helmuth, D. Kemp and S. Lewis. 2006. Chronic parrotfish grazing impedes coral recovery after bleaching. *Coral Reefs* 25: 361–368.

Sammarco, P.W. 1980. *Diadema* and its relationship to coral spat mortality: grazing, competition, and biological disturbance. *Journal of Experimental Marine Biology and Ecology* 45: 245–272.

Sammarco, P.W. 1982. Echinoid grazing as a structuring force in coral communities: whole reef manipulations. *Journal of Experimental Marine Biology and Ecology* 61: 31–55.

Sammarco, P.W. and A.H. Williams. 1982. Damselfish territoriality: Influence on *Diadema* distribution and implications for coral community structure. *Marine Ecology Progress Series* 8: 53–59.

Sánchez, J.A., M.F. Gil, L.H. Chasqui and E.M. Alvarado. 2004. Grazing dynamics on a Caribbean reef-building coral. *Coral Reefs* 23: 578–583.

Sano, M. 2000. Stability of reef fish assemblages: responses to coral recovery after catastrophic predation by *Acanthaster planci*. *Marine Ecology Progress Series* 198: 121–130.

Smith, L.D. and T.P. Hughes. 1999. An experimental assessment of survival, re-attachment and fecundity of coral fragments. *Journal of Experimental Marine Biology and Ecology* 235: 147–164.

Sonoda, K. and V.J. Paul. 1993. Effect of stony corals extracts on feeding by *Acanthaster planci*. *Marine Ecology Progress Series* 102: 161–161.

Stachowicz, J.J. and M.E. Hay. 1999. Mutualism and coral persistence: the role of herbivore resistance to algal chemical defense. *Ecology* 80: 2085–2101.

Stewart, H.L., S.J. Holbrook, R.J. Schmitt and A.J. Brooks. 2006. Symbiotic crabs maintain coral health by clearing sediments. *Coral Reefs* 25: 609–615.

Stimson, J. 1990. Stimulation of fat-body production in the polyps of the coral *Pocillopora damicornis* by the presence of mutualistic crabs of the genus *Trapezia*. *Marine Biology* 106: 211–218.

Suenaga, K. 2004. Bioorganic studies on marine natural products with bioactivity, such as antitumor activity and feeding attractance. *Bulletin of the Chemical Society of Japan* 77: 443–451.

Sussman, M., Y. Loya, M. Fine and E. Rosenberg. 2003. The marine fireworm *Hermodicecarunculata* is a winter reservoir and spring-summer vector for the coral-bleaching pathogen *Vibrio shiloi*. *Environmental Microbiology* 5: 250–255.

Szmant-Froelich, A. 1985. The effect of colony size on the reproductive ability of the Caribbean coral *Montastrea annularis* (Ellis and Solander). *Proceedings of the 5th International Coral Reef Symposium, Tahiti* pp. 295–300.

Tanner, J.E. 1997. Interspecific competition reduces fitness in scleractinian corals. *Journal of Experimental Marine Biology and Ecology* 214: 19–34.

Teruya, T., K. Suenaga, T. Koyama, Y. Nakano and D. Uemura. 2001. Arachidonic acid and alpha-linolenic acid, feeding attractants for the crown-of-thorns sea star *Acanthaster planci*, from the sea urchin *Toxopneustes pileolus*. *Journal of Experimental Marine Biology and Ecology* 266: 123–134.

Van Alstyne, K.L., C.R. Wylie and V.J. Paul. 1994. Antipredator defenses in tropical Pacific soft corals (Coelenterata: Alcyonacea) II. The relative importance of chemical and structural defenses in three species of Sinularia. *Journal of Experimental Marine Biology and Ecology* 178: 17–34.

vanVeghel, M.L.J. and R.P.M. Bak. 1994. Reproductive characteristics of the polymorphic Caribbean reef building coral *Montastrea annularis*. III. Reproduction in damaged and regenerating colonies. *Marine Ecology Progress Series* 109: 229–233.

Vine, P. 1974. Effects of algal grazing and aggressive behaviour of the fishes *Pomacentrus lividus* and *Acanthurus sohal* on coral-reef ecology. *Marine Biology* 24: 131–136.

Ward, S. 1995. The effect of damage on the growth, reproduction and storage of lipids in the scleractinian coral *Pocillopora damicornis* (Linnaeus). *Journal of Experimental Marine Biology and Ecology* 187: 193–206.

Wellington, G.M. 1980. Reversal of digestive interactions between Pacific reef corals: mediation by sweeper tentacles. *Oecologia* 47: 340–343.

Wellington, G.M. 1982. Depth zonation of corals in the Gulf of Panama: Control and facilitation by resident reef fishes. *Ecological Monographs* 52: 223–241.

West, J.M. 1997. Plasticity in the sclerites of a gorgonian coral: tests of water motion, light level, and damage cues. *Biological Bulletin* 192: 279–289.

Westneat, M.W. and J.M. Resing. 1988. Predation on coral spawn by planktivorous fish. *Coral Reefs* 7: 89–92.

Wild, C., C. Jantzen, U. Struck, O. Hoegh-Guldberg and M. Huettel. 2008. Biogeochemical responses following coral mass spawning on the Great Barrier Reef: Pelagic-benthic coupling. *Coral Reefs* 27: 123–132.

Williams, G.C. 1975. *Sex and evolution*. Monographs in Population Biology, Princeton University Press, Princeton, USA.

Williams, D.E. and M.W. Miller. 2005. Coral disease outbreak: pattern, prevalence and transmission in *Acropora cervicornis*. *Marine Ecology Progress Series* 301: 119–128.

Wise, M.J. and W.G. Abrahamson. 2007. Effects of resource availability on tolerance of herbivory: a review and assessment of three opposing models. *American Naturalist* 169: 443–454.

Witman, J.D. 1988. Effects of Predation by the Fireworm *Hermodicecarunculata* on Milleporid Hydrocorals. *Bulletin of Marine Science* 42: 446–458.

Zakai, D., O. Levy and N. Chadwick-Furman. 2000. Experimental fragmentation reduces sexual reproductive output by the reef-building coral *Pocillopora damicornis*. *Coral Reefs* 19: 185–188.

CHAPTER 8

Social Structures and Spawning Behavior of *Chaetodon* Butterflyfishes

Shinji Yabuta[1,*] and *Michael L. Berumen*[2]

INTRODUCTION

While some aspects of the biology and ecology of butterflyfishes (family Chaetodontidae) have been very well studied, especially in relation to their vulnerability to coral loss and reef degradation (e.g., Harmelin-Vivien, 1989; Harmelin-Vivien and Bouchon-Navaro, 1983; Graham, 2007; Berumen and Pratchett, 2008; Lawton et al., 2012; Wilson et al., Chapter 9), very little is known about their social structure. Social structures, defined by a whole of interactions and relationships between individuals, such as aggression, mating system, territoriality, social grouping, social bonds, care of offspring, division of labor, and so on (Hinde, 1982), are critically important in understanding the evolution and ecology of species, and have important ramifications for the structure and dynamics of populations and communities. Marine fishes are renowned for their diversity of social structures, ranging from monogamous mating between gonochoric individuals to complex polygamous mating systems often involving simultaneous or sequential hermaphrodites (Cole, 2010), both within and among families.

[1]Department of Animal Sciences, Teikyo University of Science & Technology, 2525 Yatsusawa, Uenohara, Yamanashi 409-0193, Japan.
[2]Red Sea Research Center, King Abdullah University of Science and Technology, Thuwal 23955, Kingdom of Saudi Arabia.
*Corresponding author

Chaetodon butterflyfishes exhibit a wide diversity of social structure, presumably related to a diversity of mating systems. How and why the diversity of social structure has evolved among marine fishes remains and active area of enquiry, and butterflyfishes provide a useful group to explore these questions (Fricke, 1986; Hourigan, 1989). The predominant social structure exhibited by butterflyfishes involves strong, long-term bonding between heterosexual pairs (e.g., Yabutu, 1997), considered to be indicative of monogamous mating system. Indeed, butterflyfishes (and mainly the genus *Chaetodon*) account for 31% (51/164 species) of all marine fishes considered to be monogamous (Whiteman and Côté, 2004). However, it is not clear that all species that form pairs mate exclusively within these pairs, and we need to be careful about ascribing a reproductive basis to all observations of pairing (Pratchett et al., 2006). Moreover, there are many species of butterflyfishes that do not form pairs, or are more commonly observed as solitary individuals or within larger groups.

In this chapter, we present a comprehensive overview of current knowledge relating to the social structure of butterflyfishes. We preface this by noting that information and understating of social structure within this group is limited, and this chapter more adequately identifies what we do not know rather than what we do. Importantly, there has been considerable research on the reproduction and behavior of a few distinct species (e.g., *Chaetodon lunulatus*) and almost exclusively within the genus *Chaetodon*, which will form notable case studies for this review. It is unclear however, how this information relates to other conspecifics and other genera of butterflyfishes. In this chapter, firstly we describe the mating systems of a few butterflyfish that have been well researched. Secondly, we briefly discuss the evolution on the mating systems. Thirdly, we used the information that is available (primarily group size) to allocate species to one of six broadly defined types of social structures. These six types cover the majority of readily observable behaviors in commonly studied species and represent distinct variations in traits that may be related to the mating systems. These categorizations would provide a framework for comparative study of buterflyfish behavior. Finally, we review our knowledge about the aggressive behavior and the spawning behavior of butterflyfish, which both affect the relations among individuals strongly and thus the social structures.

MONOGAMOUS SOCIAL STRUCTURES AND PAIR FORMATION

In terms of the mating systems of species, the social one is not always the same as the genetic one (Shuster and Wade, 2003; Dobson et al., 2010). Social monogamy refers to the tendency of species to exhibit heterosexual pairs that live in close spatial proximity, to cooperate in rearing offspring or defending a territory, and to have some sort of pair bond. On the other hand, genetic monogamy refers to an exclusive mating relationship between a male and a female.

Chaetodon lunulatus is amongst the best studied of all butterflyfishes and this research underlies much of our assumed knowledge on the social structures and reproductive behavior of *Chaetodon* butterflyfishes (Reese, 1975, 1991; Sutton, 1985; Yabuta, 1997, 2002; Fig. 8.1). Strong pair formation, with coordinated swimming and maintenance of very close proximity, is particularly conspicuous within this species (Reese, 1975). In the Yaeyama Islands, for example, Yabuta (2002) found that paired individuals spend 84% of their time in close proximity to their partner (see also Pratchett et al., 2006). Specific pairings of individuals are also maintained over long periods, up to seven years (Reese, 1991). For *C. lunulatus*, it has also been shown that pairs defend their territory against conspecifics (Reese, 1975; Sutton, 1985). Therefore, the species mating system is a social monogamy. In addition, it is confirmed that their mating occurs exclusively within the pair (Yabuta, 1997). Thus, this species is also genetically monogamous. A social structure closely resembling that of *C. lunulatus* has been found in

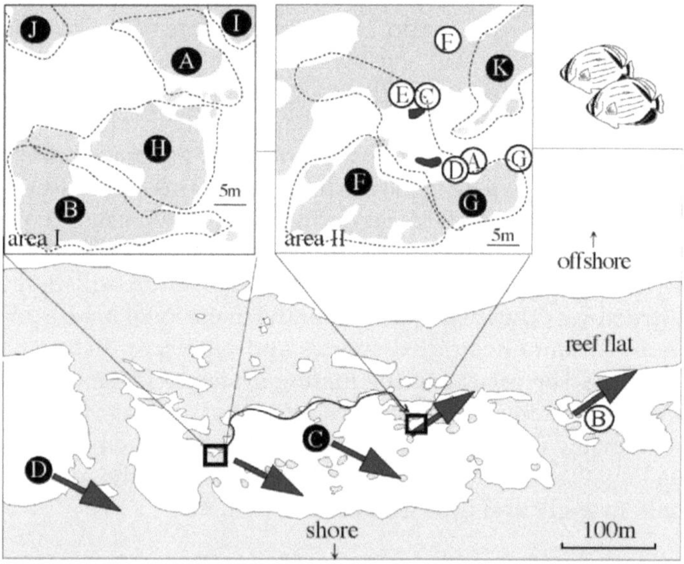

Fig. 8.1 Territories and spawning sites of *Chaetodon lunulatus* in the moat of Kuroshima Island (Yaeyama Islands, Japan) (from Yabuta, 1997). Dashed lines represent boundaries of feeding territories. Letters in solid circles represent the monogamous pairs maintaining the territories. Open circles with the pair identities indicate their respective spawning locations. Two dark gray regions in the centre of area II indicate rocks that are respectively defended as sleeping territories in the evening by pairs A or D and C or E. Arrows indicate tidal current directions at spawning time. The winding line with the arrows at both ends marks the migration route of pair A. The location of the feeding territory of pair E was unknown.

several other species, including *Chaetodon paucifasciatus* in the Gulf of Aqaba (Fricke, 1986). This species forms heterosexual pairs that are maintained for periods of up to three years. Pair members defend the territory against conspecifics and mating is restricted to within the pair (Fricke, 1986). *Chaetodon capistratus*, a Caribbean species, probably also has a monogamous and territorial social structure. Individuals usually occur in pairs and swim together in their home range to feed. Because agonistic interactions were observed across the boundaries of their home range (Neudecker and Lobel, 1982), the home range is considered a "territory" (Roberts and Ormond, 1992). In *C. capistratus*, mating was found to occur within pairs (Neudecker and Lobel, 1982). Therefore, *C. paucifasciatus*, and *C. capistratus* are socialy and genetically monogamous as well as *C. lunulatus*. These data (for *C. lunulatus*, *C. paucifasciatus*, and *C. capistratus*) have reinforced assumptions that butterflyfishes that form strong, long-term pair bonds are probably genetically monogamous: these species seem to support theories predicting that monogamous pairs are formed primarily for reproductive reasons (see Whiteman and Côté, 2004; Pratchett et al., 2006).

Other butterflyfishes with similar social structures include *Chaetodon triangulum* (Reese, 1975), *Chaetodon multicinctus* (Hourigan, 1987; Tricas, 1989), *Chaetodon ornatissimus* (Driscoll and Driscoll, 1988), *Chaetodon quadrimaculatus* (Hourigan, 1987), and *Chaetodon vagabundus* (Tanaka, 1992). All of these form heterosexual pair bonds, maintain these for long periods of time, and defend territories. Therefore, the species are socially monogamous. However, spawning behaviors have not been observed and thus it is not yet clear whether they are genetically monogamous or not. Courtship behaviors of *C. vagabundus* were observed within pairs (Yabuta, unpublished data), so although actual spawning has not been observed, this strongly suggests that the species is genetically monogamous.

POLYGAMOUS SOCIAL STRUCTURES

A different kind of social structure is found in *Chaetodon trifascialis* (Yabuta and Kawashima, 1997) and *Chaetodon fremblii* (Hourigan, 1987), which both appear to be polygynous and territorial. Each fish defends its territory against other same-sex individuals. Males have large territories encompassing two or more females' territories in a kind of a harem (Fig. 8.2). The males often visit the females within their territories, swimming together for a short time. At other times, they forage alone within their territory. The males of *C. trifascialis* spend an average of only 16% of their time with the females (Yabuta and Kawashima, 1997). When alone, they typically forage in the area outside of or at the periphery of the females' territories (Yabuta, unpublished data). Similarly, *C. fremblii* males spend only 9.1% of their time with the females (Hourigan, 1987).

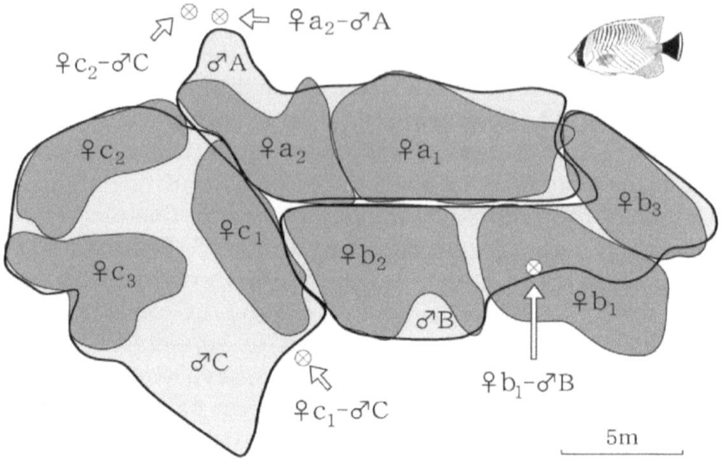

Fig. 8.2 Territories and spawning sites of *Chaetodon trifascialis* on the patch reef in Sekisei lagoon in the Yaeyama Islands (from Yabuta and Kawashima, 1997). Letters represent individual fish that defend territories while circles with crosses mark the spawning sites.

Unlike many marine fishes (e.g., Pomacentridae, Pomacanthidae, Scaridae, Serranidae), there is no recorded evidence of functional sex change among butterflyfishes. However, although non-functional, Kobayashi et al. (2010) reported hermaphroditism and sex change from female to male in immature ovaries, based on their historical observations of gonadal formation in 16 species (4 genera) of butterflyfishes. Sex changes in butteflyfishes thus warrant further investigation. Importantly, sex changes are observed in many marine fishes that show a very similar mating system to the above butterflyfishes, in which females defend territories against other females and a male defends the territory including multiple females' territories (Pinguipedidae: Stroud, 1982; Ohnishi, 1998; Labridae: Newmtzov, 1985; Malacanthidae: Baird, 1988; Balistide: Takamoto et al. 2003). These polygamous (e.g., haremic with females' territories) mating systems involve sequential hermaphroditism whereby individuals change from the subordinate to dominant sex after attaining a sufficient size or social status (Stroud, 1982; Newmtzov, 1985; Ohnishi, 1998). In these haremic fishes (e.g., *Parapercis cylindrica*) the dominant fish is a male, and the largest female changes sex and starts spawning with other females following experimental removal of the male (Stroud, 1982). If butterflyfishes do not change sex it is unclear how haremic systems are established, or what would happen following the loss of a male. Limited research on the reproductive behavior of polygamous butterflyfish is largely observational (Yabuta and Kawakami, 1997), and there is yet to be thorough experimental manipulation of these systems. Shortly before spawning (about 30 min: Yabuta, unpublished data), a male *C. trifascialis* will swim into a female's territory and accompany her until spawning. Their courtship, in which the male nudges the female's genital papilla with his snout, occurs mostly between a male and females living within his harem; only one exception has been observed. In that exception, a male nudged a female that was in the territory of another male, but was immediately chased away by the male. Spawning, therefore, was observed to occur exclusively between a male and females living within his territory (Yabuta and Kawakami, 1997). For *C. fremblii*, no spawning was documented but it was observed that the male spent almost all of his time escorting a female with a swollen abdomen. This female occupied a territory contained entirely within the male's territory. It is quite likely that other species of butterflyfishes utilize a similar polygynous mating system.

Limited evidence indicates that *C. aculeatus* uses a similar social structure in the Caribbean. Neudecker and Lobel (1982) reported that individuals foraged alone in their home range, which Roberts and Ormond (1992) considered a territory. Before spawning, the male visited the female in her home range, which was adjacent to his home range, with behaviors closely resembling that of *C. trifascialis*. In addition, Colin (1989) reported that spawning of *C. aculeatus* was observed between one male and two females. The two females spawned successively within a minute with the male in the same site, suggesting a polygynous mating system.

THE EVOLUTION OF MATING SYSTEMS IN BUTTERFLYFISH

The environmental or evolutionary factors that determine a species' social structure are not clear. All of the species discussed above are territorial benthic foragers, a common combination among butterflyfishes (Roberts and Ormond, 1992). However, some of these butterflyfishes are monogamous, whereas others are polygynous, and diet does not seem to explain the differences. Both *C. lunulatus* and *C. trifascialis* are corallivores (Roberts and Ormond, 1992); the former is monogamous while the latter is polygynous. Both *C. capistratus* and *C. fremblii* feed on sessile and sedentary invertebrates (Roberts and Ormond, 1992); the former is monogamous but the latter is polygynous. However, even among the obligate corallivores, there is variation in prey specialization, i.e., they prefer different coral species (Alwany et al., 2003; Pratchett, 2007). Furthermore, within the corallivores there are varying degrees of specialization (Berumen and Pratchett, 2008). Such dietary preferences and specializations may affect their mating systems. How food resource usage or availability influences the mating systems remains an issue worthy of further study.

Butterflyfish spawn pelagic eggs. Therefore, the place where they spawn may influence the survivorship of their eggs. Actually, it was found that a pair of monogamous *C. lunulatus* moved to a particular spawning site located away from their feeding territory. Females that had neighboring feeding territories did not spawn at the same sites (see below). In contrast, the females of *C. trifascialis,* a polygamous species, spawned within or near their feeding territory. This may make it possible for *C. trifascialis* males to monopolize spawning with multiple females. Spawning site preference and the distribution of suitable spawning sites may influence the evolution of the mating systems in butterflyfishes. At this point, however, this is speculation that will require further investigation.

One of the most apparent differences in behavior of monogamous and polygamous butterflyfishes is pair swimming. Both monogamous and polygamous butterflyfishes defend territories against conspecifics of the same sex; the territory of a single male typically includes more than one female's territory. However, in a monogamous species, a male spends most of his time swimming together with a female but a male of a polygamous species spends most of his time swimming alone. Pair swimming is a unique behavior in monogamous species. In addition, this behavior is observed in other monogamous corallivores, such as the filefish *Oxymonacanthus longirostris* (Kokita, 2002). Understanding this behavior may help us to understand an evolutionary basis for monogamy.

Why do monogamous fishes swim in a pair? Fricke (1986) argued that pair swimming may function as advertisement for possession of the territory to other conspecifics. Two brightly colored fish, swimming closely to each other should be conspicuous like two fluttering flags, to discourage conspecifics from invading the territory. At the same time, swimming alone may show other conspecifics that one owner of the territory is absent, thus potentially attracting other fish that have no territory or mate.

Pair swimming may increase the feeding rate of an individual. For territorial defense, the fish must always be vigilant of intruders. This may decrease the time available for feeding. If two fish swim together, the number of "eyes" is increased and an individual's time dedicated to territory monitoring can be reduced. Fricke (1986) carried out partner removal experiments on monogamous *C. paucifasciatus* and showed that the females' feeding rates reduced when the partners were removed. An increase of feeding rate may ultimately boost egg production. Thus there is a tangible benefit for the females and the males, if it is sure that they will reproduce with their partner.

If heterosexual pair swimming has the advantage for reproduction, we would expect the pair bond to be stronger during the reproductive season than in the non-reproductive season. This has been demonstrated in some species (Yabuta, 2007; Table 8.1). If monogamous species form pairs for reproductive purposes, we would also expect that pair formation coincides with or precedes maturation; we would not expect many solitary individuals to be mature. However, very few studies have directly investigated the

Table 8.1. Individual numbers observed in pairs, solos, and aggregations, in different seasons on the same reefs.

Species	Season	Pair	Solo	Agg.	*N*	Pair ratio	*P**
C. citrinellus	June–Aug.	78	10	0	88	0.88	NS
	Dec.	20	2	0	22	0.91	
C. lunulatus	June–Aug.	122	7	0	129	0.95	<0.001
	Dec.	28	10	0	38	0.74	
C. ornatissimus	Jun.–Aug.	46	36	0	82	0.56	<0.003
	Dec.	4	17	0	21	0.19	
C. plebeius	June–Aug.	52	11	0	63	0.82	<0.02
	Dec.	8	8	0	16	0.50	
C. vagabundus	June–Aug.	20	10	0	30	0.66	NS
	Dec.	10	6	0	16	0.63	
C. trifascialis	June–Aug.	10	33	0	44	0.22	NS
	Dec.	0	13	0	13	0	

*Significant probabilities were calculated using Fisher's exact test (2 × 2 contingency table, two-tailed). 'NS' means that the $p > 0.05$.

relationship between ontogenetic maturation and pair formation. Pratchett et al. (2006) compared maturation states of individuals within pairs and solitary individuals in *C. lunulatus* and *C. melannotus* from Kimbe Bay, Papua New Guinea. *Chaetodon lunulatus* appears to follow the prediction for monogamous fishes, with solitary individuals infrequently mature. *Chaetodon melannotus*, on the other hand, was often found as solitary individuals and sometimes as heterosexual or immature pairs.

The above advantages may have lead to the evolution of heterosexual pair swimming even in non-monogamous species. Once this behavior has evolved, a polygamous mating system could become difficult to maintain. If a male tries to enlarge his territory to cover two females' territories, the female swimming with the male will follow the male, leading to conflict between the two females. Increasing the duration and frequency of pair swimming would facilitate the evolution of monogamy. The tendency of pair swimming varies among species (Fig. 8.1), and investigating the relationship between this tendency and the type of mating system may give us some new insight to understand the evolution of mating systems.

Multiple causes may be involved in the evolution of a mating system (Shuster and Wade, 2003; Dobson et al., 2010). This is certainly the case in butterflyfishes, with many potential factors such as highly specific dietary preferences, spawning site preferences, territorial behaviors, population density, and the tendency of pair swimming. For further study, it will be necessary to examine the effect of these factors on the evolution of mating systems in butterflyfish, while also taking phylogeny into account. Again, the environmental and evolutionary mechanisms driving these systems are unclear.

SOCIAL GROUPING

For detailed understanding of social structures, it is necessary to conduct long-term studies including the identification of individuals. This requires considerable time and effort. On the other hand, it is much easier to survey social grouping. Baseline information on social grouping is fundamental to advance the comparative study of butterflyfish social systems (Reese, 1975; Hourigan, 1989; Roberts and Ormond, 1992), although this may only provide partial information on social structures. The most widely accepted method of assessing the social grouping of butterflyfish species is simply by counting the number of fish typically found swimming together. In order to compare the data of social groupings among species, it is convenient to calculate the ratio of occurrences of solitary individuals, paired individuals, and groups or aggregations (more than two fish) in relation to the total number of observed individuals. The ratios of 32 species are presented in Fig. 8.3 (Reese, 1975; Bouchon-Navaro, 1981; Fricke, 1986; Yabuta, 2007). It is readily apparent that the ratios vary among species.

For monogamous butterflyfish such as *C. lunulatus* and *C. paucifasciatus*, the pairs are usually observed swimming together continuously, maintaining a close distance (Reese, 1975; Fricke, 1986), but pair partners sometimes swim separately for short periods (Fricke, 1986; Yabuta, 2002). The pair ratios of most monogamous species are high while solitary ratios are low (Fig. 8.3). Polygynous butterflyfishes spend most of their time swimming alone in their individual territories. The males, however, visit females within their territories, swimming in a pair with the female for a short time. Consequently, polygynous species demonstrate low observed pair ratios and high solitary ratios (Fig. 8.3). Both monogamous and polygynous butterflyfish can be territorial, which may prevent the fish from aggregating. For most territorial species, such as *C. trifascialis* and *C. baronessa*, aggregation ratios are quite low (Fig. 8.3).

The social grouping ratios of some species, however, change seasonally (Table 8.1). For *Chaetodon plebeius, C. lunulatus,* and *C. ornatissimus,* the pair ratios decrease in winter, which is presumed to be the non-reproductive season in the Skisei lagoon of the Yaeyama Islands in Japan (Yabuta, 2007). As described above, *C. lunulatus* and *C. ornatissimus* have a monogamous social structure. The decreased ratio of paired fish in the winter suggests that monogamous species spend less time swimming with a partner during the non-reproductive season than during the reproductive season. Indeed, in Hawaii, *C. ornatissimus* spend only 19% of their time together with a partner during October–April, although they appear to maintain pair bonds and defend the territories (Driscoll and Driscoll, 1988).

In addition, social groupings of the same species can differ among places (Table 8.2). *Chaetodon auriga* usually occurs as solitary individuals at

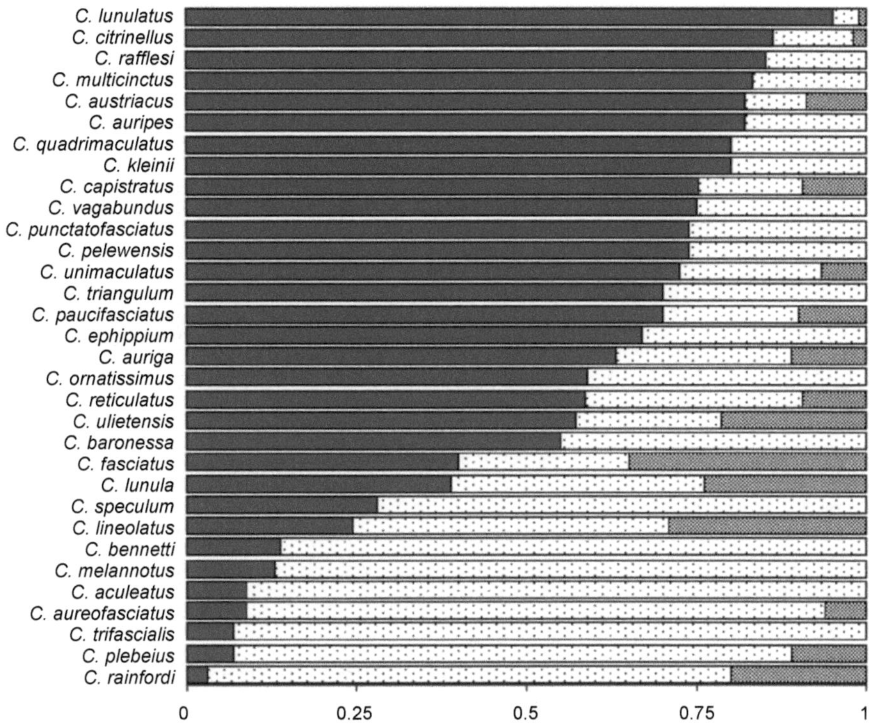

Fig. 8.3 The observed status of individuals of 32 butterflyfish species. The proportion of individuals observed in pairs, alone, or in groups (3 or more individuals) is presented for each species. The dark gray, light gray, and black bars respectively portray the paired, solitary, and aggregation proportions. This graph is drawn from the quantitative data of Reese (1975), Bouchon-Navaro (1981), Fricke (1986), and Yabuta (2007). Only the data of the largest sample size are shown when two or more samples exist for a species.

Heron Island, but they are usually paired at Eniwetok Atoll (Reese, 1975). Furthermore, *C. plebeius* occur mostly as solitary individuals at Heron Island (Reese, 1975) but usually occur in pairs in the Yaeyama Islands (Yabuta, 2007). These variations may be confounded by differences in the seasons in which the studies were conducted. In the Yaeyama Islands, the pair ratio of the species was observed to decrease in winter (Table 8.1) and the observation at Heron Island was made only in winter. However, the winter pair ratio in the Yaeyama Islands is significantly different from that at Heron Island (Fisher's exact test for a 2 × 2 contingency table; $P < 0.05$), indicating that there may be geographical differences in social behavior (Yabuta, 2007). Geographical differences in reproductive seasons even further complicate this issue. In lower latitudes, spawning may occur almost year-round, while higher latitudes demonstrate more marked peaks in summer reproductive output (Allen et al., 1981). Potential inter-annual variation in such seasonal

Table 8.2. Individual numbers of two *Chaetodon* species observed in pairs, solos, and aggregations in different locations.

Species	Season	Location*1	Pair	Solo	Agg.	N	Pair Ratio	P*2	Ref.*3
C. auriga	Mar.–Jun.	Heron	62	26	-*4	88	0.70	< 0.001	R
	unknown	Eniwetok	18	50	-*4	68	0.26		R
C. plebius	Mar.–Jun.	Heron	35	436	61	532	0.07	< 0.001	R
	Jun.–Aug.	Yaeyama	74	16	0	90	0.82		Y

*1: Heron Island is in the Great Barrier Reef; Eniwetok Atoll is in the Marshall Islands; and the Yaeyama Islands are in Okinawa.
*2: Significant probabilities were calculated using Fisher's exact test (2 × 2 or 2 × 3 contingency table, two-tailed).
*3: 'R' and 'Y' respectively signify Reese (1975) and Yabuta (2007).
*4: The individual numbers observed in aggregations in Eniwetok are not provided in the reference. Therefore, the comparison is carried out except for data of the individual numbers observed in aggregation.

trends is also unknown for butterflyfishes. Resolving these issues will require labor-intensive sampling at many locations.

CLASSIFICATION OF SOCIAL STRUCTURES

Butterflyfishes are often classified into four categories based on grouping behaviors (Hourigan, 1989; Roberts and Ormond, 1992; Yabuta, 2007): Solitary, in which the fish usually swim alone; Paired, in which the fish usually swim in pairs; Aggregated, in which fish usually swim as three or more individuals; Mixed, in which the fish exhibit no single dominant type and might be observed as single, paired, or aggregated individuals. However, this classification based solely on grouping does not capture the full diversity of butterflyfish social structures, and further distinction is necessary.

In this section, we propose a new classification based on the tendency for aggregation or territoriality as well as the tendency of pair swimming or the strength of pair bonds (Fig. 8.4). This classification is a tentative one, as results of further studies are expected to identify many useful corrections and updates. With existing data, however, social structures of six types (and four sub-types) can be inferred (Fig. 8.4).

Type 1: Monogamous and Territorial species

Chaetodon lunulatus, *C. paucifasciatus*, and *C. capistratus* are monogamous and territorial. We designate this system as type 1, with *C. lunulatus* as the typical species of this type. Probably, as described above, *C. triangulum*,

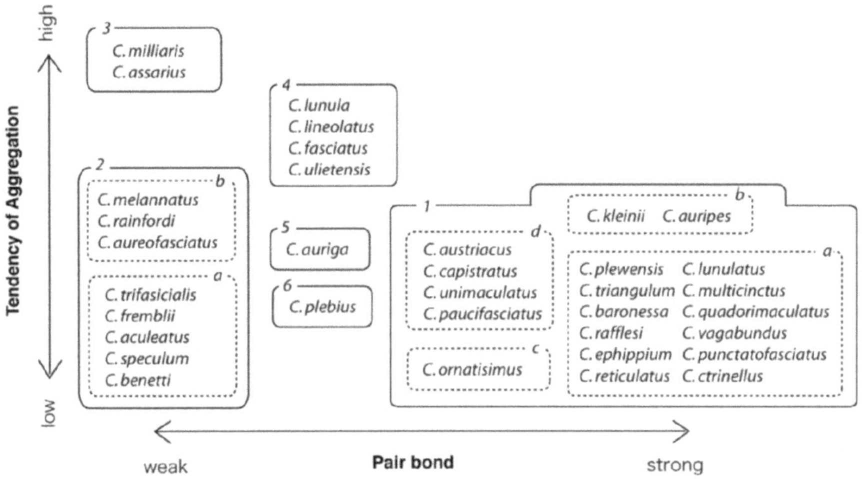

Fig. 8.4 Proposed classification of social structures of 34 butterflyfish. Species of the same social structureare enclosed within solid lines and are numbered. The species that belong to same sub-type are enclosed with dashed lines, denoted with lowercase letters. Distinct types are arranged according to axes that indicate the tendency of aggregation and the tendency of pair swimming. See text for details.

C. multicinctus, *C. ornatissimus*, *C. quadrimaculatus*, and *C. vagabundus* have the same type of social structure (and their respective sister species, where applicable). These eight species all show high pair ratios and very low aggregation ratios. The lowest pair ratio, that of *C. ornatissimus*, is 0.59; the highest aggregation rate is that of *C. paucifasciatus*: 0.10 (Fig. 8.3). Eight species aside from the eight species described above show higher pair ratios than 0.59 and lower aggregation ratios than 0.10: *Chaetodon ephippium*, *Chaetodon unimaculatus*, *Chaetodon austriacus*, *Chaetodon punctatofasciatus*, *Chaetodon pelewensis*, *Chaetodon kleinii*, *Chaetodon auripes*, *Chaetodon rafflesi*, and *Chaetodon citrinellus* (Fig. 8.3). Furthermore, *Chaetodon baronessa* and *Chaetodon reticulatus* show a pair ratio near 0.59 and a lower aggregation ratio than 0.10. These 11 species are also classifiable into type 1.

As this definition includes a large number of *Chaetodon* species, we can further subdivide this group into subtypes based on variations in their behaviors. We propose *C. lunulatus*, with the lowest aggregation ratio, as representative of sub-type 1a. Species in subgroup 1a have the strongest pair bonds are very rarely observed aggregating (Figs. 8.3 and 8.4).

Some butterflyfishes with very strong pair bonds nonetheless show notably more frequent tendencies to aggregate, and we propose classifying these into sub-type 1b. As one example of particularly interesting behavior, *C. auripes* in Wakayama, Japan, makes large diurnal aggregations during summer, swimming in the water column above a particular rock (A. Yumoto, personal communication). They are believed to be feeding on plankton while aggregating. Some other butterflyfishes, such as *C. kleinii*, form occasional feeding aggregations (Allen et al., 1998; Yabuta, personal observation). These two butterflyfish species are classifiable into sub-type 1b.

Another variation of type 1 is fishes that have slightly weaker pair bonds (sometimes seasonally) but rarely aggregate; we propose this behavior as sub-type 1c. For example, the pair ratio of *C. ornatissimus* decreases considerably in winter (Table 8.1). The value is sometimes as low as those of the Type 2 system (below). However, even when they spend less time swimming together, they maintain their pair bonds and territories (Driscoll and Driscoll, 1988). Probably, *C. ornatissimus* and *C. meyeri* are classifiable into sub-type 1c.

A final sub-type 1d, demonstrates slightly weaker pair bonds (among the Type 1 species) and also occasionally aggregate. A well-studied species typifying sub-type 1d is *C. paucifasciatus*. The aggregation ratio of *C. paucifasciatus* is slightly higher than that of *C. lunulatus* (Fig. 8.3). Frick (1986) observed that *C. paucifasciatus* formed trios during re-pairing, but no such trio formation has been observed in *C. lunulatus*. Several other species are classifiable into sub-type 1d. *Chaetodon austriacus, C. capistratus,* and *C. unimaculatus* show a slightly high aggregation ratio (Fig. 8.3). For *C. reticulatus*, the aggregation ratio is slightly high (Fig. 8.3) but the observed aggregations were mostly of juveniles (Neudecker and Lobel, 1989).

Type 2: Polygynous and Territorial Species

Several species are polygynous and territorial, which we classify as Type 2. Well-studied examples of these species include *C. trifascialis, C. fremblii,* and *C. aculeatus*. *Chaetodon trifascialis* is considered a typical species of this type. They usually occur as solitary individuals, occasionally are observed in pairs, and are rarely seen aggregating. The solitary ratios of *C. trifascialis* and *C. aculeatus* are 0.07 and 0.09 respectively, and aggregation ratios for both of them are zero. Although no quantitative data of social grouping of *C. fremblii* are available in published papers; as described above, *C. fremblii* males spend only 9.1% of their time with the females and spend the remainder of their time alone (i.e., they have a solo ratio of 0.09). Similarly, *Chaetodon bennetti, Chaetodon speculum, Chaetodon melannotus, Chaetodon aureofasciatus* and *Chaetodon rainfordi* usually occur as solitary individuals and rarely in aggregation (Fig. 8.3). *Chaetodon melannotus*, on the other hand, was often found in heterosexual or immature pairs. Additionally, a high proportion of solitary individuals were sexually mature. This suggests that *C. melannotus* may have a social structure not primarily based on reproduction. The frequent observation of Pratchett et al. (2006) of same-sex pairs seems to preclude a typical haremic system. This species is sometimes observed in large aggregations at other locations, and it is possible that these aggregations and/or pairings occur for reasons other than reproduction (e.g., foraging efficiency or territory defense). Although the pair ratio of *C. speculum* seems to be slightly high, the pair ratio of type 2 butterflyfish can become higher when the males visit the females frequently and/or spend more time with each female. Indeed, the pair ratio of *C. trifascialis* in the Yaeyama Islands is 0.25, which is as high as that of *C. speculum* (Yabuta, 2007). All of *C. bennetti*, *C. speculum*, *C. melannotus*, *C. aureofasciatus*, and *C. rainfordi* are classifiable as type 2. However, *C. rainfordi* and *C. aureofasciatus* show higher aggregation ratios than the others. Furthermore, *C. melannotus* was observed occasionally in aggregations (Allen et al., 1998; Yabuta, 2007). The territoriality of these butterflyfish might be weaker than that of *C. trifascialis.* We therefore classify *C. melannotus*, *C. rainfordi* and *C. aureofasciatus* into sub-type 2b; the remainder we classify into sub-type 2a.

Type 3: Aggregating Non-Territorial Species

Type 1 and 2 butterflyfish rarely occur in aggregations, but some butterflyfish aggregate much more frequently. For example, one of the planktivorous butterflyfish, *C. miliaris*, is known to almost always aggregate at particular sites on reefs to feed (Stimson, 1990). They are not reported to show territorial behaviors. The mating system of this species is unknown. However, Ralston (1981) reported that gonad-to-body weight ratios of *C. miliaris* were larger than those for butterflyfishes forming pairs, suggesting a promiscuous mating system. We classify species with frequent aggregation but little territorial behaviors into social structure type 3. Some other planktivorous butterflyfishes might belong to this type, such as *C. assarius* (Allen et al., 1998). For further study, it is necessary to investigate the structure of the aggregations of type 3 species, such as the sex of each member and social interactions among the members. Interestingly, some of the non-*Chaetodon* butterflyfishes, such as *Hemitaurichthys* spp. and *Heniochus acuminiatus* are planktivorous and are almost always seen in aggregations. The general tendency of planktivorous fishes to aggregate likely has evolved from foraging requirements and some species of butterflyfishes may provide interesting case studies for future research.

Type 4: Mixed Social Groups

On some occasions, *Chaetodon lunula* has been observed to form large diurnal aggregations in Hawaii (Hobson, 1974; Bouchon-Navaro, 1981; Yabuta, personal observation). In addition, *Chaetodon fasciatus* is often found foraging in large groups of up to 30 or so in the Red Sea (Ormond, 1972 cited in Roberts and Ormond, 1992; Berumen, personal observation). The social grouping data of these two species shows that they exhibit no single dominant social grouping type: the ratios of pair, solo and aggregation are all more than 0.2 (Fig. 8.3). In addition, *Chaetodon lineolatus* and *Chaetodon ulietensis* show similar social grouping data: their ratios all are more than 0.2. It is likely that the social structures of these four species are mutually similar. We call this social structure type 4.

Some butterflyfish of type 4 have occasionally been suggested to be monogamous and territorial. For example, pairs of *C. lunula* always consist of one female and one male (Reese, 1975). At least one pair of *C. fasciatus* was maintained for 6 years (Fricke, 1982). In addition, *C. fasciatus* maintains territories on the fore-reef in the Red Sea (Ormond, 1972 cited in Roberts and Ormond, 1992). Fricke (1982) reported that pairs of *C. lunula* have large territories in the Red Sea. However, the inter-individual distance between the partners in type 4 differed greatly from those of type 1 butterflyfish: type 1 pairs typically remain quite close together, while type 4 pairs are usually more than 2m apart (Fricke, 1982). In addition, as described above, type 1 butterflyfish rarely aggregate, but type 4 often do. We therefore feel there is justification in distinguishing these two types of social structures, but further work is warranted, particularly with respect to the type 4 butterflyfishes.

Types 5 and 6: Social Structures with Significant Geographic Variability

The social structures of *C. auriga* and *C. plebeius* seem to be more variable. In some locations, they are most commonly observed as solitary individuals, while in other locations they are most frequently seen in pairs (Table 8.2). Their social structure is therefore expected to be more plastic than those of other butterflyfishes. We thus tentatively classify each of the two species into separate types from other butterflyfishes: *C. auriga* into type 5 and *C. plebeius* into type 6. Furthermore, *C. auriga* shows a high pairing ratio (Fig. 8.3), which is similar to that of type 1 butterflyfish. On the other hand, both *C. auriga* and *C. plebeius* show a rather high aggregation ratio, similar to that of type 4 butterflyfish (Fig. 8.3). Rendoll (2012) reported that paired *C. auriga* individuals have a territory. However, the distance between the pair partners of *C. auriga* resembles neither those of type 1 nor type 4 (Fricke, 1986). Generally, the inter-individual distance in *C. auriga* is greater than that of type 1 but less than that of type 4. Reese (1975) described an encounter between a pair and another individual of *C. auriga*. They swam as a trio for a short time: no aggressive interaction was visible. Their territoriality might not be as strong as Types 1 and 2. The pair ratios of *C. plebeius* are very low at Heron Island similar to ratios described for fish of type 2. However, for this species, the pair ratio is high in the Yaeyama Islands where ratios are similar to those of type 1. In the Yaeyama Islands, however, the pair ratio decreases in

December (Table 8.1). This pattern is similar to that of *C. ornatissimus*, which belongs to type 1d. There may be some similarity between the social structures of *C. plebeius* and *C. ornatissimus*, but further dedicated studies will be needed to understand their social structure.

AGGRESSIVE BEHAVIOR

Most aggressive behavior in butterflyfishes appears to be related to territorial defense. Fish make a dash at a conspecific intruder entering into their territory, chasing it to drive it away. Between neighbors, a pendulum of fighting across the territorial boundary is often observed: one chases the other over the boundary into the latter's territory; then the latter chases the former. Most butterflyfish territories are considered to be feeding territories (Roberts and Ormond, 1992). However, *C. lunulatus* defends not only feeding territory but also sleeping territory when they migrate for spawning (Fig. 8.1). Such territorial behavior itself is not different from that observed when defending a feeding territory.

Type 1 butterflyfish (i.e., monogamous and territorial) often perform lateral displays during territorial disputes (*C. capistratus*: Neudecker and Lobel, 1982; *C. lunulatus*: Sutton, 1985; Yabuta, 2007; Fig. 8.5). Alternatively, an example of type 2 is *C. trifascialis*. This species does not make this kind of lateral display, although the fish might form a "carousel": two fish swimming in place, each chasing the tail of the other fish. Additionally, mouth wrestling has been observed in fighting between males of *C. trifascialis*. Mouth wrestling is a common aggressive behavior in fish: two fish lock jaws and attempt to push each other backward (Baerends and Baerends-van Room, 1950; Hurd, 1997). *Chaetodon trifascialis* is the only butterflyfish observed to exhibit this type of aggression.

Aggressive interactions related to mate guarding have been observed in some type 1 butterflyfish (*C. capistratus*: Neudecker and Lobel, 1982; *C. lunulatus*: Yabuta, 2000) and in some type 2 butterflyfish (*C. aculeatus*: Colin, 1982; *C. trifascialis*: Yabuta and Kawashima, 1997). Territorial disputes of butterflyfish usually do not last long and rarely involve serious injury. However, fighting between males over a female can be prolonged. Yabuta (2000) described an example of a long aggressive interaction between males in *C. lunulatus,* a type 1 fish, with a serious outcome. Before the aggression, the two males each had a territory. The female, whose territory overlapped both males' territories, was defending her territory against other females. During one daily observation, one male intruded into the other male's territory. The ordinary aggressive lateral display was shown, but without an immediate resolution. It was repeated, but the intruding male never retreated. Escalated fighting with physical attacks ensued (Video of this encounter can be viewed online (see

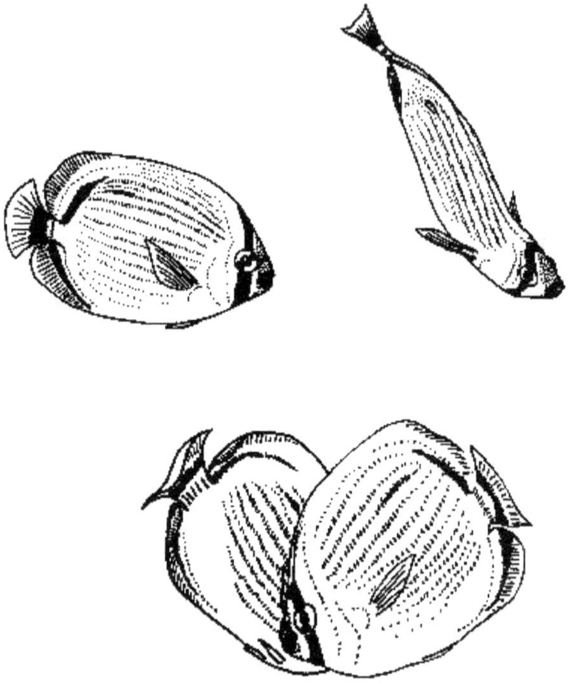

Fig. 8.5 The tail-up display of *Chaetodon lunulatus* is a kind of lateral display. The fish respond to the other fish approaching to it with the display (top). The approaching fish often responds to the display with the same display, resulting in a mutual display (bottom). The display is used in territorial disputes and upon meeting with the partner. The display to the partner is considered to reduce the risk of inappropriate attacks between partners (from Yabuta, 2000).

Video References section)). During the fight, physical attacks occurred many times, with both males weakened by injury. During observations on the following day, the intruding male was no longer present, and two days later, the other defending male had also disappeared. It is likely that the two males did not survive their injuries. This suggests that aggressive interactions of butterflyfish are potentially dangerous, introducing the possibility for serious consequences. Colin (1982) reported less violent but prolonged fighting between males over a female in *C. aculeatus* (type 2). Neudecker and Lobel (1978) reported that *C. aculeatus* males visited a female before spawning; the pair performed a coordinate display when they met.

Type 1 butterflyfish are monogamous and territorial. They defend their territory aggressively against conspecifics, but appear to recognize their partner somewhat reliably. There are documented cases where partner recognition seems to have failed and actually led to attacks against the partner (e.g., *C. lunulatus*, Yabuta, 2002). This apparent uncertainty may be the cause of lateral displays or tail-up displays preceding territorial disputes

(Yabuta, 1999, 2002; Fig. 8.5). The display performed for the partner can reduce the risk of attack between partners: it has been estimated that 70% of high-risk attacks toward partners were prevented by this display (Yabuta, 2002). Similarly to *C. lunulatus,* the lateral display of *C. vagabundus* is used in in territorial disputes and also in meetings between partners (S. Yabuta and T. Saito, unpublished data). These aggressive interactions and partner displays are considered to have evolved because they serve to prevent inappropriate attacks against partners (Yabuta, 2008).

Meeting displays, however, have not been described in the polygynous butterflyfish. *Chaetodon trifascialis* males frequently enter females' territories to visit. When visiting, the male swims a few meters above the bottom. Females often respond to the male in the same manner as against an intruder: they make a dash at the male. The female usually stops short of an actual attack, however, no other display is used. Video of this visiting behavior can be viewed online (see Video References section). It remains unclear whether this behavior is used also in aggressive interactions or not.

At Lizard Island on Australia's Great Barrier Reef, Berumen and Pratchett (2006) recorded all aggressive interactions of butterflyfishes during 780 individual feeding observations. Notably, levels of aggression seemed to be linked to the availability of preferred resources. In areas where highly preferred food was abundant, more aggressive interactions were observed than in areas with lower quality foods (Berumen et al., 2004; Berumen and Pratchett, 2006). As aggressive interactions are potentially costly, it is likely that such behavior is only worthwhile if suitable resources are at stake or if access to key spawning partners is threatened.

PAIR SPAWNING AND SPAWNING AGGREGATIONS

Type 1 butterflyfishes spawn mostly in pairs, before which no elaborate courtship has been observed (Lobel, 1978; Neudecker and Lobel, 1982; Fricke, 1986; Colin, 1989; Yabuta, 1997). The male typically follows the female closely, keeping his snout against the female's ventral flank or anal fin area. This behavior is also often observed even when actual spawning does not occur. When spawning does occur, the pair ascends slowly from the bottom, maintaining the tandem position described above. At the top of the ascent, the female releases her eggs; immediately after, the male release his sperm. At the gametes' release, the pair darts back to the bottom. Video of such pair spawning behavior in *C. lunulatus* is available online (see Video References section). This type of broadcast spawning introduces the possibility for non-partner males to attempt to fertilize a female's eggs. Although this "sneaking" of males has been reported in other families (e.g., wrasses), it is rarely reported in butterflyfishes. Lobel (1989), however, reported sneaking in *C. multicinctus*. It is likely that other males followed a spawning pair and attempted to release gametes at the same instant that the observed pair spawned.

Pair spawning is also observed in type 2 butterflyfishes, although no elaborate courtship display was observed before spawning (Lobel, 1978; Yabuta and Kawashima, 1997). For example, in *C. trifascialis*, it appears that the male typically will only make very brief visits to a female's territory during the day. However, the evening of a spawning event, the male will make an extended stay until the actual spawning event (Yabuta and Kawashima, 1997). Similar male visiting behavior and pair spawning has been reported in *C. aculeatus* (Neudecker and Lobel, 1982). Although some reports describe spawning involving three or more individuals (see below), it is possible that many of these observed groups consisted of a spawning pair with an additional male or males (perhaps 'sneaking' as described above). Pair spawning seems to be typical for *Chaetodon* butterflyfishes, irrespective of their social structure.

Group spawning seems most likely to occur among type 3 butterflyfishes. This group contains planktivorous species and they are frequently swimming in aggregation. Moyer and Nakamura (1994) reported that *C. ephippium*, type 4, made a large offshore group (>100 fish) in the evening at Lizard Island in the Great Barrier Reef. Because all spawning of butterflyfish has been observed in the evening, it is possible that this grouping is related to reproduction. However, it is far more common to find *C. ephippium* in heterosexual pairs. Alternatively, the large group may be some kind of feeding aggregation. It is unclear what the purpose of such aggregations is, although anecdotal reports are common, including aggregations of multiple species at once (e.g., *C. vagabundus, C. auriga, C. ephippium, C. lunula,* and *C. lunulatus* at Lizard Island, M Berumen pers. obs.). We are not aware of any specific reports of butterflyfishes in such aggregations spawning.

TIMING AND SEASONALITY OF SPAWNING

All observations indicate that spawning of butterflyfish occurs at around sunset, irrespective of the social structure (Lobel, 1978, 1989; Suzuki et al., 1980; Neudecker and Lobel, 1982; Fricke, 1986; Colin, 1989; Yabuta, 1997; Yabuta and Kawashima, 1997). Dusk spawning would benefit fish spawning pelagic eggs because it reduces the risk of predation by planktivores, which are less active at dusk. Dusk spawning appears to be commonplace among *Chaetodon* butterflyfishes and many other reef fishes. Although they spawn at dusk, the precise time of spawning varies among species. In the Caribbean, spawning by *C. capistratus* occasionally occurs at the same sites as those used by *C. aculeatus*, but it takes place about five minutes after the latter (Colin, 1989). In the Yaeyama Islands, spawning of *C. lunulatus* occurs from 20 minutes before sunset to 30 minutes after sunset whereas *C. trifascialis* spawning occurs up to one hour after sunset. Observing this nocturnal spawning presents a major challenge, as it is so dark under the water that observers must carefully follow the focal fish to prevent losing them. Fortunately, *C. trifascialis* adopts a nocturnal coloration pattern (two white spots surrounded by a dark area dorsolaterally on the body) that is visible even at dusk, aiding observers in recognizing individual fish. In contrast, *C. lunulatus*, which spawns earlier in the evening, transforms to nocturnal coloration only after spawning. The coloration seems cryptic rather than distinctive, perhaps serving to reduce night time predation risk. The distinctive nocturnal colorations in spawning, such as that of *C. trifascialis*, may help the fish to recognize each other in the dark (Lobel, 1978).

Lunar periodicity of spawning has been reported for some butterflyfishes. In the Yaeyama Islands, 11 identified pairs of *C. lunulatus* were observed in the field for 101 days: during 27 May–7 August, 25 August–3 September, and 21 October–11 November (Yabuta, 1997). In all, 14 spawnings were observed: all occurred within five days of either a full or new moon (Fig. 8.6). For *C. trifascialis*, eight female and three male individuals were observed for 32 days every evening: during 6 May–7 July (Yabuta and Kawashima, 1997). Four spawning events were observed within 1–3 days after a full moon and also on the day of the new moon. Eight courtship behaviors without actual spawning were observed to occur on the day of new moon and the five days following, as well as on the day of new moon.

Some other species do not show clear lunar periodicity of spawning. Neudecker and Lobel (1982) reported *C. aculeatus* at St. Croix spawning on two evenings (two days apart) during a new moon but not on three evenings during the moon's first quarter. In the Bahamas, *C. aculeatus* spawned at the new moon and three days before the new moon (Colin, 1989), but at 8 and 9 days before the full moon in St. Croix (Neudecker and Lobel, 1982). In Puerto Rico, *C. aculeatus* were observed to spawn from the night of the full moon to 8 days after the full moon. This suggests little to no lunar periodicity of spawning for this species.

During the new and full moon periods, the tide level changes greatly, thereby producing a strong current. For that reason, spawning during these periods might reduce the risk of predation on pelagic eggs because these can be carried away rapidly from the reef and reef-resident egg predators. Butterflyfish spawning does not always accompany lunar periodicity, which seems to be less common than dusk spawning. The degree of advantage of spawning accompanying the lunar cycle can vary among locations, and thus may determine whether lunar periodicity is observed or not. Further studies are necessary to elucidate what triggers butterflyfish spawning. In

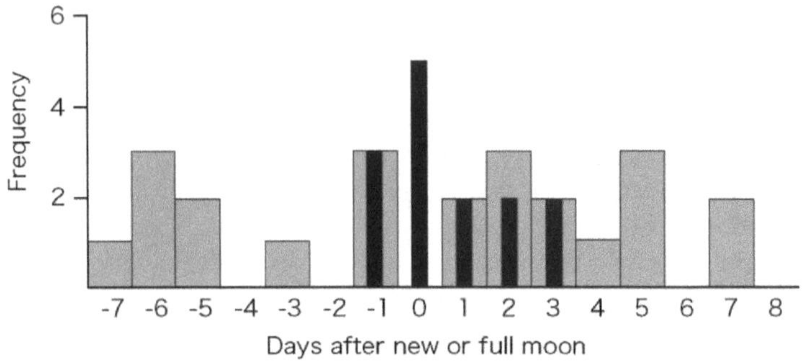

Fig. 8.6 Lunar periodicity of spawning (black bar) and the migrations without spawning (gray column,) in *Chaetodon lunulatus* (from Yabuta, 1997).

these cases, it is to be noted that neither the occurrence of courtship nor the ripe abdomen of females always represents the occurrence of actual spawning. For example, the courtship of *C. lunulatus* can occur independent of the lunar cycle; in such a case, the abdomens of the females are swollen (Yabuta, 1997).

In Caribbean butterflyfishes, there seems to be a spring spawning peak during February–April. Both *C. aculeatus* and *C. capistratus* were observed to spawn in April (Colin and Clavijo, 1988) and February (Neudecker and Lobel, 1982). The water temperature was about 25°C. In addition, in Hawaii, spawning of *C. fremblii* and *C. unimaculatus* was observed during February–March (Lobel, 1978). For *C. multicinctus*, the first spawning of the year was observed in January; the respective peaks of courtship activity and spawning were observed from March until mid-July (Lobel, 1989). In the Sinai peninsula of the Red Sea, *C. paucifasciatus* was observed to spawn in June and July (Fricke, 1986). Although it is a warmer season, the water temperature is not as high then: 24–26°C (Fishelson et al., 1987). Colin (1989) argued that butterflyfishes might not spawn actively at temperatures greater than about 26°C. Comparative observations in warmer areas of the Red Sea, such as the Central Red Sea, would be useful to test this hypothesis.

In the Yaeyama Islands of Okinawa, in contrast, *C. lunulatus* and *C. trifascialis* were observed to spawn when the water temperature was higher than 26°C. In fact, *C. lunulatus* spawned during May–July and September (Yabuta, 1997; Yabuta, unpublished data); *C. trifascialis* spawned during May–June (Yabuta and Kawashima, 1997). The average sea surface temperatures during this season were all higher than 26°C: 26.0°C in May, 27.8°C in June, 29.4°C in July, and 28.5°C in September (1986–1990, Ishigaki Meteorological Observatory). In the moat, where the spawning of *C. lunulatus* occurred, the water temperature on spawning days ranged from 25.6°C to 30.9°C (Yabuta, 1997). However, these butterflyfishes seem not to have spawned when the water temperature was highest. An upper threshold may exist, above which spawning does not occur. For example, *C. lunulatus* was not observed to spawn in August when the average sea surface temperature is 29.4°C (1986–1990, Ishigaki Meteorological Observatory). Spawning and courtship activities of *C. trifascialis* were observed during 6 May to 8 June, but not between June 9 and July 7. The water temperature of the former period was 25.1–27.6°C, which was lower than that of the latter period 26.2–29.3°C (Yaeyama Marine Park Research Station, per. com.).

Some systematic observations for spawning of *C. lunulatus* were conducted during November–December. However, no spawning was observed (Yabuta, unpublished data). It was most likely that *C. lunulatus* did not spawn during those periods. The average surface sea temperature was 22.6°C in December (1986–1990, Ishigaki Meteorological Observatory). The butterflyfish appear not to spawn actively when the water temperature is too low or too high. Further study is necessary to reveal the precise relationship between water temperature and spawning of butterflyfishes. In addition, the seasonal change of ocean current direction might affect the seasonality of spawning (Lobel, 1978). The timing by which other organisms living in the same area release their pelagic eggs might affect the butterflyfish spawning behavior also (Lobel, 1978).

In general, butterflyfishes appear to have few limitations to spawning unless the water temperature is too low and too high. Undoubtedly there is also local adaptation and this optimal window changes geographically within and among species. In some areas the potential reproductive season could extend through a large part of the year. It is not known how many times individuals spawn throughout the season, although a single pair of *C. lunulatus* spawned five times in a 101-day period (Yabuta, 1997). A better understanding of reproductive capacity (combining reproductive season, individual output, and local output) would greatly improve our overall understanding of butterflyfish population dynamics.

SPAWNING SITE PREFERENCE AND SPAWNING MIGRATION

Most spawning that has been observed in butterflyfishes has occurred within or near feeding territories (Lobel, 1978; Neudecker and Lobel, 1982; Fricke, 1986; Colin, 1989; Fig. 8.1; Fig. 8.2). However, in some species, spawning migrations have been reported. Some pairs of *C. lunulatus* migrated from their territories to specific spawning sites 100–400 m away (e.g., pairs A, B, C, and D in Fig. 8.1). The spawning sites experience an offshore current, as opposed to the inshore current predominant in the feeding territories of these specific individuals. The pairs of those in feeding territories in the area of offshore current did not migrate; they spawned near their territories (pair F and G in Fig. 8.1). These results suggest that the pairs preferred the areas of offshore current for spawning. Many coral reef fishes that spawn pelagic eggs migrate offshore or to the leeward edges of reefs for spawning (e.g., Johannes, 1978; Robertson, 1983). Spawning at such sites might reduce the risk of predation on the pelagic eggs because these would be carried away rapidly from the reef and reef-resident egg predators (Johannes, 1978).

Spawning migration has not been reported for any other butterflyfish. It is likely that spawning migration is not unique to *C. lunulatus*. At the fringing reefs of Sesoko Island of Okinawa, pairs of *C. vagabundus* migrate considerable distances every evening from their feeding territories to their sleeping sites (Tanaka, 1992). Courtship is sometimes observed at the sleeping sites after sunset, although spawning was not confirmed (Tanaka, 1992). Similarly, one of the authors (SY) observed a pair of *C. unimaculatus*, the female's abdomen noticeably swollen, migrating down current along a reef for more than 200 m one evening at Kuroshima Island. Colin (1989) reported that a pair of *Chaetodon striatus* swam approximately 50 m along a reef before spawning, although the position of their feeding territory was not reported. So it is possible that this behavior is more common, but rarely documented.

Spawning within or near the feeding territories is likely to have occurred because the butterflyfish had their feeding territories in areas of offshore current, thereby rendering migration unnecessary, as was the case for pairs of *C. lunulatus* that did not migrate to spawn. Butterflyfish might migrate only when their feeding territories are not located in areas that are suitable for spawning. Such intraspecific variation in spawning migration has been reported for some other coral reef fishes (e.g., acanthurids: Robertson, 1983).

CONCLUSION

The social structures of butterflyfishes remain poorly studied with the exception of a few species. While there is little data available for the majority of known species, assessment of existing data does indicate that there may be distinct social structures present in the family. These social structures are primarily distinguishable by the strength of pair bonds and the tendency to aggregate. As with many other reef fishes, these social structures are likely evolved to facilitate feeding or reproduction. Monogamous social structures are among the most prevalent in the family, although species range from aggregating to haremic systems. Butterflyfishes display varying degrees of aggression intra- and inter-specifically; in some cases this appears to be motivated by feeding territories or access to mates. While some species appear to have spawning events associated with lunar phases, other species do not. Likewise, there is some evidence of seasonality in butterflyfish spawning, although in some locations, species may be able to spawn year-round where temperatures are suitable. It is possible that species have an upper and lower thermal bound and can spawn in any month within these bounds. Given the strong link between social structures and reproductive biology, further study in this family could greatly improve our understanding of their population dynamics and general ecology.

REFERENCES

Allen, G.R., R. Steene and M. Allen. 1998. *A guide to angelfishes and butterflyfishes*. Odyssey Publishing, Perth.

Alwany, M., E. Thaler and M. Stachowitsch. 2003. Food Selection in Two Corallivorous Butterflyfishes, *Chaetodon austriacus* and *C. trifascialis*, in the Northern Red Sea. *Marine Ecology* 24: 165–177.

Barird, T.A. 1988. Female and male territoriality and mating system of the sand tilefish, *Malacanthus plumieri*. *Environmental Biology of Fishes* 22: 101–116.

Berumen, M.L. and M.S. Pratchett. 2008. Trade-offs associated with dietary specialization in corallivorous butterflyfishes (Chaetodontidae: *Chaetodon*). *Behavioral Ecology and Sociobiology* 62: 989–994.

Bouchon-Navaro, Y. 1981. Quantitative distribution of the Chaetodontidae on a fringing reef of Moorea Island (French Polynesia). *Journal of Experimental Marine Biology and Ecology* 55: 145–157.

Colin, P.L. 1989. Aspects of the spawning of western Atlantic butterflyfishes (Pisces: Chaetodontidae). *Environmental Biology of Fishes* 25: 131–141.

Davies, N. B. 1991. Mating Systems In: *Behavioural Ecology*, J.R. Krebs and N.B. Davies (Eds.) 3rd Edition. Blackwell, Oxford, pp. 263–294.

Dobson, F.S., B.M. Way and C. Baudoin. 2010. Spatial dynamics and the evolution of social monogamy in mammals. *Behavioral Ecology* 21: 742–752.

Driscoll, J.W. and J.L. Driscoll. 1988. Pair behavior and spacing in butterflyfishes (Chaetodontidae). *Environmental Biology of Fishes* 22: 29–37.

Fricke, H.W. 1986. Pair swimming and mutual partner guarding in monagomous butterflyfish (Pisces, Chaetodontidae): a joint advertisement for territory. *Ethology* 73: 307–333.

Gore, M.A. 1983. The effect of a flexible spacing system on the social organization of a coral reef fish, *Chaetodon capistratus*. *Behavior* 85: 118–145.

Hinde. 1982. Ethology: its nature and relations with other sciences. Oxford University Press. New York/Oxford.

Hourigan, T.F. 1987. The behavioral ecology of three species of butterflyfish (family Chaetodontidae). Ph. D. Dissertation, University of Hawaii, Honolulu.

Hourigan, T.F. 1989. Environmental determinants of butterflyfish social structures. *Environmental Biology of Fishes* 25: 61–78.

Hurd, P.L. 1997. Cooperative signalling between opponents in fish fights. *Animal Behaviour* 54: 1309–15.

Johannes, R.E. 1978. Reproductive strategies of coastal marine fishes in the tropics. *Environmental Biology of Fishes* 3: 65–84.

Kobayashi, K., K. Suzuki and S. Hioki. 2007. Early gonadal formation in *Chaetodon auripes* and hermaphroditism in sixteen Japanese butterflyfishes (Chaetodontidae). *Japanese Journal of Ichthyology*. 54: 21–40 (in Japanese with English abstract).

Kokita, T. 2002. The role of female behavior in maintaining of a coral-reef filefish. *Ethology* 108: 157–168.

Lobel, P.S. 1989. Spawning behavior of *Chaetodon multicinctus* (Chaetodontidae); pairs and intruders. *Environmental Biology of Fishes* 25: 125–130.

Moyer, T., and K. Nakamura. 1994. *The world of fish: social behavior and spawning ecology*. Toukai University Press, Tokyo (In Japanese).

Nemtzov, S.C. 1985. Social control of sex change in the Red Sea razorfish, Xyrichthys pentadactylus (Teleostes, Labridae). *Environmental Biology of Fishes* 14: 199–211.
Neudecker, S. and P.S. Lobel. 1982. Mating systems of Chaetodontid and pomacanthid fishes at St. Croix. *Zeitschrift für Tierpsychologie* 59: 299–318.
Ohnishi, N. 1998. Studies on the life history and the reproductive ecology of the polygynous sandperch, *Parapercis snyderi*. PhD dissertation, Osaka City University.
Pratchett, M.S. 2005. Dietary overlap among coral-feeding butterflyfishes (Chaetodontidae) at Lizard Island, northern Great Barrier Reef. *Marine Biology* 148: 373–382.
Pratchett, M.S. 2007. Dietary selection by coral-feeding butterflyfishes (Chaetodontidae) on the Great Barrier Reef, Australia. *The Raffles Bulletin of Zoology* 14: 171–176.
Ralston, S. 1981. Aspect of reproductive biology and feeding ecology of *Chaetodon miliaris*, a Hawaiian endemic butterflyfish. *Environmental Biology of Fishes* 6: 167–176.
Reese, E.S. 1975. A comparative field study of the social behavior and related ecology of reef fishes of the family Chaetodontidae. *Zeitschrift für Tierpsychologie* 37: 37–61.
Reese, E.S. 1991. How behavior influences community structure of butterflyfishes (family Chaetodontidae) on Pacific coral reefs. *Ecology International Bulletin* 19: 29–41.
Roberts, C.M. and R.F.G. Ormond. 1992. Butterflyfish social behavior, with special reference to the incidence of territoriality: a review. *Environmental Biology of Fishes* 34: 79–93.
Robertson, D.R. 1983. On the spawning behavior and spawning cycles of eight surgeonfishes (Acanthuridae) from the Indo-Pacific. *Environmental Biology of Fishes* 9: 193–223.
Sano, M. 1989. Feeding habits of Japanese butterflyfishes (Chaetodontidae). *Environmental Biology of Fishes* 25: 195–203.
Shuster, S.M. and M.J. Wade. 2003. Mating systems and strategies. Princeton University Press.
Stimson, J.S. 1990. Density dependent recruitment in the reef fish *Chaetodon miliaris*. *Environmental Biology of Fishes* 29: 1–13.
Stroud G.J. 1982. The taxonomy and biology of fishes of the genus *Parapercis* (Teleostei: Mugilididae) in Great Barrier Reef waters. PhD dissertation, James Cook University.
Sutton, M. 1985. Patterns of spacing in a coral reef fish in two habitats on the Great Barrier Reef. *Animal Behavior* 33: 1332–1337.
Suzuki, K., Y. Tanaka and S. Hioki. 1980. Spawning behavior, eggs, and larvae of the butterflyfish, *Chaetodon nippon*, in an aquarium. *Japanese Journal of Ichthyology* 26: 334–341.
Takamoto, G., S. Seki, Y. Nakashima, K. Karino and T. Kuwamura. 2003. Protogynous sex change in the haremic triggerfish Sufflamen chrysopterus (Tetraodontiformes). *Ichthyological Research* 50: 281–283.
Tanaka, N. 1992. Pair territory and diurnal migration of the vagabond butterflyfish *Chaetodon vagabundus*. *Galaxea* 11: 66.
Tricas, T.C. 1989. Determinants of feeding territory size in the corallivorous butterflyfish, *Chaetodon multicinctus*. *Animal Behavior* 37: 830–841.
Yabuta, S. 1997. Spawning migrations in the monogamous butterflyfish, *Chaetodon trifasciatus*. *Ichthyological Research* 44: 177–182.
Yabuta, S. 1999. Behavioral rules and tail-up display in extra- and intra-pair interactions of the butterflyfish, *Chaetodon lunulatus*. *Journal of Ethology* 17: 79–86.
Yabuta, S. 2000. Behaviors in agnostic interaction of the butterflyfish (*Chaetodon lunulatus*). *Journal of Ethology* 18: 11–15.
Yabuta, S. 2002. Uncertainty in partner recognition and the tail-up display in a monogamous butterflyfish. *Animal Behavior* 63: 165–173.
Yabuta, S. 2007. Social groupings in 18 species of butterflyfish and pair bond weakening during the nonreproductive season." *Ichthyological Research* 54: 207–210.
Yabuta, S. 2008a. Signal function of tail-up posture in monogamous butterflyfish (*Chaetodon lunulatus*): a dummy experiments in the field. *Ichthyological Research* (in press).
Yabuta, S. 2008b. Evolution of cross-contextual displays: the role of risk of inappropriate attacks to non-opponents, such as partners. *Animal Behavior* (in press).

Yabuta, S. and M. Kawashima. 1997. Spawning behavior and haremic mating system in the corallivorous butterflyfish, *Chaetodon trifascialis*, at Kuroshima Island, Okinawa. *Ichthyological Research* 44: 183–188.

VIDEO REFERENCES

The videos of some *Chaetodon* behaviors discussed in this chapter are available on the Movie Archives of Animal Behavior website (http://www.momo-p.com), using the following data numbers as search keys:

1) Territorial disputes in *C. lunulatus*: momo010711cl01b
2) Meeting display between pair partners in *C. lunulatus*: momo041211cl01a
3) Fatal fighting in *C. lunulatus*: momo010711cl02a
4) Encounter of a male visiting the territory of a female in polygynous butterflyfish, *C. trifascialis*: momo011001ct02b
5) Mouth wrestling between males of *C. trifascialis*: momo011001ct01b
6) Spawning behavior of *C. lunulatus*: momo010316cl01a

CHAPTER 9

Susceptibility of Butterflyfish to Habitat Disturbance: Do 'chaets' Ever Prosper?

Shaun K. Wilson,[1,*] *Nicholas A.J. Graham*[2,a] and *Morgan S. Pratchett*[2,b]

INTRODUCTION

Disturbance is an integral component of ecology, responsible for re-setting successional clocks and preventing the dominance of climax communities (Grime, 1973; Connell, 1978). However, ecologists, environmentalists and resource managers are concerned that the diversity, frequency, intensity, and spatial extent of disturbances is increasing (Hoegh-Guldberg, 1999; Wooldridge et al., 2005), and communities are perpetually in the early stages of succession, or have 'flipped' to alternate stable states (Scheffer et al., 2001).

The impact of disturbance on butterflyfishes (family Chaetodontidae) has been exclusively studied on coral reefs, reflecting their predominantly tropical distribution and the close association of many species with reef habitat (Allen et al., 1998). Coral reefs are an ecosystem in which

[1]Marine Science Program, Department of Environment and Conservation 7 Dick Perry Ave, Kensington, WA 6151, Australia.
E-mail: shaun.wilson@dec.wa.gov.au
[2]ARC Centre of Excellence for Coral Reef Studies, James Cook University, Townsville QLD 4811, Australia.
[a]E-mail: nick.graham@jcu.edu.au
[b]E-mail: morgan.pratchett@jcu.edu.au
*Corresponding author

environmental and ecological anomalies are an inherent and important part of the system. However, disturbances associated with global warming such as coral bleaching (Glynn, 1993; Brown, 1997; Hoegh-Guldberg, 1999), storms (Emanuel, 2005; Webster et al., 2005) and disease (Harvell et al., 1999, 2002; Bruno et al., 2007), combined with changes in sea levels (Rahmstorf, 2007) and water chemistry (Pelejero et al., 2005; Hoegh-Guldberg et al., 2007) threaten the long term viability of reef habitats. At the same time, exploitation is removing large bodied predatory species (Russ, 2002), altering top down processes and resulting in trophic cascades (Pauly et al., 1998). This may be beneficial for many small fish, such as chaetodontids, as removal of predators can result in the release of prey species (Graham et al., 2003). However, removal of invertebrate feeding species through fishing has also been associated with increased abundance of the coral feeding seastar, *Acanthaster planci* (Dulvy et al., 2004; Sweatman, 2008), and bio-eroding urchins (McClanahan and Shafir, 1990), which could have detrimental consequences for some chaetodontids, through loss of their preferred food or habitat.

Eutrophication, caused by release of sewerage at sea, agriculture associated land runoff and pollution may alter bottom up processes on coral reefs (Grigg, 1994; Lapointe, 1997). Increased nutrients in the water encourage rapid growth of algae (Schaffelke, 1999), which compete with corals for space on reefs. These bottom up processes may be particularly important following disturbances that result in extensive coral mortality, as space vacated by corals is quickly occupied by algae (Diaz-Pulido and McCook, 2002) and overgrowth by fleshy macroalgae may prevent recovery of coral communities.

In both the Caribbean (Gardner et al., 2003) and Indo-Pacific (Bruno and Selig, 2007) habitat disturbances have resulted in regional declines of coral cover over the past 30 years. The impact that coral loss has had on reef associated communities is an area of intense interest, as these communities provide many of the economic and ecological services humans derive from coral reefs (Moberg and Folke, 1999). Much of the work to date has focussed on the effects of disturbance on reef fishes (see Wilson et al., 2006), in particular butterflyfish from the genus *Chaetodon*. The chaetodontids are typically diurnally active species with bright body colour patterns that make them conspicuous to humans and therefore easy to survey. In addition, more than half of chaetodont species rely on coral for food (Cole et al., 2008) or shelter (Pratchett et al., 2008a) and early studies on reef disturbances identified members of the Chaetodontidae as highly susceptible to the effects of coral loss (Bouchon-Navaro et al., 1985; Williams, 1986). Butterflyfishes are an ideal family for studying the intricacies of habitat loss on fish because response to habitat disturbance varies among species and locations (e.g.,

Bouchon-Navaro et al., 1985; Shibuno, 1999; Syms, 1998; Pratchett et al., 2006; Graham, 2007).

The loss of coral may affect chaetodontid fish by changing the availability of food resources, refugia and settlement sites for juveniles. All three of these factors have the potential to influence survival and abundance of chaetodontid fish, however their relative importance will vary between species. Thus, whilst disturbance is expected to have a detrimental effect on some species, a change in benthic composition and community structure may be beneficial to others (e.g., Berumen and Pratchett, 2006a). In this chapter we consider how different chaetodontid species are influenced by different types of disturbances and the ecological processes that allow some "chaets" to prosper, whilst others decline in abundance.

INTERSPECIFIC VARIATION IN RESPONSE TO DISTURBANCE

To assess the influence of coral depleting disturbance on different species of chaetodontid we carried out a meta-analysis of 17 studies that had recorded abundance of chaetodontid fish and percent coral cover both before and after disturbance (Table 9.1). As small disturbances can have negligible impact on fish communities (e.g., Cheal et al., 2002) we only considered studies where total coral cover had declined by >10%. The responses of individual species to coral declines were calculated by dividing the proportional change in the abundance of each fish species by the proportional decline in coral cover within the local environment. Changes in fish abundance and coral cover (Δ) were transformed (Δ^T), prior to calculating the final metric value using the equation.

$$\Delta^T = \log_e(1 + \Delta/101)$$

This transformation distributes metric values about zero and accounts for the right skew in data, which is often associated with % difference values (Kaiser et al., 2006). The mean metric values are presented for each fish species and 95% confidence intervals about the mean value have been used to determine if species consistently increase or decline following disturbances. Finally, each chaetodontid species was classified as an obligate corallivore, facultative corallivore or non-coral feeder based on feeding observations and gut content analyses, following Cole et al. (2008).

Results from the meta-analysis indicate that coral depleting disturbances tend to have a detrimental effect on the abundance of obligate corallivores, whilst there is a tendency for abundance of facultative species to increase, although this response is highly variable (Fig. 9.1). That disturbance has a greater effect on species that feed exclusively on live coral than those that

Table 9.1 Studies used in meta-analysis to assess the impact of coral loss on chaetodontid abundance.

		% Coral cover		
Location	**Disturbance**	**Before**	**After**	**Reference**
Iriomote Island, Japan	Bleaching	95	0	Sano (2004)
Okinawa Island, Japan	Experiment	90	0	Sano et al. (1984)
Capricorn bunker, GBR, Australia	Storm	80	10	Halford et al. (2004)
Kona, Hawaii	Storm	78	37	Walsh (1983)
Lau islands, Fiji	Multiple	64-44	18-24	Wilson et al. (2008b)
Seychelles	Bleaching	64-29	10-3	Spalding and Jarvis (2002)
Palm Islands, GBR, Australia	Bleaching	63-39	45-9	Thompson and Malcolm (1999)
Lizard Island, GBR, Australia	Experiment	55	43-34	Syms (1998)
Seychelles	Bleaching	50-23	17-0	Jennings (unpublished)
Ishigaki Island, Japan	Bleaching	39	3	Shibuno et al. (1999)
Seychelles	Bleaching	37-19	9-5	Graham et al. (2006, 2007)
Cairns sector, GBR, Australia	Storm	36-21	23-9	Cheal et al. (2002)
Moorea, French Polynesia	COTS	36	16	Bouchon-Navaro et al. (1985)
Trunk Reef, GBR, Australia	Bleaching	33	15	Pratchett et al. (2004)
Trunk Reef, GBR, Australia	Multiple	33	3	Pratchett et al. (2006)
Lizard Island, GBR, Australia	COTS	31-24	18-14	Pratchett (2001)
Chagos	Bleaching	30-41	19-26	Graham and Spalding (unpublished)

do not rely solely on coral for food reflects differences in the ecological versatility of these two feeding groups. Facultative coral feeders can switch or supplement coral diets with algae and motile invertebrates, which are likely to be more common following coral mortality. Previous studies of chaetodontid response to habitat disturbance on the Great Barrier Reef (Pratchett et al., 2006), Seychelles (Graham, 2007) and French Polynesia (Bouchon-Navaro et al., 1985; Berumen and Pratchett, 2006) have also demonstrated that the decline of obligate corallivores is greater than facultative counterparts after coral loss. However meta-analysis findings suggest that there is considerable variation in the response of both obligate and facultative coral feeders (Fig. 9.1). Abundance of at least one facultative species, *Chaetodon kleinii*, consistently declines, whilst abundance of some species (e.g., *C. vagabundus* and *C. ephippium*), often increases after disturbances. Similarly, whilst obligate corallivores typically decline in

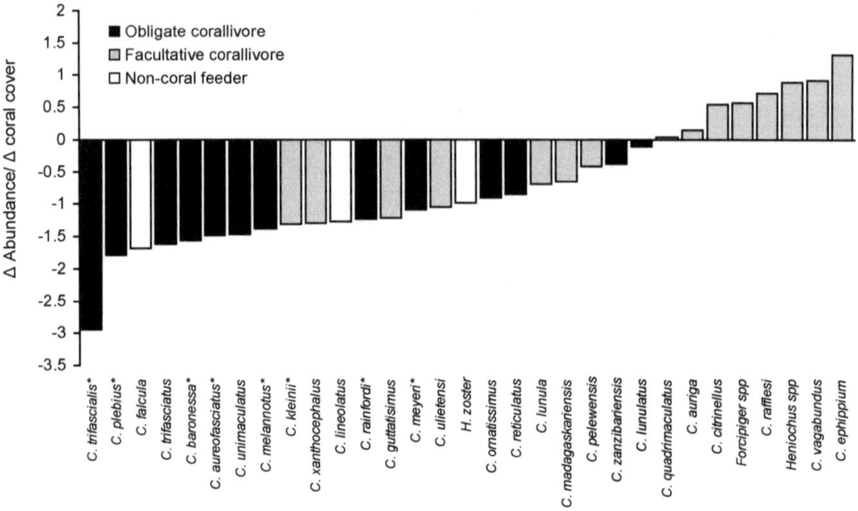

Fig. 9.1 The effects of coral loss on 31 butterflyfish species. Response for each species are average values, calculated from two or more independent observations. * species where mean response value and 95% confidence interval failed to intersect 0, indicating the response was consistent among study locations. Negative values are indicative of abundance decline.

abundance after coral loss, the extent of response is not equivalent among species, and only 7 of the 13 species consistently exhibit population declines. Indeed studies have found abundance of obligate coral feeders may remain stable or increase, even after extensive coral mortality (e.g., Bouchon-Navaro et al., 1985; Pratchett et al., 2004). The remainder of this chapter will examine variation in the response of chaetodontids to disturbance, dealing specifically with the influence of different disturbance types, ecological versatility and recruitment.

Different Types of Disturbances

Disturbances are discrete events that disrupt ecosystems altering the structure of communities. The cause of disturbance will determine which components of the environment are most severely affected and how communities respond. On coral reefs habitat disturbances can be divided into two broad categories: i) physical and ii) biological (Wilson et al., 2006). Physical disturbances are those that result in coral mortality and loss of physical reef structure (e.g., storms) and biological disturbances are those that cause coral mortality while structural complexity of the reef is not immediately affected (e.g., outbreaks of crown-of-thorns starfish and coral bleaching). Physical disturbances are expected to have a much greater impact on abundance of fish, as they reduce availability of dietary resources and structural complexity (Wilson et al., 2006). A loss of structural complexity and associated shelter may be particularly important for chaetodontids, as many species are small bodied, with maximum total lengths < 20cm (Allen et al., 1998), making them highly susceptible to predation (Munday and Jones, 1998). Using data from studies listed in Table 9.1, and meta-analysis metrics described above, we compared the effects of physical and biological disturbances on seven species of chaetodontid.

Changes in abundance of chaetodontid species due to physical and biological disturbances were very similar (Fig. 9.2), and although there were significant differences in response to disturbance between species (ANOVA $F_{6,55}$= 8.82, P = 0.023), there was no significant response due to disturbance type (ANOVA $F_{1,55}$ = 2.97, P = 0.344) or the interaction between disturbance type and species (ANOVA $F_{6,55}$ = 0.85, P = 0.953). These results are in contrast to a similar analysis that showed physical disturbance has a greater impact on coral reef fish (Wilson et al., 2006). Importantly, six of the species included in the current analysis are obligate corallivores and it is therefore not surprising that declines in availability of their only food source have a similar impact to declines in habitat structure. Consequently, among obligate coral feeding chaetodontids, coral mortality is the primary cause of population declines and the impact of structural complexity loss associated with coral mortality is insignificant.

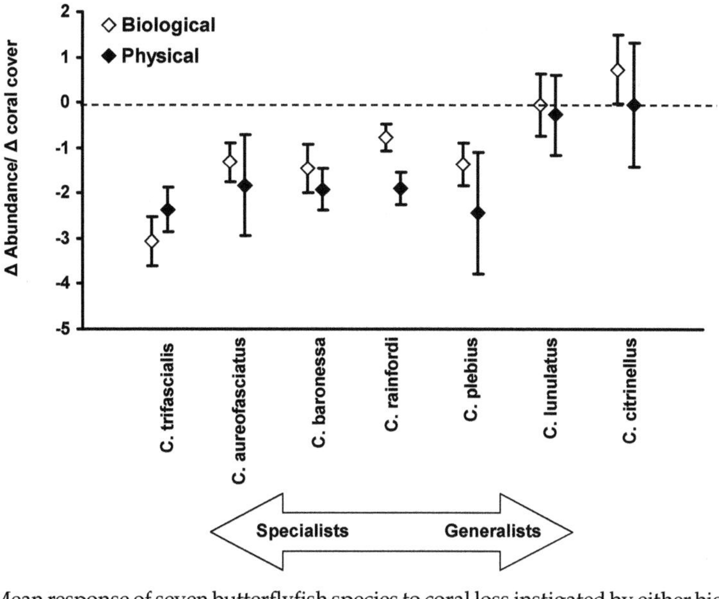

Fig. 9.2 Mean response of seven butterflyfish species to coral loss instigated by either biological (e.g., COTS, coral bleaching) or physical (e.g., severe storms) disturbances. Error bars are standard errors calculated from two or more observations. Species are arranged left to right from the most to least specialised based on feeding observations (Pratchett, 2005, 2007).

Physical disturbances may however have a more obvious and damaging impact on facultative coral feeding chaetodontids than biological disturbances. In the Granitic Seychelles extensive coral bleaching in 1998 caused > 90% mortality of corals (Sheppard, 2003). Subsequent erosion of coral skeletons led to loss of structural complexity (Engelhardt, 2004), although the extent of structural collapse was spatially variable (Graham et al., 2006). At sites where structural complexity had been retained, abundance of facultative corallivores had remained stable, whereas where complexity had been lost their abundance had declined (Graham et al., 2009). Within the same sites, change in coral cover had no apparent influence on numbers of facultative coral feeders, but could be positively correlated to changes in abundance of obligate corallivores (Graham et al., 2009). Facultative coral feeders are known to shelter overnight within the reef matrix, returning to specific locations nightly (Ehrlich et al., 1977). Disturbances that ultimately result in loss of structural complexity may therefore have a broad impact on chaetodontid communities, with the potential to affect species reliant on refuge provided by coral skeletons as well as coral feeding specialists.

The delayed breakdown of reef structure observed on Seychelles reefs also highlights the time scale over which different disturbances can act. The full impact of biological disturbances may not be felt until 5–10 years after coral mortality (Pratchett et al., 2009) and may have a more extensive impact on fish communities than the initial loss of coral. As coral bleaching can be a much more geographically extensive disturbance than storms, which are typically patchy in their impact (Bythell et al., 2000), the long-term impacts of bleaching, where structure is lost, can be more substantial and longer lasting than the impacts of storms.

The time taken for coral loss to occur may also affect butterflyfish communities. Outbreaks of crown-of-thorns starfish are protracted disturbances where coral declines occur gradually over several years, whilst coral bleaching can cause extensive mortality over large areas within 2–3 weeks (Wilson et al., 2009). However, a comparison of disturbances caused by coral bleaching versus outbreaks of crown-of-thorns starfish failed to detect a significant difference ($F_{1,96} = 0.46$, $P = 0.34$) in how these disturbances affect obligate and facultative feeding chaetodontids (Fig. 9.3).

Response of butterflyfishes to outbreaks of crown-of-thorns starfish and coral bleaching may however vary among species. Different disturbance types can affect different components of the coral community, which can influence fish response to coral loss. For example, different corals vary greatly in their susceptibility to coral bleaching and outbreaks of crown-of-thorns starfish (Pratchett et al., 2008a). As a consequence, the composition of coral communities can be largely dependant on the nature of preceding disturbance types (McClanahan et al., 2007). Since some chaetodontids

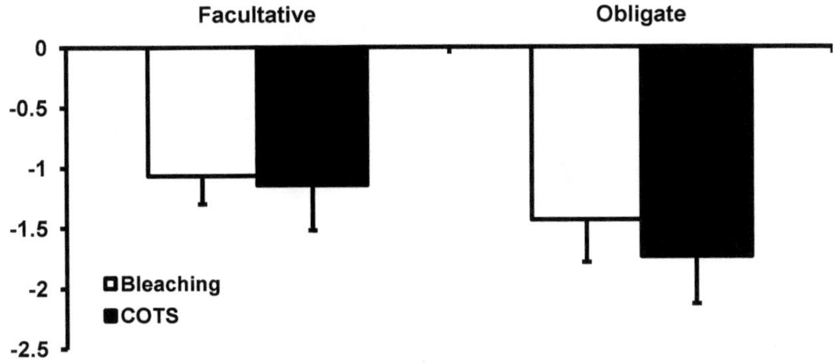

Fig. 9.3 Response of facultative and obligate coral feeding butterflyfish to coral decline instigated by coral bleaching or COTS outbreaks. Standard errors calculated from response of eight or more species.

preferentially feed on specific corals (Pratchett, 2005; Pratchett, Chapter 6), a shift in the structure of coral communities is expected to influence the composition of chaetodontid communities. Such a shift was observed on Moorean reefs, where 20 years after COTS had reduced coral cover by ~60%, coral coverage had returned to pre-disturbance levels. Post disturbance communities were however, characterised by *Pocillopora* rather than *Acropora* species (Berumen and Pratchett, 2006). Correspondingly, overall abundance of chaetodontids after coral recovery was comparable to pre-disturbance levels, however among obligate corallivores there was a decline in species that preferentially feed on *Acropora* and an increase in species that feed on *Pocillopora* (Berumen and Pratchett, 2006).

Recovery of coral communities following disturbances may also vary spatially, giving rise to differing fish communities. In the Lau islands, Fiji, outbreaks of crown-of-thorns starfish combined with localised coral bleaching between 2000 and 2002 (Cumming et al., 2002; Dulvy et al., 2004) reduced coral cover by up to 70%, although the extent of loss and subsequent recovery varied among islands (Wilson et al., 2008a). Prior to disturbance, coral communities on four of the islands were similar and dominated by corals from the families Acroporidae, Mussidae and Faviidae (Fig. 9.4). Chaetodontid communities on these reefs were characterised by a high abundance of obligate coral feeders before disturbance and facultative or non-coral feeders after coral loss. The island of Kabara was already affected by outbreaks of crown-of-thorns starfish during initial surveys and coral cover was <1% at the time of initial surveys. Six years later, coral cover at Kabara had increased and benthic communities and structural complexity were similar to those at Totoya Island (Fig. 9.4). Chaetodontid communities at both these islands were characterised by facultative corallivores, yet the species composition at the two islands were markedly different (Fig. 9.4).

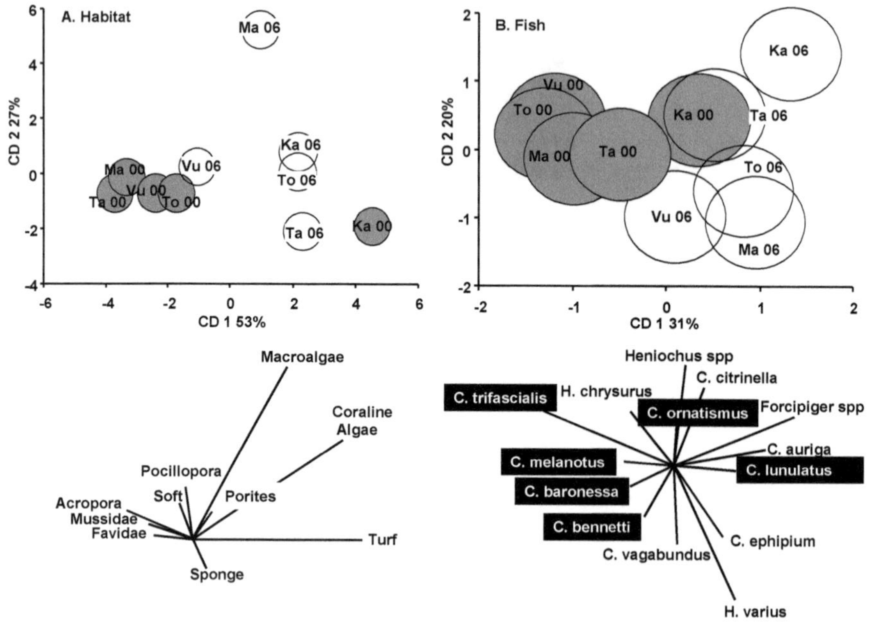

Fig. 9.4 Canonical discriminate analysis (CDA), showing change in A. benthic habitat and B. chaetodontid communities following coral decline at five Fijian islands; Ka Kabara, Ma Matuku, Ta Tavunasici, To Totoya, Vu Vuqava. Benthic and fish data were collected from 18 150m^2 areas at each island. Ellipses are 95% confidence intervals around the group centroid for data collected in 2000 (grey) and 2006 (white). Vectors for the significant benthic and fish variables are displayed below each CDA. Highlighted chaetodontid species are obligate corallivores.

At Kabara, chaetodontid communities in 2006 were dominated by *Forcipiger* spp., *Chaetodon citrinellus* and the obligate corallivore *Chaetodon ornatissimus*, whilst at Totoya communities were characterised by *Heniochus varius*, *Chaetodon ephippium* and the obligate corallivore *Chaetodon lunulatus*. This is consistent with the findings of Anderson et al. (1981), who demonstrated chaetodontids occupy specific niches and geographical replacement occurs between species of the same niche.

Similarly, at the island of Vuaqava, total coral cover had not changed six years after the disturbance, however the coral community was now dominated by *Pocillopora*, rather than *Acropora* corals (Fig. 9.4). Accordingly, the previously abundant *Acropora* feeding specialist, *Chaetodon trifascialis*, was not observed during 2006 surveys of Vuaqava, and the coral feeding generalist *Chaetodon lunulatus* was the most prevalent obligate corallivore. These results demonstrate that recovery patterns of butterflyfish may be governed by recovery patterns of corals, but may also be driven by factors other than habitat composition.

Whilst most of the current literature focuses on how disturbances indirectly affect chaetodontid fish via their impact on habitat, disturbances can also have a direct effect on fish. Global warming is causing long term changes in the marine environment, increasing water temperature and decreasing pH. Changes in water temperature and chemistry have the potential to influence fish growth, physiology and reproduction, which may affect future abundance and distribution patterns (Munday et al., 2008).

Disturbance to communities caused by fishing may also have a direct affect on some chaetodontid species. Trading of ornamental coral reef fishes is an international industry, with an estimated wholesale value of up to 30 million USD per annum (Wood, 2001). Chaetodontids are among the most popular and profitable fishes within the aquarium trade (Chan and Sadovy, 1998; Wood, 2001), with collectors targeting species with high market value (Sadovy and Vincent, 2002). By targeting specific chaetodontid species, collectors can reduce local abundance and alter community composition (Lawton et al., Chapter 11). For example, a Hawaiian study showed populations of three chaetodontid target species were significantly lower in areas open to collecting, whilst population levels of a non-target species was similar in areas open and closed to ornamental fishing (Tissot and Hallacher, 2003).

The removal of fish for human consumption may also affect butterflyfishes (Lawton et al., Chapter 11). However, fishers typically target large bodied predatory fish (Pauly et al., 1998), which can alter top down processes, resulting in the predatory release of small bodied prey species (Graham et al., 2003). Moreover, the presence of predatory fish can significantly reduce recruitment of butterflyfishes (Almany, 2003). Raymundo et al. (2009) suggested that increased prevalence of coral diseases in areas subject to overfishing in the Philippines, resulted from increased abundance of corallivorous butterflyfishes following removal of piscivorous fishes. However, Raymundo et al. (2009) did not actually document increased abundance of butterflyfishes in areas subject to overfishing compared to near by areas with limited fishing. There is actually very little evidence to suggest trophic cascades instigated by changed fishing pressure influence chaetodontid abundance. A comparison of fishing and habitat effects on three chaetodontid species, all obligate corallivores, determined the indirect effects of fishing on fish abundance were negligible in comparison to changes in coral cover (Wilson et al., 2008a). The response may, however, vary for species with lower dependence on live coral.

Ecological Versatility and Chaetodontid Response to Disturbance

Species capable of using a broad array of dietary and habitat resources are expected to be more adept in dealing with changes in their environment. Thus, it is hypothesised that generalist species are less affected by disturbances than specialised species (Vazquez and Simberloff, 2002) and that highly specialised species are more susceptible to extinction (McKinney, 1997). Greater resilience to changes in resource availability has been observed among several families of coral reef fish, including coral dwelling gobies (Munday, 2004), damselfish (Wilson et al., 2008b) and chaetodontids (Graham, 2007; Pratchett et al., 2008a). Among the chaetodontids it is clear that coral loss has a greater impact on species that are obligate coral feeders (Fig. 9.1), however within this group there is considerable variation in both response to disturbance and dietary specialisation (Pratchett, Chapter 6).

Our analyses found differences in disturbance response by obligate corallivore species was related to their level of dietary specialisation (Fig. 9.2). Many of the obligate coral feeders preferentially feed on particular genera, or even species of coral (Tricas, 1989a; Pratchett, 2007; Graham, 2007). *Chaetodon trifascialis* has perhaps the most selective diet, feeding almost exclusively on a single species of coral, *Acropora hyacinthus* (Irons, 1989; Pratchett, 2005). Accordingly, meta-analysis of disturbance impact on butterflyfish indicates that *C. trifascialis* consistently declines in abundance following coral decline and that populations decline more severely than other chaetodontid species (Fig. 9.1). In contrast, *Chaetodon lunulatus* feeds from more than 50 different coral species (Pratchett, 2005), and is the obligate corallivore least affected by coral loss (Figs. 9.1 and 9.2). Indeed density of *C. lunulatus* on reefs is often unaffected following coral decline (Syms, 1998; Cheal et al., 2002; Pratchett et al., 2004), emphasising the resilience of this generalist feeder to changes in dietary resources.

Changes in resource availability, may, however, have a sub-lethal affect on fish, allowing them to persist under duress. For example, Pratchett et al. (2004) demonstrated that although coral loss of >50% did not affect abundance of *C. lunulatus*, lipid reserves in the liver had been significantly depleted. Lipids are crucial energy reserves for fish (Cowey and Sargent, 1977), and when reserves are critically low fish health and reproductive output are compromised (Adams, 1999). It is unclear if there are interspecific differences in chaetodontid ability to store lipids, although spatial variation in the type of dietary resources clearly influences liver lipid content (Berumen et al., 2005). Consequently, fish with pre-existing high liver lipid content may be less susceptible to declines in food resources, although long term prognosis for the survival of these individuals is dire unless coral resources are restored.

Response to disturbance may also manifest in the form of changed growth rates rather than changes in abundance. A decline in coral cover was associated with reduced growth in the coral feeding leather jacket, *Oxymonacanthus longirostris* (Kokita and Nakazono, 2001), whilst increased algal cover following COTS outbreaks is associated with increased growth of herbivorous fish (Hart and Russ, 1996). Species composition and availability of coral can influence growth rates of chaetodontids; growth of both *Chaetodon plebeius* and *C. trifascialis* being dependant on the species of coral they feed upon (Berumen and Pratchett, 2008). Disturbances that alter the composition of coral communities may therefore have a detrimental effect on growth of chaetodontids when coverage of preferred dietary corals is reduced.

Finally, reduced coral cover following disturbance may alter the foraging behaviour of chaetodontids. The feeding territories of corallivores like *C. trifascialis* are inversely proportional to the density of coral (Irons, 1989) and experimental reduction of coral cover results in the expansion of *Chaetodon multicinctus* territories (Tricas, 1989b). If coral loss is extensive, feeding territories can break down and chaetodontids may roam over large areas in search of food (Samways, 2005). Foraging over wider and unfamiliar areas may expend more energy and increase susceptibility to predation (Tricas, 1985), although, as with changes in body condition and growth, modified feeding behaviour may allow fish to persist under sub-optimal conditions.

Habitat Disturbance and Recruitment of Chaetodontids

The supply of larval fish from plankton is an important determinant of distribution and abundance patterns of most coral reef fish. For species with very specific habitat or dietary requirements, availability of these resources governs recruitment and adult population densities (Munday and Jones, 1998). Abundance of adult chaetodontids can be positively correlated with coral cover (Bell et al., 1985; Bouchon-Navaro and Bouchon, 1989; Bozec et al., 2005), prompting suggestions that chaetodontid density is resource limited. The strength of this relationship is, however, spatially and temporally variable and often weak or non-existent when coral coverage is high (Findley and Findley, 1985; Fowler, 1990; Cox, 1994; Cadoret et al., 1999). Moreover recruitment rates of chaetodontids are typically low (Pratchett et al., 2008b). Consequently, when coral cover is high, populations of coral dependant chaetodontids are more likely to be governed by supply rate of recruits. Similarly, space vacated by corals following disturbance is rapidly colonised by algae (Diaz-Pulido and McCook, 2002), making habitat conditions favourable for chaetodontid species that feed on algae and motile invertebrates. When disturbance is patchy or only affects specific reef zones, increase in invertivore and herbivore density may relate to migration from unaffected areas. However, when coral loss is extensive, increased density of fish over large spatial scales will be dependant on recruitment of new fish into the population. Supply of recruits from the plankton is temporally and spatially variable (Doherty, 1991), which partially explains the variable response of chaetodontids expected to benefit from coral loss.

Ontogenetic differences in diet and habitat requirements may also compromise the ability of some fish to repopulate reefs after disturbance. Some coral reef fish have strong affiliations with coral as juveniles, yet adult conspecifics have no apparent coral association (Booth and Beretta, 1994; Gutiérrez, 1998; Feary et al., 2007). For these fish species, reduced coral cover may restrict recruitment, limiting adult population sizes (Halpern et al., 2005). Thus a decline in adult populations of non-coral feeding chaetodontids may occur following coral depletion if species recruit to live coral. This proposal has gained support from findings at Kimbe Bay, Papua New Guinea, where abundance of 75% of fish species declined following >90% coral loss (Jones et al., 2004). Many of these fish species have no affinity with live coral as adults and their decline was largely attributed to reduced availability of coral for fish recruitment (Jones et al., 2004).

To determine if there are any ontogenetic shifts in resource requirements of chaetodontids we compared coral dependence of 10 adult and juvenile species. There is a strong, positive, relationship ($R^2 = 0.95$) between the contribution of coral to adult chaetodontid diets and the proportion of juvenile conspecifics recruiting to live coral (Fig. 9.5). Chaetodontid species that feed predominantly on corals as adults, recruit into corals preferentially fed upon by adults (Pratchett et al., 2008b). This is most likely because recruits of coral feeding chaetodontids have diets very similar to those

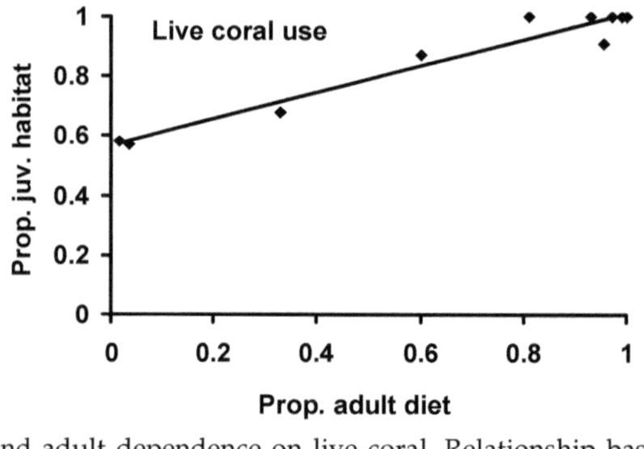

Fig. 9.5 Juvenile and adult dependence on live coral. Relationship based on comparison between juvenile habitat use and diet of adult conspecifics. Based on data from Pratchett (2005) and Pratchett et al. (2009).

of adults, and start feeding on corals as soon as they settle (Harmelin-Vivien, 1989). Conversely, species with limited reliance on coral for food, e.g., *Chaetodonauriga* and *C. vagabundus*, are more closely associated with non-coral habitats as juveniles (Pratchett et al., 2008b). These results infer that reduced availability of coral following disturbance will only affect recruitment of coral-feeding butterflyfishes.

Habitat disturbances may, however, affect recruitment of a broader range of chaetodontid species if either: i) the presence of coral on reefs provides chemical cues which attract planktonic larvae to reefs from oceanic waters; ii) post-disturbance erosion and collapse of reef structure occur at a spatial scale relevant to a broad range of small bodied juvenile fishes, increasing their susceptibility to predation (Graham et al., 2007); iii) disturbance reduces the availability of non-coral habitats, that are essential for juveniles (e.g., mangroves and seagrass; Munday et al., 2008). Under these scenarios adult stocks of species with no apparent dependence on coral will be affected by disturbance. Accordingly, several studies have documented declines in the abundance of non-corallivorous butterflyfishes (e.g., *C. auriga*) following extensive coral loss (e.g., Williams, 1986), but are at a loss to explain this response.

CONCLUSION

Disturbance represents an important and natural driver of many chaetodontid populations, as many species feed exclusively on coral and habitat disturbances on reefs typically remove live coral. The most severely affected species are those with highly specialised diets that selectively feed from only a few coral species. Long-term persistence of these species is threatened by increasing disturbance frequency and intensity, which prevent recovery of coral communities. Chaetodontids with generalist diets, that feed on a broad array of coral species, as well as algae and motile invertebrates may benefit from disturbance instigated changes in benthic communities. However, if there is no coral recovery, structural complexity of the reef is expected to collapse, which will affect those species that shelter among coral skeletons. Consequently, disturbances that reduce the structural complexity of the reef (e.g., severe storms or the long-term impacts of bleaching or outbreaks of crown-of-thorns starfish where recovery is absent) can have a much greater impact on chaetodontid communities, as they decrease availability of both coral and shelter. The physical and chemical changes to the environment associated with global warming may also influence the supply of chaetodontid recruits, reproduction, growth and physiology of adults, and this is an area that requires further research.

Fishing may also influence butterflyfish communities in some locations (Lawton et al., Chapter 11), although impacts attributed to fishing are expected to be less than that of habitat loss. Direct effects of fishing on butterflyfishes are restricted to those species targeted by ornamental collectors, or limited small-scale artisanal fisheries, and there is limited evidence to suggest butterflyfishes will be affected by trophic cascades instigated by increased fishing pressure (but see Raymundo et al., 2009). Thus, dynamics of butterflyfish assemblages can be directly attributed to changes in benthic composition, although supply of new fish recruits will determine the rate of change and species composition. Importantly, the diversity of life history traits within the family Chaetodontidae ensures that environmental conditions will naturally favour some species and that these 'chaets' may prosper.

ACKNOWLEDGEMENTS

Comments from the anonymous reviewers improved the content of this chapter. Financial support for work carried out in Fiji and the western Indian Ocean was provided by National Geographic and the Leverhulme Trust.

REFERENCES

Adams, S.M. 1999. Ecological role of lipids in the health and success of fish populations. In *The ecological role of lipids in freshwater ecosystems*, M.T. Arts and B.C. Wainman (eds.). Springer-Verlag, New York, USA pp. 132–160.

Allen, G.R., R. Steene and M. Allen. 1998. *A guide to angelfishes and butterflyfishes*, Odyssey Publishing, Perth, Australia.

Almany, G.R. 2003. Priority effects in coral reef fish communities. *Ecology* 84: 1920–1935.

Anderson, G., A. Ehrlich, P. Ehrlich, J. Roughgarden, B. Russell and F. Talbot. 1981. The community structure of coral reef fishes. *American Naturalist* 117: 476–495.

Bell, J.D., M.L. Harmelin-Vivien and R. Galzin. 1985. Large scale spatial variation in abundance of butterflyfshes (Chaetodontidae) on Polynesian reefs. *Proceedings of the 5th International Coral Reef Symposium, Tahiti* pp. 421–426.

Berumen, M.L. and M.S. Pratchett. 2006. Recovery without resilience: Persistent disturbance and long-term shifts in the structure of fish and coral communities at Tiahura Reef, Moorea. *Coral Reefs* 25: 647–653.

Berumen, M.L. and M.S. Pratchett. 2008. Trade-offs associated with dietary specialisation in corallivorous butterflyfishes (Chaetodontidae: *Chaetodon*). *Behavioral Ecology and Sociobiology* 62: 989–994.

Berumen, M.L., M.S. Pratchett and M.I. McCormick. 2005. Within-reef differences in diet and body condition of coral-feeding butterflyfishes (Chaetoodontidae). *Marine Ecology Progress Series* 287: 217–227.

Booth, D.J. and G.A. Beretta. 1994. Seasonal recruitment, habitat associations and survival of pomacentrid reef fish in the US Virgin Islands. *Coral Reefs* 13: 81–89.

Bouchon-Navaro, Y. and C. Bouchon. 1989. Correlations between chaetodontid fishes and coral communities of the Gulf of Aqaba (Red Sea). *Environmental Biology of Fishes* 25: 47–60.

Bouchon-Navaro, Y., C. Bouchon and M.L. Harmelin-Vivien. 1985. Impact of coral degradation on a chaetodontid fish assemblage (Moorea, French Polynesia). *Proceedings of the 5th International Coral Reef Symposium, Tahiti* pp. 427–432.

Bozec, Y.M., S. Doledec and M. Kulbicki. 2005. An analysis of fish habitat associations on disturbed coral reefs: chaetodontid fishes in New Caledonia. *Journal of Fish Biology* 66: 966–982.

Brown, B.E. 1997. Coral bleaching: causes and consequences. *Coral Reefs* 16: 129–138.

Bruno, J.F. and E.R. Selig. 2007. Regional Decline of Coral Cover in the Indo-Pacific: Timing, Extent, and Subregional Comparisons. *PLoS ONE* 2: e711.

Bruno, J.F., E.R. Selig, K.S. Casey, C.A. Page, B.L. Willis, C.D. Harvell, H. Sweatman and A.M. Melendy. 2007. Thermal stress and coral cover as drivers of coral disease outbreaks. *PLoS Biology* 5: 1220–1227.

Bythell, J.C., Z.M. Hillis-Starr and C.S. Rogers. 2000. Local variability but landscape stability in coral reef communities following repeated hurricane impacts. *Marine Ecology Progress Series* 204: 93–100.

Cadoret, L., M. Adjeroud and M. Tsuchiya. 1999. Spatial distribution of chaetodontid fish in coral reefs of the Ryukyu Islands, southern Japan. *Journal of the Marine Biological Association of the* UK 79: 725–735.

Chan, T.T.C. and Y.J. Sadovy. 1998. Profile of the Marine Aquarium Fish Trade in Hong Kong. *Aquarium Sciences and Conservation* 2: 197–213.

Cheal, A.J., G. Coleman, S. Delean, I. Miller, K. Osborne and H. Sweatman. 2002. Responses of coral and fish assemblages to a severe but short-lived tropical cyclone on the Great Barrier Reef, Australia. *Coral Reefs* 21: 131–142.

Cole, A.J., M.S. Pratchett and G.P. Jones. 2008. Diversity and functional importance of coral-feeding fishes on tropical coral reefs. *Fish and Fisheries* 9: 286–307.

Connell, J.H. 1978. Diversity in Tropical Rain Forests and Coral Reefs. *Science* 199: 1302–1310.

Cowey, C.B. and J.R. Sargent. 1977. Lipid nutrition in fish. *Comparative Biochemistry and Physiology* 57: 269–273.

Cox, E.F. 1994. Resource use by corallivorous butterflyfishes (Family Chaetodontidae) in Hawaii. *Bulletin of Marine Science* 54: 535–545.

Cumming, R.L., M.A. Toscano, E.R. Lovell, B.A. Carlson, N.K. Dulvy, A. Hughes, J.F. Koven, N.J. Quinn, H.R. Sykes, O.J.S. Taylor and D. Vaughan. 2000. Mass coral bleaching in the Fiji Islands, 2000. *Proceedings of the 9th International Coral Reef Symposium, Bali* pp. 1161–1169.

Diaz-Pulido, G. and L.J. McCook. 2002. The fate of bleached corals: patterns and dynamics of algal recruitment. *Marine Ecology Progress Series* 232: 115–128.

Doherty, P.J. 1991. Spatial and temporal patterns in recruitment. In *The Ecology of Fishes on Coral Reefs*, P.F. Sale (ed.). Academic Press, Inc., San Diego, USA pp. 261–292.

Dulvy, N.K., R.P. Freckleton and N.V.C. Polunin. 2004. Coral reef cascades and the indirect effects of predator removal by exploitation. *Ecology Letters* 7: 410–416.

Ehrlich, P.R., F.H. Talbot, B.C. Russell and G.R.V. Anderson. 1977. The behaviour of chaetodontid fishes with special reference to Lorenz' "poster colouration" hypothesis. *Journal of Zoology* 183: 213–228.

Emanuel, K. 2005. Increasing destructiveness of tropical cyclones over the past 30 years. *Nature* 436: 686–688.

Engelhardt, U. 2004. The status of scleractinian coral and reef-associated fish communities 6 years after the 1998 mass coral bleaching event. Final report. Seychelles Marine Ecosystem Management Project, Reefcare International, Townsville, Australia.

Feary, D.A., G.R. Almany, M.I. McCormick and G.P. Jones. 2007. Habitat choice, recruitment and the response of coral reef fishes to coral degradation. *Oecologia* 153: 727–737.

Findley, J.S. and M.T. Findley. 1985. A search for pattern in butterflyfish communities. *American Naturalist* 126: 800–816.

Fowler, A.J. 1990. Spatial and temporal patterns of distribution and abundance of chaetodontid fishes at One Tree Reef, southern GBR. *Marine Ecology Progress Series* 64: 39–53.

Gardner, T.A., I.M. Côté, J.A. Gill, A. Grant and A.R. Watkinson. 2003. Long-Term Region-Wide Declines in Caribbean Corals. *Science* 301: 958–960.

Glynn, P.W. 1993. Coral reef bleaching: ecological perspectives. *Coral Reefs* 12: 1–17.

Graham, N.A.J. 2007. Ecological versatility and the decline of coral feeding fishes following climate driven coral mortality. *Marine Biology* 153: 119–127.

Graham, N.A.J., R.D. Evans and G.R. Russ. 2003. The effects of marine reserve protection on the trophic relationships of reef fishes on the Great Barrier Reef. *Environmental Conservation* 30: 200–208.

Graham, N.A.J., S.K. Wilson, S. Jennings, N.V.C. Polunin, J.P. Bijoux and J. Robinson. 2006. Dynamic fragility of oceanic coral reef ecosystems. *Proceedings of the National Academy of Sciences of the United States of America* 103: 8425–8429.

Graham, N.A.J., S.K. Wilson, S. Jennings, N.V.C. Polunin, J. Robinson, J.P. Bijoux and T.M. Daw. 2007. Lag Effects in the Impacts of Mass Coral Bleaching on Coral Reef Fish, Fisheries, and Ecosystems. *Conservation Biology* 21: 1291–1300.

Graham, N.A.J., S.K. Wilson, M.S. Pratchett, N.V.C. Polunin and M.D. Spalding. 2009. Coral mortality versus structural collapse as drivers of corallivorous butterflyfish decline. *Biodiversity and Conservation* 18: 3325–3336.

Grigg, R.W. 1994. Effects of sewage discharge, fishing pressure and habitat complexity on coral ecosystems and reef fishes in Hawaii. *Marine Ecology Progress Series* 103: 25–34.

Grime, J.P. 1973. Competitive exclusion in herbaceous vegetation. *Nature* 242: 344–347.

Gutiérrez, L. 1998. Habitat selection by recruits establishes local patterns of adult distribution in two species of damselfishes: *Stegastesdorsopunicans* and *S. planifrons*. *Oecologia* 115: 268–277.

Halford, A., A.J. Cheal, D. Ryan and D.M. Williams. 2004. Resilience to Large-Scale Disturbance in Coral and Fish Assemblages on the Great Barrier Reef. *Ecology* 85: 1892–1905.

Halpern, B.S., S.D. Gaines and R.R. Warner. 2005. Habitat size, recruitment, and longevity as factors limiting population size in stage-structured species. *American Naturalist* 165: 82–94.

Harmelin-Vivien, M.L. 1989. Implications of feeding specialization on the recruitment processes and community structure of butterflyfishes. *Environmental Biology of Fishes* 25: 101–110.

Hart, A.M. and G.R. Russ. 1996. Response of herbivorous fishes to crown-of-thorns starfish *Acanthaster planci* outbreaks. III. Age, growth, mortality and maturity indices of *Acanthurusnigrofuscus*. *Marine Ecology Progress Series* 136: 25–35.

Harvell, C.D., K. Kim, J.M. Burkholder, R.R. Colwell, P.R. Epstein, D.J. Grimes, E.E. Hofmann, E.K. Lipp, A.D. Osterhaus, R.M. Overstreet, J.W. Porter, G.W. Smith and G.R. Vasta. 1999. Emerging Marine Diseases—Climate Links and Anthropogenic Factors. *Science* 285: 1505–1510.

Harvell, C.D., C.E. Mitchell, J.R. Ward, S. Altizer, A.P. Dobson, R.S. Ostfeld and M.D. Samuel. 2002. Climate Warming and Disease Risks for Terrestrial and Marine Biota. *Science* 296: 2158–2162.

Hoegh-Guldberg, O. 1999. Climate change, coral bleaching and the future of the world's coral reefs. *Marine and Freshwater Research* 50: 839–866.

Hoegh-Guldberg, O., P.J. Mumby, A.J. Hooten, R.S. Steneck, P. Greenfield, E. Gomez, C.D. Harvell, P.F. Sale, A.J. Edwards, K. Caldeira, N. Knowlton, C.M. Eakin, R. Iglesias-Prieto, N. Muthiga, R.H. Bradbury, A. Dubi and M.E. Hatziolos. 2007. Coral Reefs Under Rapid Climate Change and Ocean Acidification. *Science* 318: 1737–1742.

Irons, D.K. 1989.Temporal and areal feeding behavior of the butterflyfish, *Chaetodon trifascialis*, at Johnston Atoll. *Environmental Biology of Fishes* 25: 187–193.

Jones, G.P., M.I. McCormick, M. Srinivasan and J.V. Eagle. 2004. Coral decline threatens fish biodiversity in marine reserves. *Proceedings of the National Academy of Sciences of the United States of America* 101: 8251–8253.

Kaiser, M.J., K.R. Clarke, H. Hinz, M.C.V. Austen, P.J. Somerfield and I. Karakakkis. 2006. Global analysis of response and recovery of benthic biota to fishing. *Marine Ecology Progress Series* 311: 1–14.

Kokita, T. and A. Nakazono. 2001. Rapid response of an obligately corallivorous filefish *Oxymonacanthuslongirostris* (Monacanthidae) to a mass coral bleaching event. *Coral Reefs* 20: 155–158.

Lapointe, B.E. 1997. Nutrient Thresholds for Bottom-Up Control of Macroalgal Blooms on Coral Reefs in Jamaica and Southeast Florida. *Limnology and Oceanography* 1119–1131.

McClanahan, T.R. and S.H. Shafir. 1990. Causes and consequences of sea urchin abundance and diversity in Kenyan coral reef lagoons. *Oecologia* 83: 362–370.

McClanahan, T.R., M. Ateweberhan, N.A.J. Graham, S.K. Wilson, C.R. Sebastian, M.M.M. Guillaume and J.H. Bruggemann. 2007. Western Indian Ocean coral communities: bleaching responses and susceptibility to extinction. *Marine Ecology Progress Series* 337: 1–13.

McKinney, M.L. 1997. Extinction vulnerability and selectivity: combining ecological and paleontological views. *Annual Review of Ecology Evolution and Systematics* 495–516.

Moberg, F. and C. Folke. 1999. Ecological goods and services of coral reef ecosystems. *Ecological Economics* 29: 215–233.

Munday, P.L. 2004. Habitat loss, resource specialization, and extinction on coral reefs. *Global Change Biology* 10: 1642–1647.

Munday, P.L. and G.P. Jones. 1998. The ecological implications of small body size among coral-reef fishes. *Oceanography and Marine Biology: An Annual Review* 36: 373–411.

Munday, P.L., G.P. Jones, M.S. Pratchett and A.J. Williams. 2008. Climate change and the future for coral reef fishes. *Fish and Fisheries* 9: 261–285.

Pauly, D., V. Christensen, J. Dalsgaard, R. Froese and F. Torres Jr. 1998. Fishing down marine food webs. *Science* 279: 860–863.

Pelejero, C., E. Calvo, M.T. McCulloch, J.F. Marshall, M.K. Gagan, J.M. Lough and B.N. Opdyke. 2005. Preindustrial to modern interdecadal variability in coral reef pH. *Science* 309: 2204–2207.

Pratchett, M.S. 2001. Influence of coral symbionts on feeding preferences of crown-of-thorns starfish *Acanthaster planci* in the western Pacific. *Marine Ecology Progress Series* 214: 111–119.

Pratchett, M.S. 2005. Dietry overlap among coral-feeding butterflyfishes (Chaetodontidae) at Lizard Island, northern Great Barrier Reef. *Marine Biology* 148: 373–382.

Pratchett, M.S. 2007. Dietary selection by coral-feeding butterflyfishes (Chaetodontidae) on the Great Barrier Reef, Australia. *The Raffles Bulletin of Zoology* Supplement No. 14: 171–176.

Pratchett, M.S., S.K. Wilson, M.L. Berumen and M.I. McCormick. 2004. Sublethal effects of coral bleaching on an obligate coral feeding butterflyfish. *Coral Reefs* 23: 352–356.

Pratchett, M.S., S.K. Wilson and A.H. Baird. 2006. Declines in the abundance of *Chaetodon* butterflyfishes following extensive coral depletion. *Journal of Fish Biology* 69: 1269–1280.

Pratchett, M.S., P.L. Munday, S.K. Wilson, N.A.J. Graham, J.E. Cinner, D.R. Bellwood, G.P. Jones, N.V.C. Polunin and T.R. Mcclanahan. 2008a. Effects of climate-induced coral bleaching on coral-reef fishes—Ecological and economic consequences.*Oceanography and Marine Biology: An Annual Review* 46: 251–296.

Pratchett, M.S., M.L. Berumen, M.J. Marnane, J.V. Eagle and D.J. Pratchett. 2008b. Habitat associations of juvenile versus adult butterflyfishes. *Coral Reefs* 27: 541–551.

Pratchett, M.S., S.K. Wilson, N.A.J. Graham, P.L. Munday, G.P. Jones and N.V.C. Polunin. 2009. Multi-scale temporal effects of climate-induced coral bleaching on motile reef organisms. In *Coral Bleaching: patterns and processes, causes and consequences*, M. van Oppen and J. Lough (eds.). Springer, Berlin, Germany pp. 139–158.

Rahmstorf, S. 2007. A semi-empirical approach to projecting future sea-level rise. *Science* 315: 368–370.

Raymundo, L.J., A.R., Halford, A.P. Maypa and A.M. Kerr. 2009. Functionally diverse reef-fish communities ameliorate coral disease. *Proceedings of the National Academy of Sciences of the United States of America* 106: 17067–17070.

Russ, G.R. 2002. Yet another review of marine reserves as reef fishery management tools. In *Coral reef fishes: dynamics and diversity in a complex ecosystem*, P.F. Sale (ed.). Academic Press, New York, USA pp. 421–444.

Sadovy, Y.J. and A.C.J. Vincent. 2002. Ecological issues and the trades in live reef fishes. In *Coral reef fishes: dynamics and diversity in a complex ecosystem*, P.F. Sale (ed.). Academic Press, New York, USA pp. 391–340.

Samways, M.J. 2005. Breakdown of butterflyfish (Chaetodontidae) territories associated with the onset of a mass coral bleaching event. *Aquatic Conservation: Marine and Freshwater Ecosystems* 15: S101–S107.

Sano, M. 2004. Short-term effects of a mass coral bleaching event on a reef fish assemblage at Iriomote Island, Japan. *Fisheries Science* 70: 41–46.

Sano, M., M. Shimizu and Y. Nose. 1984. Changes in structure of coral reef fish communities by destruction of hermatypic corals: Observational and experimental views. *Pacific Science* 38: 51–80.

Schaffelke, B. 1999. Short-term nutrient pulses as tools to assess responses of coral reef macroalgae to enhanced nutrient availability. *Marine Ecology Progress Series* 182: 305–310.

Scheffer, M., S. Carpenter, J.A. Foley, C. Folke and B. Walker. 2001. Catastrophic shifts in ecosystems. *Nature* 413: 591–596.

Sheppard, C.R.C. 2003. Predicted recurrences of mass coral mortality in the Indian Ocean. *Nature* 425: 294–297.

Shibuno, T., K. Hashimoto, O. Abe and Y. Takada. 1999. Short-term changes in the structure of a fish community following coral bleaching at lshigaki Island, Japan. *Galaxea* 51–58.

Spalding, M.D. and G.E. Jarvis. 2002. The impact of the 1998 coral mortality on reef fish communities in the Seychelles. *Marine Pollution Bulletin* 44: 309–321.

Sweatman, H. 2008. No-take reserves protect coral reefs from predatory starfish. *Current Biology* 18: R598–R599.

Syms, C. 1998. Disturbance and the structure of coral reef fish communities on the reef slope. *Journal of Experimental Marine Biology and Ecology* 230: 151–167.

Thompson, A. and H. Malcolm. 1999. Benthic and fish monitoring of fringing reefs in the Brook, Palm and Rattlesnake Island groups: Status post 1998 coral bleaching event. Technical report. Queensland Parks and Wildlife Service, Townsville, Australia. pp. 26.

Tissot, B.N. and L.E. Hallacher. 2003. Effects of aquarium collectors on coral reef fishes in Kona, Hawaii. *Conservation Biology* 17: 1759–1768.

Tricas, T.C. 1985. The economics of foraging in coral-feeding butterflyfishes of Hawaii. *Proceedings of the 5th International Coral Reef Symposium, Tahiti* pp. 409–414.

Tricas, T.C. 1989a. Prey selection by coral-feeding butterflyfishes: strategies to maximize the profit. *Environmental Biology of Fishes* 25: 171–185.

Tricas, T.C. 1989b. Determinants of feeding territory size in the corallivorous butterflyfish, *Chaetodonmulticinctus*. *Animal Behaviour* 37: 830–841.

Vázquez, D.P. and D. Simberloff. 2002. Ecological specialization and susceptibility to disturbance: conjectures and refutations. *American Naturalist* 159: 606–623.

Walsh, W.J. 1983. Stability of a coral reef fish community following a catastrophic storm. *Coral Reefs* 2: 49–63.

Webster, P.J., G.J. Holland, J.A. Curry and H.R. Chang. 2005. Changes in tropical cyclone number, duration, and intensity in a warming environment. *Science* 309: 1844–1846.

Williams, D.M. 1986. Temporal variation in the structure of reef slope fish communities (central Great Barrier Reef): short-term effects of *Acanthaster planci* infestation. *Marine Ecology Progress Series* 28: 157–164.

Wilson, S.K., N.A.J. Graham, M.S. Pratchett, G.P. Jones and N.V.C. Polunin. 2006. Multiple disturbances and the global degradation of coral reefs: are reef fishes at risk or resilient? *Global Change Biology* 12: 2220–2234.

Wilson, S.K., R. Fisher, M.S. Pratchett, N.A.J. Graham, N.K. Dulvy, R.A. Turner, A. Cakacaka, N.V.C. Polunin and S.P. Rushton. 2008a. Exploitation and habitat degradation as agents of change within coral reef fish communities. *Global Change Biology* 14: 2796–2809.

Wilson, S.K., S.C. Burgess, A.J. Cheal, M. Emslie, R. Fisher, I. Miller, N.V.C. Polunin and H.P.A. Sweatman. 2008b. Habitat utilization by coral reef fish: implications for specialists vs. generalists in a changing environment. *Journal of Animal Ecology* 77: 220–228.

Wilson, S., A. Dolman, A. Cheal, M. Emslie, M. Pratchett and H.P.A. Sweatman. 2009. Maintenance of fish diversity on disturbed coral reefs. *Coral Reefs* 28: 3–14.

Wood, E.M. 2001. Collection of coral reef fish for aquaria: global trade, conservation issues and management strategies. Marine Conservation Society, Ross-on-Wye, UK pp. 80.

Wooldridge, S., T. Done, R. Berkelmans, R. Jones and P. Marshall. 2005. Precursors for resilience in coral communities in a warming climate: a belief network approach. *Marine Ecology Progress Series* 295: 157–169.

CHAPTER 10

Corallivorous Butterflyfishes as Ambassadors of Coral Reefs

Michael P. Crosby,[1,*] *Ernst S. Reese*[2] and *Michael L. Berumen*[3]

INTRODUCTION OF THE BUTTERFLYFISH INDICATOR METHOD

The concept of using one or a small group of species to indicate the condition of a larger community or ecosystem is not new. Hunting and gathering peoples, both past and present, intuitively apply the idea (Lee and Devore, 1968), and similarly early naturalists/conservationists like Aldo Leopold (1966) recognized the usefulness of the concept. A more recent example of the utility of the concept in the terrestrial environment is the finding that the decline of migratory songbird populations is an indication of degraded and hostile forests across the continental United States (Askins, 1995). To our knowledge, the concept of indicator species was first applied formally to marine organisms by the celebrated British marine biologist F.S. Russell in the 1930s (Fraser, 1962; Raymont, 1963). In a series of papers, Russell showed that certain species of zooplankton were indicative of water masses that contained commercially important species of fish. Russell emphasized

[1]Mote Marine Laboratory, 1600 Ken Thompson Parkway Sarasota, FL 34236, USA.
E-mail: mcrosby@mote.org
[2]Department of Zoology and the Hawaii Institute of Marine Biology, The University of Hawaii, Honolulu, Hawaii, USA.
E-mail: ereese@hawaii.edu
[3]Red Sea Research Center, King Abdullah University of Science and Technology, Thuwal, 23955, Saudi Arabia.
E-mail: michael.berumen@kaust.edu.sa
*Corresponding author

that a good indicator species must be easily recognized and its distribution must characterize a particular mass of seawater. Two edited volumes address the use of marine organisms as indicators for detecting ecological impacts, usually from human activities, on marine ecosystems (Soule and Kleppel, 1988; Schmitt and Osenberg, 1996), with chapters on coral reefs (Hourigan et al., 1988; Jones and Kaly, 1996).

The seminal paper by Odum and Odum (1953) clearly established coral reefs as communities and, with the addition of energy flow, the concept of the coral reef ecosystem was born. Surprisingly the idea of co-evolution, in the broad sense, between corals and coral feeding fishes was not widely recognized until much later (Reese, 1977), nor was the possibility recognized that this relationship might prove useful for the conservation and management of coral reef ecosystems (Reese, 1981). The idea that the distribution, abundance, and behaviour of coral feeding fishes of the family Chaetodontidae, the butterflyfishes, could be used as indicators of environmental stress on coral reefs was established by Hourigan et al. (1988).

Corallivorous butterflyfishes have been considered, studied, and debated as potential indicator species for coral reef ecosystems for many years. It is worth pointing out that this debate is predicated on the assumption that more hard coral cover is a desired and/or 'healthier' state of the system than a state with decreased coral cover and increased algal cover, for example (see Bellwood et al., 2004). Herein we present a closer look and summary of some of the ideas and potential applications underlying the usefulness of some butterflyfishes as indicator species in this context. In the simplest case, assessing coral cover on a particular reef is perhaps best achieved by learning to properly survey benthic cover and directly assessing the coral communities as opposed to using a proxy (i.e., surveying an indicator species). While we certainly agree that learning to identify a relatively small number of conspicuous fish is easier than distinguishing among a taxonomically problematic and challenging group (e.g., corals), there are also some circumstances in which a proxy can be somewhat beneficial. For example, butterflyfishes may be useful as indicator species for detecting more subtle and ongoing changes in the 'health' of a coral reef ecosystem, as we will describe below.

RELATIONSHIP OF OBLIGATE CORALLIVORE BUTTERFLYFISH TO CORAL

Fishes of the family Chaetodontidae are found in all tropical seas of the world. There are about 130 species in 11 genera with 90 of the species in the genus *Chaetodon*. The systematics of the family is well studied using classical systematic methods (Burgess, 1978), cladistic analysis (Blum, 1989), and molecular phylogenetics (Fesslerand Westneat, 2007; Bellwood et al., 2010; Bellwood and Pratchett, Chapter 1). Butterflyfishes are characterized as diurnally active, brightly coloured inhabitants of coral reefs. They can be classified into several feeding guilds: corallivores (sometimes grouped by varying degrees of obligation), benthic omnivores, and planktivores. The coral feeders are especially interesting because they are closely associated with the living coral reef for both food and shelter. Coral feeding species tend to live in monogamous pairs and, as a general rule, are home ranging or territorial (see Yabuta and Berumen, Chapter 8).

Because of the close association of Chaetodontidae with coral communities (Reese, 1981; Bouchon-Navaro et al., 1985; Hourigan et al., 1988; Bouchon-Navaro and Bouchon, 1989), their distinctive appearance, diurnal behaviour cycle and territoriality, it has been proposed that obligate corallivore butterflyfish should be excellent indicators of subtle, sub-lethal change in coral reefs (Reese, 1977, 1981; Crosby and Reese, 1996; Reese and Crosby, 1999). Obligate corallivores essentially feed exclusively on living coral (Cole and Pratchett, Chapter 5) and require extensive access to live corals for food. Facultative corallivore butterflyfish derive a significant portion of their diet from living coral, but occasionally may feed on non-coral organisms, especially when coral is not abundant (Reese, 1977, 1981; Cole et al., 2008; Chapter 5, Cole and Pratchett).

Several morphological, life history, behavioural, and ecological characteristics make obligate corallivore butterflyfishes candidates to be indicator species for ecological conditions on coral reefs. In many species, pairs of chaetodontids show a very high degree of site fidelity with the same individual pairs being found on the same territories for seven to eight years (Reese, 1991) and these territories may be quite small, often less than 100m^2 (Crosby and Reese, 2005; Berumen and Pratchett, 2006). Their life span is 6 to 20 years depending on the species (Berumen, 2005; Zekeria et al., 2006; Berumen et al., 2012; Yabuta and Berumen, Chapter 8). Predation on adults is minimal (see St. John, 1999) but may play a significant role in limiting population sizes during the larval dispersal stage and also in the early post-settlement stage (Webster, 2002). Their striking colour patterns make them ideal for behavioural observation in the field. Not only can species be readily identified from a distance and inter-specific interactions recorded (e.g., Berumen and Pratchett, 2006), but variation within each species permits individual recognition in the field, facilitating study of behaviour such as territoriality (e.g., Reese, 1973).

Not all chaetodontids are candidates for indicator species of change in coral reefs. The planktivores in particular, hovering above the reef facing into the current to intercept plankton, are not sensitive to the corals on the reef beneath them. Likewise, the more omnivorous species, feeding on algae and benthic invertebrates other than corals, tend to be opportunistic and they may be able to feed on prey in proportion to their abundance. As their food source changes, they may change their diets and are not, therefore, as useful for indicating changes occurring in the ecosystem.

In contrast, corallivorous chaetodontids make ideal indicators because they feed directly on the corals. Many species are obligate corallivores and do not feed on anything else (e.g., Pratchett, 2005). Furthermore, they show strong preferences for certain species of corals (e.g., Berumen et al., 2005; Pratchett, 2007) that provides a further dimension of sensitivity to the system, but caution must be applied since feeding electivity and prey use may be flexible in some species (Pratchett, Chapter 6). Since they are territorial, strongly site-attached, and live for many years, they provide a temporal component to the system that has great value. Even if changes occur very slowly in the ecosystem, the same individual pairs of butterflyfishes will be present to experience the change. A habitat must provide an animal with food, shelter, and the opportunity to reproduce. If a specific habitat does not provide these minimum requirements, the animal must either move between habitats on a diurnal, seasonal, or permanent basis, or it will not survive. If a species is frequently moving between habitats, its behavioural ecology becomes more complex, making it a poor indicator species candidate for the focal habitat of interest. The territorial and home-ranging behaviour of many corallivorous butterflyfishes therefore minimizes this complexity.

EVOLUTION OF THE BUTTERFLYFISH INDICATOR METHOD

Just the Numbers

There are a number of studies where entire assemblages of chaetodontid fishes are correlated with coral cover (e.g., Sano et al., 1984, 1987; Bell et al., 1985; Bouchon-Navaro et al., 1985; Findley and Findley, 1989; Williams, 1986; Roberts and Ormond, 1987; Roberts et al., 1988; White, 1988; Cadoret et al., 1999). Although generally there is a good correlation between chaetodontid abundance and coral cover, the results are sometimes mixed and do not always support the hypothesis that chaetodontid fishes can be used as indicators of coral cover. Poor correlations between chaetodontid abundance and coral cover are often due to entire assemblages of butterflyfishes being counted rather than just the corallivorous species (see Pratchett and Berumen, 2008). In general, indicator approaches aim to determine overall hard coral cover. When attempting to address questions of coral community composition, greater depth of analysis is likely required as discussed below.

Although Roberts et al. (1988) are critical of the abundance approach for a butterflyfish indicator methodology, when they plotted the abundance of three species of obligate corallivores in the Red Sea against hard coral cover, they obtained a correlation of 0.8. In addition, they agree that repeated censuses at the same sites over time would be useful in assessing changes in coral reef communities. White (1988) provides further support for abundance approaches in a study where 34 species of Chaetodontid fishes were counted at eight locations. Only about a third of these species can be classed as coral feeders, both facultative and obligate, but nevertheless when the number of species was plotted against hard (scleractinian) coral cover, the correlation was 0.71. Jones and Kaly (1996) showed graphically a high correlation between the obligate corallivores *Chaetodon trifascialis* (Fig. 10.1a) and *C. lunulatus* and total percent coral cover in the Funafuti lagoon, Tuvalu. Interestingly, they found the same good correlation between algal cover and *Chaetodon citrinellus* and *Chaetodon auriga*, which illustrates the danger in counting all species of chaetodontids (Fig 10.1b).

Another problem when relying on 'just the numbers' arises due to problems related to sublethal effects and responses of butterflyfish species to changes in coral availability. For example, Pratchett et al. (2004) documented the effect of a major coral bleaching event on Trunk Reef (on Australia's Great Barrier Reef) in 2000. *Chaetodon lunulatus*, an obligate corallivore, was studied before (May 2000) and after (March 2002) the bleaching event on this reef. Although live coral cover decreased by more than 55%, the abundance of *C. lunulatus* did not change significantly. However, further investigation revealed that *C. lunulatus* had switched its diet to consume less preferred species of corals, which happened to be less impacted by the bleaching event than preferred prey species. Analysis of the lipid storage in the livers revealed that fish in 2002 had far lower energy stores than in 2000. Liver lipid content provides a very sensitive measure of the physiological condition and subsequent fitness for individual fish (see Cowey and Sargent, 1977; Black and Love, 1986; Green and McCormick, 1999). This study demonstrated one of the potential pitfalls of using corallivorous butterflyfishes as indicator species—some species possess flexibility in their diets and a capacity to tolerate sublethal effects of degradation in coral communities. Assessment of Trunk Reef based solely on the abundance of *C. lunulatus* would not have indicated any change in the state of the reef.

Two similar studies on Lizard Island (Great Barrier Reef), Berumen et al. (2005) and Berumen and Pratchett (2006), explored populations of obligate corallivores living in different habitats around the island. The most specialized fish studied, *Chaetodon baronessa*, had no differences in densities between exposed fore-reef habitats and sheltered back-reef habitats despite marked differences in coral communities. Interestingly, however, there were notable differences in the lipid stores in the livers of the fish between these

A

B

Fig. 10.1 Good and bad candidates for indicator species. A) *Chaetodon trifascialis* is an obligate coral feeding butterflyfishes that exhibits strong association with specific coral types, mainly *Acropora*, Photo by Luiz Rocha; B) *Chaetodon auriga* is a generalist feeder, which sometimes feeds on corals, but may also be abundant on reefs dominated by macro-algae. Photo by Michael Berumen.

Colour image of this figure appears in the colour plate section at the end of the book.

habitats, with fish from exposed fore-reefs having much greater energy stores. Corresponding with these differences were several other interesting results: 1) *C. baronessa* in fore-reef areas was far more selective in its feeding behaviour than in back-reef areas, 2) *C. baronessa* was far more aggressive in intra- and interspecific interactions on fore-reefs than on back-reefs, 3) mean territory sizes of *C. baronessa* were twice as large on back-reefs than on fore-reefs. Taken together, these studies suggest that characteristics other than just abundance of populations will be indicative of the 'quality' of the reef that they are associated with.

On the extremes, obligate corallivorous butterflyfishes are obvious choices: only a reef with very high coral cover can sustain the highest densities of obligate corallivores, while a complete loss of corals will quickly lead to the loss of these fishes from a reef (e.g., Pratchett et al., 2006). However, it is the changes intermediate in magnitude and time that are less clear, and at these scales, there are many studies that find support for the concept and many studies that do not. The widespread disagreement most likely means that there will never be general acceptance for an approach based solely on single assessments of butterflyfish community assemblages. This leads us, in this chapter, to explore a more nuanced approach.

Behaviour

The general butterflyfish indicator hypothesis as introduced by Crosby and Reese (1996) and Reese and Crosby (1999) predicted that percentage coral cover would be inversely related with territory size and agonistic encounters for obligate corallivore butterflyfishes, while positively related with size and number of adult fishes.

Because butterflyfishes are conspicuous inhabitants of coral reefs throughout the Indo-Pacific region and can be easily and inexpensively surveyed, these small fish are ideally suited to help in efforts to protect and preserve the world's coral reefs. Specifically, many Chaetodontidae species are obligate coral feeders, depending exclusively on the live tissue of corals for their food. Because their metabolic or energetic demand is so intimately linked to the existence and overall condition or 'health' of the coral substrate, it is believed that these species of butterflyfish are excellent candidates as 'indicators' of changes in conditions on the coral reef (Hourigan et al., 1988). The conceptual ideas are simple. Corallivorous butterflyfishes have co-evolved with, and are intimately related to, the corals on which they feed (Reese, 1977, 1981). In addition, the distribution and abundance of these fish should be directly correlated with that of the corals. Therefore, if the corals are adversely affected by stressful environmental conditions (such as chronic low levels of pollution) their health would deteriorate and they will be less nutritious to the butterflyfish. Coral-feeding butterflyfish should respond to declines in coral quality or abundance by behavioural and spatial adjustments that can be easily and rapidly observed (i.e., changes in feeding rates, territory size and other related behaviours). Thus, counting the density and behavioural traits of the brightly coloured fish is a relatively simple and inexpensive way to assess the general health of a coral reef (e.g.,

Crosby and Reese, 2005). Divers can be taught in a few hours to recognize and census butterflyfish along a transect (Crosby and Reese, 1996). This is especially important considering the fact that both chronic low levels of pollutants on the reef and the slowly deteriorating condition of the corals are difficult to detect by conventional methods.

Crosby and Reese (1996) first published a manual and accompanying training video describing a proposed methodology for the butterflyfish indicator monitoring concept. That manual described a non-invasive and 'low-tech' approach to assessing and monitoring coral reef habitats. Its purpose was to serve as a reference for individuals with limited technical science background and expertise. Contents included step-by-step instructions for observing coral-feeding butterflyfish behaviour and how to relate changes in these traits to changes in coral habitat condition, as well as basic techniques for assessing coral and fish biodiversity and abundance. Crosby and Reese (1996) felt it had become evident that new and innovative techniques were needed that could provide an 'early warning' of stress within coral habitats so that investigations to determine causes of the stress and implementation of counteractive management strategies to protect these critical habitats could be initiated. A critical need also existed for relatively low cost, simple (low technology) methods for assessing the overall stress levels ('health') of coral reef habitats throughout the world. This need continues to be especially high in regions that do not have readily available equipment and expertise for conducting comprehensive 'high-tech' ecological studies. Even within the more 'developed' regions of the world, the availability of low cost methods for coral reef assessments that could be conducted by persons without extensive technological training and education would be of immense value to monitoring the stress in these ecosystems.

Erdmann (1997) offers a thoughtful critique of the butterflyfish indicator method as developed by Crosby and Reese (1996) and points out a number of areas where caution is needed. We agree but believe them to have been addressed in several subsequent reports (discussed in the following section). As an alternative to butterflyfishes, stomatopods are suggested as bio-indicators of marine pollution stress on coral reefs (Erdmann and Caldwell, 1997) but most of the criticisms of the butterflyfish indicator approach are equally applicable to stomatopods. Moreover, F.S. Russell has noted that 'it is easier to count brightly coloured relatively large butterflyfishes than cryptic, cavity-dwelling stomatopods'. The choice is up to the researcher, public agency, environmental group or individual volunteer as to what methods most appropriately address their needs.

Subsequently, Reese and Crosby (1999) outlined an assortment of advantages and disadvantages for this methodology. Conventional monitoring programs are often expensive, time consuming, require scientifically trained personnel, involve complex sampling schedules, and require the processing, storage and analysis of samples. Photographic sampling methods, which at first glance appear to be an excellent choice, are expensive and analysis is extremely time consuming (Leujak and Ormond, 2007). Furthermore, permanent transecting lines or pins may not be permitted in preserves and sanctuaries.

In contrast, the use of indicator species has many advantages. It is less expensive and does not require scientifically trained personnel. It eliminates the costs in time and money of establishing a conventional sampling plan, including the archiving, storage and analysis of samples. It is not necessary to leave permanent markers in the field. The use of indicator species to monitor coral reefs may be the method of choice when gradual change is suspected (e.g., due to low but chronic input of pollutants into a reef area), where previous baseline data do not exist, when funds are limited, or where scientifically trained personnel are not available. These conditions exist in many developing nations where coral reefs are an important resource and there is a need for conservation and management.

Choosing the best one or two species to serve as indicators is extremely important. It necessitates some knowledge of the life history and behavioural ecology of species, preferably species that are abundant, conspicuous, diurnal, long lived, site attached, and dependent on the ecosystem to be monitored for their food and shelter. Fortunately this information often exists for one area where the species was studied and can be applied effectively elsewhere. Thus, for example, it is possible to recommend species of butterflyfishes as potential indicator species for many coral reefs in the Red Sea, Indian and Pacific Oceans based on studies in Hawai'i, Guam, Jordan, Israel, Egypt, Iran, Australia, the Marshall Islands, and French Polynesia, among others.

Samways (2005) reports the reaction of three species of Seychelles obligate corallivorous butterflyfishes to the onset of the 1998 mass-bleaching event, which caused between 70% and 99% estimated mortality of Seychelles hard corals (Goreau et al., 2000). A few months prior to this large-scale bleaching event, the reefs on Cousine Island, Seychelles, were being assessed using different comparative methodologies (Samways and Hatton, 2001), although the authors at the time obviously were not aware of the impending bleaching event. Data on butterflyfish behaviour were, therefore, gathered serendipitously at the onset of the bleaching event.

The baseline data of Reese (1989) and Irons (1989) make comparisons meaningful, with the assumption that behaviour is similar in the Pacific and Indian Oceans. In view of the findings of Reese (1975, 1981) and Sutton (1985) this is probably a fair assumption. What is very clear is that both *C. trifascialis* and *C. trifasciatus* were ranging so widely when surveyed in the Seychelles that they were frequently observed over non-food substrates, although this is not unusual for *C. trifascialis* even on healthy reefs (Bouchon-Navaro and Bouchon, 1989). Morevoer, the linear distances that that these fishes were swimming in the Seychelles were an order of magnitude higher than has been recorded previously (e.g., Reese, 1989). Any territorial mosaic, either intra- or inter-specific, was not detectable. It appeared that the whole of this small reef was each individual's or pair's foraging area. During this roaming, there were frequent and aggressive encounters with damselfishes chasing the butterflyfishes. Such wide-ranging behaviour exists in the normal behavioural repertoire, when individuals make excursions to areas beyond their home territories (Reese, 1989). The impression was that on Cousine the whole corallivorous butterflyfish assemblage had moved into excursion behaviour. This is very different from the small (510m^2) territories held by *C. trifascialis* on *Acropora*-rich reefs in the Pacific (Irons, 1989). It may be that adverse conditions, including stress from coral bleaching, trigger a more open population situation when otherwise, under salubrious conditions, the population is relatively closed. Mean territory sizes of an obligate corallivore (*C. baronessa*) were approximately 53m^2 in areas with high abundances of preferred coral prey species, and about 120m^2 in back-reef locations with reduced availability of preferred prey (Berumen and Pratchett, 2006). It is possible that subtle changes in the 'quality' of coral prey occur as a result of stress on the coral, leading to a change in behaviour of the fish as they attempt to maintain a level of nutritional intake. However, such behavioural change due to sublethal changes in corals has not yet been documented, and would be a worthwhile mechanism to investigate further, especially in the context of using butterflyfish as potential indicators. It should be noted that determining territory sizes requires a nontrivial investment of in-water observation time. It may take 50 minutes of observation to accurately delineate each territory (Crosby and Reese, 1996) and additional time to record the area delineated.

Crosby and Reese (2005) examined intra-specific behavioural manifestations, as measured by territory size and conspecific agonistic encounters, population size and recruitment of the obligate corallivore butterflyfish *Chaetodon multicinctus* (Family: Chaetodontidae) over a 5-year period at three Hawaiian coral reef sites, each receiving different levels of non-point-source terrigenous sediment runoff. All three sites indicated significant correlations for percentage coral cover with territory size, fish size, and number of adult *C. multicinctus*, and for agonistic encounters with number of juvenile *C. multicinctus*. However, the site most impacted by sediment runoff indicated no significant correlations for percentage coral cover with territory size, fish length, number of adult *C. multicinctus*, number of agonistic encounters, or *C. multicinctus* juveniles; nor for number of agonistic encounters with number of juveniles. The site moderately to minimally impacted by sediment runoff exhibited significant correlations for percentage coral cover with number of adults, while the site with no sedimentation impacts exhibited significant correlations for percentage coral cover with fish size, number of adults and number of juveniles. Significant differences were found to exist between all three sites for agonistic encounters, territory size, fish size, number of *C. multicinctus* juveniles and percentage coral cover. The most highly impacted site exhibited a significantly higher number of adult *C. multicinctus* that were significantly smaller in size than either the minimally impacted and non-impacted sites. The study design purposely selected high coral cover habitats (means ranged from 81.4% to 96.5%) at each site. Within this limited range of relatively high coral cover habitat, relative habitat stability resulting from a range of historic sediment inputs between sites appears to be the stronger forcing function for observed intra-specific behavioural manifestations, population size and recruitment than percentage coral cover. Study results indicate that obligate corallivore butterflyfish behavioural manifestations, population size and recruitment may be used as a benchmark for changes in habitat stability for coral reef communities, and support the general premise of the butterflyfish indicator species hypothesis.

Clearly the indicator method is not useful or appropriate where large episodic events like storms, oil spills, or crown-of-thorns starfish (*Acanthaster planci*) outbreaks damage coral reefs, as the method is most effective in revealing subtle gradual changes. In addition, for the method to be applied effectively, coral cover should be relatively high (at least about 40%) in areas to be studied. For example, Crosby and Reese (2005) reported that at their Kuheia study site on the Hawaiian Island of Kaho'olawe, coral growth is good on the ridges or spurs, but there are no corals whatsoever in the sediment clogged depressions or grooves. As a result, their transect study lines were established on the ridges where both corals and fish are present. There should also be a good assemblage of coral feeding butterflyfishes, and at least one of the species should be relatively common and territorial. Since behavioural observations play an important role in the method, water clarity should be reasonably good; perhaps not less than 10 m, and it must not be too rough. Fishes stop 'behaving' (at least in an observable manner) when it is rough and visibility is poor.

Another disadvantage to the indicator species method is that it only provides data on the abundance, distribution and diversity of coral feeding species of butterflyfishes and of the more common species of corals, if species are being recorded, or the coral growth forms, if they are being used. Therefore, only within these limitations can diversity estimates be made. Furthermore, it is necessary to know to what extent collections of reef fishes occur for the aquarium trade and whether cyanide and other chemicals are in use to collect target species for the fine dining market.

1. The indicator species method for monitoring the occurrence of change on coral reefs may be the method of choice; (a) when change is suspected to occur slowly but persistently, (b) when no prior baseline data exist, (c) when funds and trained personnel are limited, (d) when volunteers are available, and, of course, (e) when corals are sufficiently abundant to support an existing population of one or more potential indicator species of coral feeding butterflyfish.
2. The indicator species monitoring method is inexpensive, easy to learn, does not involve complicated, expensive, and time consuming sampling protocols with the concomitant problems of sample storage, analysis, and data reduction. With the indicator species method, quantitative data are ready for entry into a database literally as soon as the diver is out of the water.
3. The method can be used in a step-wise incremental application with each step adding increasingly valuable information. The analysis of behaviour of the indicator species provides valuable insights into the dynamics of the reef community and may provide a useful 'early warning' of the degradation of the coral reef. An added bonus of the method is that volunteers learn to appreciate the subtle workings and interactions of the coral reef ecosystem. This is a benefit of great value in terms of public awareness, involvement and support of conservation and management programs for coral reefs.

TESTING THE BUTTERFLYFISH INDICATOR METHOD

Due to its 'low technology' approach, the butterflyfish indicator method also has the advantage of a high benefit-to-cost ratio. The Crosby and Reese (1996) butterflyfish indicator method, in varying forms, has been employed in numerous countries. In some of these locations, some elements of the method were employed even if not directly as part of an indicator species program. These locations include: Saipan, Guam, American Samoa and Hawaii (Crosby et al., 2000; Crosby and Reese, 2005) the Bunaken Marine Park, North Sulawesi, Indonesia since 1998 (Massimo Boyer, personal communication), elsewhere in Indonesia (Bawole and Boli, 2000; U. Killguss, personal communication), Australia (Pratchett et al., 2004; Berumen et al., 2005; Berumen and Pratchett, 2006), Andaman and Nicobar Islands, Indian Ocean (R. Jeyabhaskaran, personal communication), Seychelles and South Africa (Samways et al., personal communication), Eritrea (Zekeria and Videler, 2000), Saudi Arabia (Joseph and Sy, 2000) and the Gulf of Aqaba (Khalaf and Crosby, 2005). We will review a few case studies below. Many of these studies explicitly incorporated behaviour as a component of the study (e.g., Crosby et al., 2005; Crosby and Reese, 2005).

The Fifth Indo Pacific Fish Conference was held in Noumea, New Caledonia, in 1997. A field trip took the participants to Amedee Island, which is on the barrier reef south of Noumea and marks the entrance to the harbour. Visitors to the island are greeted by an information board which states, 'Around Amedee Islet, fish communities are rich: 97 different species have been recorded, 24 of which are butterflyfishes. The presence of species of that family (Chaetodontidae) is an indicator of a healthy coral environment.' This statement would seem to be based on the research of many scientists studying corals and fish in French Oceania and the Red Sea (e.g., Harmelin-Vivien and Bouchon-Navaro, 1983).

When coral feeding species are studied at one location over time, the results become clearer. For example, at Moorea, French Polynesia, the obligate corallivores *C. trifascialis* and *C. trifasciatus* were abundant on the reefs in 1979, but disappeared completely after intense predation by *Acanthaster planci*, in 1981. They were replaced (as numerically dominant) by the algal browser *C. citrinellus* in a survey conducted in 1983. Berumen and Pratchett (2006) resurveyed the same reef communities in 2003. The findings of this study, 20 years later, revealed that the abundance of butterflyfishes was equivalent to the predisturbance levels of 1979. Surveys of the coral communities also revealed that coral cover at the site was equivalent to predisturbance levels. At this point, one would be tempted to add this to the list of case studies in support of butterflyfish as general indicators of reef health. However, more careful analysis revealed some very important and perhaps more useful points. The most abundant butterflyfish in 1979 was *Chaetodon pelewensis*, whereas in 2003 it was *Chaetodon reticulatus*. Both are obligate corallivores, but they appeared to have different level of dietary specialisation. Investigation of the coral community structure indicated that there had been a fundamental shift in the makeup of the coral communities (specifically, a change from an *Acropora* dominated community to a *Pocillipora* dominated community). Another interesting finding from the study was that an extreme *Acropora* specialist, *C. trifascialis*, was present before the *A. planci* outbreak, disappeared locally immediately after the outbreak, and was extremely rare in the 2003 study. This suggests that *C. trifascialis* may be susceptible to local extinction given a loss of its preferred (almost exclusive) prey species, *Acropora hyacinthus*.

When considered as a whole, the Moorea case study can tell us two things. First, butterflyfish communities may be useful as indicators of subtle changes on a reef, such as gradual shifts in dominance or makeup of coral communities. Notably, this conclusion arose from a careful simultaneous assessment of both butterflyfish and coral community composition. Such relationships may need to be established at each site where a similar approach would be enacted. Another important caveat to this point is that it does not clarify whether or not one reef community is 'healthier' or 'better' than another (i.e., is an *Acropora* dominated reef 'better' than a *Pocillipora* dominated reef?). Second, if the aim is to have an indicator of a single coral species, then the extreme specialization of *C. trifascialis* (Fig. 10.1) may be a suitable indicator for *A. hyacinthus* or similar congeners.

In a thorough study of the behaviour, ecology, population dynamics and life history of the butterflyfishes of the Southern Red Sea, Zekeria (2003) provides strong evidence that *Chaetodon larvatus* is a candidate for an indicator species for ecological conditions of coral reefs in the region. *C. larvatus* feeds during daylight hours on corals, particularly branching *Montipora* corals, lives in heterosexual pairs in relatively small territories (mean territory size ranges from 33 to 45m^2) which it defends against con-specifics and *Chaetodon semilarvatus*, another coral feeding species. *C. larvatus* is the most abundant butterflyfish in the Southern Red Sea, and its abundance correlates with the abundance of corals, especially branching *Montipora*.

Shokri et al. (2005) report that in the Persian Gulf waters of Iran, occurrence of *Chaetodon nigropunctatus* and *Chaetodon melapterus* was always in association with coral formations. *C. melapterus* and *C. nigropunctatus* are reported to feed exclusively or primarily on coral polyps (Lieske and Myers, 1994; Carpenter et al., 1997). However, correlation analyses of butterflyfish with percent coral cover yielded mixed results for Shokri et al. (2005). Their results indicate that *C. nigropunctatus* may be an excellent candidate for such an indicator species in Iranian waters of the northern Persian Gulf, but further studies are required to confirm this. Importantly, Pratchett et al. (2013) showed that *C. nigropunctatus* is a facultative (not obligate) corallivore, at least in the southern Persian Gulf. We encourage more extensive surveys in the northern Persian Gulf to provide more convincing support for the use of *C. nigropunctatus* as an indicator of changing conditions in coral reefs.

Khalaf and Crosby (2005) concluded that density of *Chaetodon paucifasciatus* is the single best indicator species for proportion of benthic habitat covered by hard coral, and, when combined with *Chaetodon austriacus*, explains the majority of variability in percentage hard coral coverage. Their results support the butterflyfish indicator species concept that specific coral-feeding butterflyfish species, and not total butterflyfish abundance, may be a valuable indicator of changes to coral reef ecosystems. However, additional studies are needed to elucidate the behavioural traits (i.e., feeding rates, territory size and agonistic encounters) of the candidate species *Chaetodon melannotus, C. austriacus* and *C. paucifasciatus*. Such studies should be included as part of implementation of a broader, regional long-term butterflyfish monitoring programme.

BUTTERFLYFISH INDICATOR METHOD AS A VEHICLE FOR EDUCATION AND BUILDING BRIDGES

Volunteer Monitoring and Education

Incorporating volunteers into reef assessment and monitoring programs is increasingly popular and studies have shown that with proper training, volunteers can play a productive role in detecting trends of coral reef processes (Crosby and Reese, 1996; Reese and Crosby, 1999; Brown, 1998; Hallacher and Tissot, 1999; McManus and Vergara, 1999; Hodgson and Stepath, 1999; Stepath, 1999). Participants gain a new sense of appreciation for the complexities of reef ecosystems and learn principles of stewardship. The alarming decline of coral reef communities, coupled with the limited funds and expertise available for studying their demise, requires reliance on alternative methods for gathering information for resource management. The key for successful volunteer monitoring programs is to ensure appropriate levels of quality control and data management.

Special efforts should be made by volunteer monitoring programs to interest and involve school teachers in using the butterflyfish indicator concept in science education. In addition, tourists at coastal resorts and recreational divers are often seeking opportunities to contribute to positive environmental efforts, learn new and novel techniques, and have an educational experience. By training hotel staff and local/regional recreational diving operators in the butterflyfish monitoring method, programs will also be promoting sound eco-tourism opportunities while allowing for businesses to advertise unique opportunities for tourists and expand their offerings. Training workshop activities will not only provide additional information on the overall condition of coral reefs, but will serve to educate participants and enhance environmental awareness regarding the ecological relationships inherent between the organisms of coral reefs. A significant focus of workshops on volunteer butterflyfish monitoring and data management should be on translation and transfer of technical information into user-friendly data visualization products and 'lessons learned' recommendations for managers, policy-makers, educators, local community groups and non-government organisations.

Learning to recognise one or two brightly coloured butterflyfishes and to observe their behaviour, and to learn to recognize growth forms of corals are easy tasks. It is not necessary to learn scientific names or to have scientifically trained personnel. It is easy to train volunteers in one day to learn to collect the necessary data. In fact we encourage the use of volunteers because the training requirements are minimal, the data collection is easy, and therefore there is less chance for error or inexperienced volunteer bias than in more complicated, more demanding, and more learning-intensive monitoring techniques.

The necessary equipment, as described in the manual (Crosby and Reese, 1996), includes one or more 30 m by 1 cm brightly coloured polypropylene transect lines with coloured surveyor's tape tied through the weave at 1 m intervals, and inexpensive clipboards or writing slates with regular soft lead No. 1 pencils. Writing slates or plastic overlays of the data sheets may be copied on any copier, erased and reused, or underwater writing paper, such as Xerox Never-Tear plastic paper, may be used, but this will add to the cost (a pack of 100 sheets costs about US$50). This plastic paper is very durable and can be erased and reused and goes through most copying machines. Simple carpenter's nails and surveyor's tape are needed to mark the territory boundaries, and fiberglass measuring tapes of 30–50 m (cost about $35–45), needed to measure the territory size, are available at most hardware stores. It is essential to have an underwater watch to time the various activities and brands like Casio are reliable and inexpensive (about $35).

Sample data sheets are shown in the manual and can be modified to meet the requirements of the particular monitoring site to depict the indicator species being used and other species of corallivorous butterflyfishes as well as the most common coral growth forms. It is necessary to have access to a computer and appropriate statistical packages. We use data spread sheets created using the Microsoft Excel program. Details of the data management procedures we use are available in our manual (Crosby and Reese, 1996).

Another advantage of the method is that it can be used in an incremental, stepwise manner with each step providing more data and increasing accuracy and sensitivity to the predictive strength of the method. In the simplest first step, only the fish and coral counts are made along the transects. The second step is marking the territory boundaries and recording agonistic chasing events (it is convenient to do these two parameters at the same time as described in the manual). The third step is measuring the sizes of the territories. The final fourth step is measuring feeding rates. The complete data set provides a wealth of quantitative data about the behavioural ecology of the indicator species and the overall condition of the coral reef.

Objectives of a butterflyfish indicator volunteer monitoring program may be to promote:

- community participation, education, training and understanding, and NGO involvement in resource management issues;
- improved linkages between scientist, managers, environment groups and the public;
- environmental management capacity building;

- eco-tourism and positive interactions between different socio-economic groups;
- a new non-governmental focus for conservation and management of coral reef resources;
- potential for cultural cooperation, exchange and understanding across a region or between countries based on the shared objective of coral reef conservation and management;
- linkages with international environmental instruments (i.e., the International Coral Reef Initiative and the Global Coral Reef Monitoring Network) and marine protected areas.

General approach of a butterflyfish indicator volunteer monitoring program could focus on:

- community education, outreach and training;
- coordinated and joint trans-boundary professional and volunteer coral reef assessments;
- coordinated and joint trans-boundary seminars, meetings and workshops;
- mutual exchange of scientific, resource management and technological information.

Focal Point for Regional Marine Science Partnerships

Marine science has demonstrated its clear utility for providing an improved knowledge base on which informed marine and coastal resource management decisions can and should be based. However, marine science partnerships and cooperation also have great potential to serve as a vehicle for improving international relations (Crosby, 2007), and the United Nations Convention on Law of the Sea encourages signatories to promote marine science capacity building in developing countries. Crosby (2007) reported a number of successful international marine science partnerships and specific models that should be considered for implementation in other regions of the world. Often, such partnerships developed between nations are built upon a long history of close ties and cooperative relationships. However, marine science partnerships have also been demonstrated to not only support the conduct of excellence in cooperative research, but also to serve as vehicles for improving relationships between nations.

As a specific example of the utility of butterflyfish to serve as trans-boundary ambassadors for coral reef conservation, education and resource management, there was clear agreement and strong support amongst a diverse group of participants at the *Middle East Regional Science Symposium and Workshop: Butterflyfish (Family Chaetodontidae) Research and Monitoring* (Khalaf and Crosby, 2005) for the creation of a regionally coordinated effort to develop an initial baseline set of shared data on a spatial range from the Gulf of Aqaba, through the Red Sea, around the Arabian Peninsula, and up into the Persian Gulf. There was a feeling amongst participants that there probably exist other limited regional databases from studies and monitoring programmes that, although not specifically designed to focus on obligate corallivore butterflyfish, could be 'mined' for useful information. Some participants suggested that such a regional butterflyfish database could be built on the existing Gulf of Aqaba data-management programme of the Aqaba Special Economic Zone Authority in Aqaba, Jordan. Such an effort would need to allow for remote accessing of data by all participants and standardized data management techniques for quality control and quality assurances.

Workshop participants were unanimous in their support for the value of establishing a long-term reef monitoring programme for the Middle East region using butterflyfishes, in addition to other international coral monitoring techniques, aimed at helping marine and coastal resource management authorities in decision-making about the future of country-specific and regionally connected coral reef systems. At the close of the workshop, participants also recommended that a subsequent regional conference be held in one or two years to discuss the results of their efforts to implement such monitoring programmes, and the potential for establishing a coordinated Middle East regional butterflyfish monitoring data-management programme to share data helpful for developing a regional perspective of trends and status of relationships between butterflyfish and coral reefs. In this sense, there was a general feeling that butterflyfishes could indeed serve as ambassadors at large for coral reef health in the Middle East region.

PRIORITIES FOR FUTURE BUTTERFLYFISH RESEARCH

Important outcomes of a conference held on butterflyfish as indicator species (Khalaf and Crosby, 2005) included a general consensus on the unique value of obligate corallivore butterflyfish as indicators of changing conditions in coral reefs in the Middle East region. Collectively, workshop participants felt their combined experiences in the region supported the general premise of the butterflyfish indicator species hypothesis (Crosby and Reese, 1996; Reese and Crosby, 1999). However, they also recommended that further regional specific research (i.e., relationship of physiological stress levels in coral with butterflyfish population dynamics) is needed to confirm the utility of any specific obligate corallivore butterflyfish as an 'early warning system' for stress in corals. Several specific areas of focus were recommended.

Specific suggestions were made to initiate more extensive studies to:

- better document and elucidate the behavioural traits (i.e., feeding rates, territory size and agonistic encounters) of the Middle East regional candidate indicator chaetodontid species *C. austriacus*, *C. melannotus*, *C. paucifasciatus* and *C. nigropunctatus*;
- determine chaetodontid biomass estimates as part of monitoring programmes in order to determine whether carrying capacity of the habitat has a stronger impact on biomass than on the number of individual butterflyfish;
- examine whether combined use of butterflyfish species richness and abundance is also promising as proxies of total reef fish density and biomass estimates; and
- determine the pre- and post-settlement process and factors responsible for controlling the structure of butterflyfish assemblages.

Studies such as those recommended by the workshop should include data gathered on corallivore butterflyfish behaviour on both stressed reefs and healthy, unstressed reefs. It is the deviation from 'normal' behaviour that workshop participants felt would be indicative of potential change to the entire coral reef in question. For instance, deviation from unstressed baseline characteristics, along with observation of early bleaching corals such as *Montipora* spp. and branching *Acropora* spp., may be highly likely to indicate the coming of a major coral bleaching event.

Although it is well established that after a reef is severely degraded or destroyed, populations of coral feeding butterflyfishes are reduced or eliminated altogether (Sano et al., 1984, 1987; Bell et al., 1985; Harmelin-Vivien and Galzin, 1985; Williams, 1986), the process of this event remains unknown. As the corals become moribund and eventually die, what do the fishes do? Do the larger or more aggressive pairs of butterflyfish enlarge their territories to compensate for the decreasing amount of food at the expense of the smaller or less aggressive pairs of fish, which then must either starve or emigrate? Do these strongly territorial and site-attached fish stay and die as their food supply is depleted or do they emigrate from the reef when some food threshold value is reached? Our Kaho'olawe research did not enable us to track these events or provide answers to these questions. Some experimental evidence suggests that the fish will continue to feed even when no nutritional benefit seems to be conveyed from a poor food source (Berumen and Pratchett, 2008), but it is unclear if this would apply to a reef-wide scale. The effort to understand the mechanisms acting on corallivores during a decline of coral cover, however, is continuing and we are alert to any opportunity to observe the behaviour *in situ* of an indicator species on a reef undergoing a process of gradual degradation.

REFERENCES

Askins R.A. 1995. Hostile landscapes and the decline of migratory songbirds. *Science* 267: 1956–1957.

Bawole, R. and P. Boli. 2000. (presentation/abstract). Chaetodontid-fish association with coral lifeform at Lemon Island, Manokwari-Irian Jaya. *9th International Coral Reef Symposium, Bali, Indonesia.*

Bell, J., M. Harmelin-Vivien and R. Galzin. 1985. Large scale spatial variation in abundance of butterflyfishes (Chaetodontidae) on Polynesian reefs. *Proceedings of the Fifth International Coral Reef Congress, Tahiti* 5: 421–426.

Bellwood, D.R., L. van Herwerden and N. Konow. 2004. 'Evolution and biogeography of marine angelfishes.' *Molecular Phylogenetics and Evolution* 33: 140–155.

Bellwood, D.R., S. Klanten, P.F. Cowman, M.S. Pratchett, N. Konow and L. van Herwerden. 2010. Evolutionary history of the butterflyfishes (f: Chaetodontidae) and the rise of coral feeding fishes. *Journal of Evolutionary Biology* 23(2): 335–349.

Berumen, M.L. 2005. The importance of juveniles in modelling growth: butterflyfish at Lizard Island. *Env. Biol. Fishes* 72: 409–413.

Berumen, M.L. and M.S. Pratchett. 2008. Trade-offs associated with dietary specialization for corallivorous butterflyfishes. *Behavioural Ecology and Sociobiology* 62: 989–994.

Berumen, M.L., E.L. Trip, J.H. Choat and M.S. Pratchett. 2012. Differences in demographic traits of four butterflyfish species between two reefs of the Great Barrier Reef separated by 1,200 km. *Coral Reefs* 31: 169–177.

Berumen, M.L. and M.S. Pratchett. 2006. Recovery without resilience: persistent disturbance and long-termshifts in the structure of fish and coral communities at Tiahura Reef, Moorea. *Coral Reefs* 25(4): 647–653.

Berumen, M.L., M.S. Pratchett and M.I. McCormick. 2005. Within-reef differences in diet and body condition of coral-feeding butterflyfishes (Chaetodontidae). *Marine Ecology Progress Series* 287: 217–227.

Black, D. and R.M. Love. 1986. The sequential mobilization and restoration of energy reserves in tissues of Atlantic cod during starvation and re-feeding. *Journal of Comparative Physiology and Biology* 156: 469–479.

Blum, S.D. 1989. Biogeography of the Chaetodontidae: an analysis of allopatry among closely related species. *Environmental Biology of Fishes* 25: 9–31.

Bouchon-Navaro, Y. and C. Bouchon. 1989. Correlations between Chaetodontid fishes and coral communities of the Gulf ofAqaba (Red Sea). *In:* P.J. Mootta (ed.). *The Butterflyfishes: Success on the Coral Reef.* Kluwer: Dordrecht. pp. 47–60.

Bouchon-Navaro, Y., C. Bouchon and M.L. Harmelin-Vivien. 1985. Impact of coral degradation on a Chaetodontid fish assemblage assemblage (Moorea, French Polynesia). *Proceedings of the 5th International Coral Reef Symposium* 5: 427–432

Brown, E. 1999. Long term monitoring of coral reefs on Maui, Hawai'i and the applicability of volunteers. *In:* J.E. Maragos and R. Grober-Dunsmore (eds.). *Proceedings of the Hawai'i Coral Reef Monitoring Workshop, June 9-11, 1998, Honolulu, Hawai'i.* Division of Aquatic Resources, Department of Land and Natural Resources, Hawaii State Government. pp. 131–146.

Burgess, W.E. 1978. *Butterflyfishes of the World.* T.F.H. Publications, Neptune City, N.J. 832 pp.

Cadoret, L., M. Adjeroud and M. Tsuchiya. 1999. Spatial distribution of chaetodontid fish in coral reefs of the Ryukyu Islands, southern Japan. *Journal of the Marine Biological Association of the U.K.* 79: 725–735.

Carpenter K.E., P.L. Harrison, G. Hodgson, A.H. Alsaffar and S.H. Alhazeem. 1997. *The corals and coral reef fishes of Kuwait.* Kuwait Institute for Scientific Research, Kuwait. 166 pp.

Cole, A.J., M.S. Pratchett and G.P. Jones. 2008. Diversity and functional importance of coral-feeding fishes on tropical coral reefs. *Fish and Fisheries* 9: 286–307.

Cowey, C.B. and J.R. Sargent. 1977. Lipid nutrition in fish. *Comparative Biochemistry and Physiology - Part B* 57(B): 269–273.
Crosby, M.P. 2007. Improving International Relations Through Marine Science Partnerships. *In: Law, Science & Ocean Management*, M.H. Nordquist, R. Long, T.H. Heidar and J. Norton Moore, eds. MartinusNijhoff Publishers, Leiden/Boston. pp. 271–293.
Crosby, M.P. and E.S. Reese. 2005. Relationship of habitat stability and intra-specific population dynamics of an obligate corallivore butterflyfish. *Aquatic Conservation: Marine and Freshwater Ecosytems* 15: 13–25.
Crosby, M.P., R. Bohne and K. Geenen. 2000. *Alternative access management strategies for marine and coastal protected areas: A reference manual for their development and assessment.* U.S. Man and the Biosphere Program.Washington, DC. 164 pp.
Crosby, M.P. and E.S. Reese. 1996. *A Manual for Monitoring Coral Reefs With Indicator Species: Butterflyfishes as Indicators of Change on Indo-Pacific Reefs.* Office of Ocean and Coastal Resource Management, National Oceanic and Atmospheric Administration, Silver Spring, MD. 45 pp.
Erdmann, M.V. and R.L. Caldwell. 1997. Stomatopod crustaceans as indicators of marine pollution stress on coral reefs. *In:* H.A. Lessios and I.G. Macintyre (eds.). *Proceedings of the 8th International Coral Reef Symposium Vol. 2. Smithsonian Tropical Research Institute, Panama*. pp. 1521–1526.
Erdmann, MV. 1997. The butterflyfish bioindicator hypothesis. *Reef Encounter* 21: 8–9.
Fessler, J.L. and M.W. Westneat. 2007. Molecular phylogenetics of the butterflyfishes (Chaetodontidae): Taxonomy and biogeography of a global coral reef fish family. *Molecular Phylogenetics and Evolution* 45: 50–68.
Findley J.S. and M.T. Findley. 1989. Circumtropical patterns in butterflyfish communities. *Env. Biol. Fishes* 25: 33–46.
Fraser, J. 1962. *Nature adrift*. G.T. Foulis and Co. Ltd., London. 178 pp.
Goreau, T., T. McClanahan, R. Hayes and A. Strong. 2000. Conservation of coral reefs after the 1998 global bleaching event. *Conservation Biology* 14: 5–15.
Green, B.S. and M.I. McCormick. 1999. Influence of larval feeding history on the body condition of *Amphiprion melanopus*. *Journal of Fish Biology* 55: 1273–1289.
Hallacher, L. and B. Tissot. 1999. Quantitative underwater ecological surveying techniques: A coral reef monitoring workshop. *In:* J.E. Maragos and R. Grober-Dunsmore (eds.).*Proceedings of the Hawai'i Coral Reef Monitoring Workshop, June 9-11, 1998, Honolulu, Hawai'i.*Division of Aquatic Resources, Department of Land and Natural Resources, Hawaii State Government. pp. 147–158.
Harmelin-Vivien, M.L. and Y. Bouchon-Navaro. 1983. Feeding diets and significance of coral feeding among Chaetodontid fishes in Moorea (French Polynesia). *Coral Reefs* 2: 119–127.
Hodgson, G. and C.M. Stepath. 1999. Using Reef Check for long-term coral reef monitoring in Hawai'i. *In:* J.E. Maragos and R. Grober-Dunsmore (eds.). *Proceedings of the Hawai'i Coral Reef Monitoring Workshop, June 9-11, 1998, Honolulu, Hawai'i.* Division of Aquatic Resources, Department of Land and Natural Resources, Hawaii State Government pp. 173–184.
Hourigan, T.F., Tricas, T.C. and E.S. Reese. 1988. Coral reef fishes as indicators of environmental stress in coral reefs, *In: Marine Organisms as Indicators*, D.F. Soule and G.S. Kleppel (eds.). Springer Verlag, New York pp. 107–135.
Irons, D.K. 1989. Temporal and areal feeding behaviour of the butterflyfish, *Chaetodon trifascialis*, at Johnston Atoll. *Environmental Biology of Fishes* 25: 187–193.
Jones, G.P. and U.L. Kaly. 1996. Criteria for selecting marine organisms in biomonitoring studies. *In:* Schmitt, R.J., Osenberg, C.W. (eds.). *Detecting ecological impacts*. Academic Press. New York, USA pp. 29–46.
Joseph, P.S. and J.C. Sy. 2000. (presentation/abstract). Butterflyfish population diversity in Yanbu coral. *9th International Coral Reef Symposium, Bali, Indonesia.*

Khalaf, M. and M.P. Crosby. 2005. Assemblage structure of Butterflyfishes and their use as indicators of Gulf of Aqaba benthic habitat in Jordan *Aquatic Conservation: Marine and Freshwater Ecosystems* 15: 27–43.

Lee, R.B. and I. Devore. 1968. *Man the hunter*. Aldine, Chicago, USA 415 pp.

Leopold, A. 1966. *A Sand County almanac with other essays on conservation from Round river*. Oxford Univ Press New York, USA. 269 pp.

Leujak, W. and R.F.G. Ormond. 2007. Comparative accuracy and efficiency of six coral community survey methods. *J. Exp. Mar. Biol. Ecol.* 351: 168–187.

Lieske, E. and R. Myers. 1994. *Collins pocket guide: Coral Reef Fishes, Indo-Pacific and Caribbean Including the Red Sea*. Harper Collins, London. 400 pp.

McManus, J.W. and S.G. Vergara. 1999. ReefBase 3.0 and the ReefBase aquanaut system. *In:* J.E. Maragos and R. Grober-Dunsmore (eds.). *Proceedings of the Hawai'i Coral Reef Monitoring Workshop, June 9-11, 1998, Honolulu, Hawai'i*. Division of Aquatic Resources, Department of Land and Natural Resources, Hawaii State Government pp. 159–172.

Odum, H.T. and E.P. Odum. 1955. Trophic structure and productivity of a windward coral reef community on Eniwetok Atoll. *Ecological Monographs* 25: 291–320.

Pratchett, M.S. 2007. Dietary selection by coral-feeding butterflyfishes (Chaetodontidae) on the Great Barrier Reef, Australia.' *Raffles Bulletin of Zoology* S14: 171–176.

Pratchett, M.S. and M.L. Berumen. 2008. Inter-specific variation in distributions and diets of coral reef butterflyfishes (Teleostei: Chaetodontidae). *Journal of Fish Biology* 73: 1730–1747.

Pratchett, M.S., S.K. Wilson, M.L. Berumen and M.I. McCormick. 2004. Sub-lethal effects of coral bleaching on an obligate coral-feedingbutterflyfish. *Coral Reefs* 23: 352–356.

Pratchett, M.S., S.K. Wilson and A.H. Baird. 2006. Declines in the abundance of Chaetodon butterflyfishes following extensive coral depletion. *Journal of Fish Biology* 69: 1269–1280.

Pratchett, M.S., S.K. Wilson, M.L. Berumen and M.I. McCormick. 2004. Sublethal effects of coral bleaching on an obligate coral feeding butterflyfish. *Coral Reefs* 24: 352–356.

Pratchett, M.S. 2005. Dietary overlap among coral-feeding butterflyfishes (Chaetodontidae) at Lizard Island, northern Great Barrier Reef. *Marine Biology* 148: 373–382.

Pratchett, M.S., A.S. Hoey, D.A. Feary, A.G. Bauman, J.A. Burt and B. Riegl. 2012. Functional composition of Chaetodon butterflyfishes at a peripheral and extreme coral reef location, the Persian Gulf. *Marine Pollution Bulletin* DOI: 10.1016/j.marpolbul.2012.10.013.

Raymont, J.E.G. 1963. *Plankton and productivity in the oceans*. Pergamon Press, New York USA. 660 pp.

Reese, E.S. 1973. Duration of residence by coral reef fishes on 'home' reefs. *Copeia* 1973: 145–149.

Reese, E.S. 1975. A comparative field study of the social behaviour and related ecology of reef fishes of the family Chaetodontidae. *Zeitschrift fur Tierpyschologie* 37: 37–61.

Reese, E.S. 1989. Orientation behaviour of butterflyfishes (family Chaetodontidae) on coral reefs: spatial learning of route specific landmarks. *In:* P.J. Motta (ed.). *The Butterflyfishes: Success on the Coral Reef*. Kluwer: Dordrecht 79–86.

Reese, E.S. 1977. Coevolution of corals and coral eating fishes. *Proceedings of the 3rd International Coral Reef Symposium* 1: 267–274.

Reese, E.S. 1981. Predation on corals by fishes of the family Chaetodontidae: implications for conservation and management of coral reef ecosystems. *Bulletin of Marine Science* 31: 594–604.

Reese, E.S. and M.P. Crosby. 1999. The use of indicator species for coral reef monitoring. *In:* J.E. Maragos and R. Grober-Dunsmore (eds.). *Proceedings of the Hawai'i Coral Reef Monitoring Workshop, June 9-11, 1998, Honolulu, Hawai'i*. Division of Aquatic Resources, Department of Land and Natural Resources, Hawaii State Government pp. 121–128.

Roberts, C.M., R.F.G. Ormond and A.R.D. Shepherd. 1988. the usefulness of butterflyfishes as environmental indicators on coral reefs. *Proceedings of the 6th International Coral Reef Symposium* 2: 331–336.

Roberts CM. and R.F.G. Ormond. 1987. Habitat complexity and coral reef fish diversity and abundance on Red Sea fringing reefs. *Marine Ecology Progress Series* 41: 1–8.

Samways, M.J. 2005. Breakdown of butterflyfish (Chaetodontidae) territories associated with the onset of a mass coral bleaching event. *Aquatic Conservation: Marine and Freshwater Ecosystems* 15: S101–S107.

Samways, M.J. and M.J. Hatton. 2001. An appraisal of two coral reef monitoring manuals for gathering baseline data. *Bulletin of Marine Science* 69: 471–485.

Sano, M., M. Shimizu and Y. Nose. 1984. Changes in structure of coral reefs fish communities by destruction of hermatypic corals: observational and experimental views. *Pacific Science* 38: 51–79.

Sano, M., M. Shimizu and Y. Nose. 1987. Long-term effects of destruction of hermatypic corals by *Acanthaster planci* infestation of reef fish communities at Iriomote Island, Japan. *Marine Ecology Progress Series* 37: 191–199.

Schmitt, R.J. and C.W. Osenberg. 1996. *Detecting ecological impacts*. Academic Press. New York USA. 403 pp.

Shokri, M.R., S.M.R. Fatemi and M.P. Crosby. 2005. The Status of Butterflyfishes (Chaetodontidae) in the Northern Persian Gulf, I.R. Iran. *Aquatic Conservation: Marine and Freshwater Ecosystems* 15: 91–99.

Soule, D.F. and G.S. Kleppel. 1988. *Marine Organisms as Indicators*. Springer-Verlag, New York. 342 pp.

St. John, J. 1999. Ontogenetic changes in the diet of the coral reef grouper *Plectropomus leopardus* (Serranidae): patterns in taxa, size and habitat of prey. *Marine Ecology Progress Series* 180: 233–246.

Stepath. C.M. 1999. Does Community-based coral reef monitoring promote awareness? *In:* J.E. Maragos and R. Grober-Dunsmore (eds.). *Proceedings of the Hawai'i Coral Reef Monitoring Workshop, June 9-11, 1998, Honolulu, Hawai'i.* Division of Aquatic Resources, Department of Land and Natural Resources, Hawaii State Government pp. 33–38.

Sutton, M. 1985. Patterns of spacing in a coral reef fish in two habitats on the Great Barrier Reef. *Animal Behaviour.* 33: 1322–1337.

Webster, M.I. 2002. Role of predators in the early post-settlement demography of coral reef fishes. *Oecologia.* 131(1): 52–60.

White, A.T. 1988. Chaetodon occurrence relative to coral reef habitats in the Philippines with implications for reef assessment. *Proceedings of the 6th International Coral Reef Symposium* 2: 427–432.

Willams, D.M. 1986. Temporal variation in the structure of reef slope communities (central Great Barrier Reef): short-term effects of *Acanthaster planci* infestation. *Marine Ecology Progress Series* 28: 157–164.

Yabuta and Berumen, Chapter 8 of this book.

Zekeria, Z.A. 2003. *Butterflyfishes of the Southern Red Sea: ecology and population dynamics.* Ph.D. thesis. Netherlands University of Groningen. 120 pp.

Zekeria, Z.A. and J.J. Videler. 2000. Correlation between the abundance of butterflyfishes and coral communities in the southern Red Sea. *In: Proceedings of the Ninth International Coral Reef Symposium, Bali.* 1: 487–492.

Zekeria, Z.A., S. Weertman, B. Samuel, T. Kale-ab and J.J. Videler. 2006. Growth of *Chaetodon larvatus* (Chaetodontidae: Pisces) in the southern Red Sea. Marine Biology 148(5): 1113–1122.

CHAPTER 11

Harvesting of Butterflyfishes for Aquarium and Artisanal Fisheries

Rebecca J. Lawton,[1,*] *Morgan S. Pratchett*[2] and *J. Charles Delbeek*[3]

INTRODUCTION

Direct harvesting of fishes and other coral reef organisms represents the most significant and pervasive anthropogenic impact on coral reef ecosystems (e.g., Roberts, 1995; Jackson et al., 2001; Myers and Worm, 2003), causing a range of direct and indirect effects on populations, communities, and ecosystems. Extensive harvesting of very large and previously abundant reef-associated organisms has caused massive changes in coral reef ecosystems throughout the last two centuries (Jackson et al., 2001), and contemporary fisheries continue to overexploit many reef species (e.g., Morris et al., 2000). Coral reef fisheries account for a relatively small proportion of global fisheries yield (<1%), but are fundamental in providing food and livelihoods in many tropical nations (e.g., Bell et al., 2009). Also, coral reef fishes are probably more vulnerable to exploitation compared

[1]School of Marine and Tropical Biology, James Cook University, Townsville, Australia.
E-mail: rebecca.lawton@jcu.edu.au
[2]ARC Centre of Excellence for Coral Reef Studies, James Cook University, Townsville, Australia.
E-mail: morgan.pratchett@jcu.edu.au
[3]SteinhardAquarium, California Academy of Sciences, San Francisco, California, USA.
E-mail: CDelbeek@calacademy.org
*Corresponding author

to many other fisheries stocks (Russ, 1991), and overfishing of reef fishes with critical functional roles, such as herbivorous fishes, has contributed to extensive and accelerating degradation of coral reef habitats (e.g., Hughes, 1994, 1996) and greatly increases vulnerability to subsequent disturbances (Bellwood et al., 2003; Hughes et al., 2007).

Most reef-based fisheries tend to target large piscivorous species (mostly Serranidae, Lutjanidae, and Lethrinidae), which are the predominant target of extensive hook-and-line fisheries (Russ, 1991) and are preferred food fishes, at least in most industrialised countries (Sadovy and Vincent, 2002). However, many other families of coral reef fishes, including butterflyfishes (Chaetodontidae) may be subject to fishing-pressure from relatively non-selective artisanal fisheries (Mangi and Roberts, 2006) or highly specific fisheries that service extensive aquarium or curio markets (Grey et al., 2005). Modernisation and globalisation are also increasing the efficiency and diversity of fishes that are harvested within coral reef ecosystems (Berkeset al., 2006). For example, the rapid growth of international markets for marine aquarium fishes (mostly since the 1970's), combined with improved transport routes, has created entirely new fisheries in relatively isolated developing countries (e.g., Philippines), based on species that were not otherwise harvested because they were not regarded as an important food source (Wood, 2001; Bruckner, 2005; Sadovy and Vincent, 2002).

This review focuses on potential effects of wild harvesting of coral reef butterflyfishes (family Chaetodonitdae). While butterflyfishes would not generally be considered to be subject to excessive fishing pressure, these small iconic coral reef fishes are being increasingly targeted by both artisanal fisheries and marine aquarium collectors (Fig. 11.1). In terms of the marine aquarium industry, butterflyfishes (along with damselfishes and angelfishes) are among the most important families of coral reef fishes (e.g., Green, 2003). For artisanal fisheries, extensive depletion of large piscivorous fishes is increasing pressure on smaller benthic feeding fishes, which include butterflyfishes (e.g., Friedlander and Parrish, 1997; Mangi and Roberts, 2006). On the whole, wild harvesting of butterflyfishes is probably fairly localised and somewhat limited, but we must also recognise that direct harvesting compounds upon other indirect pressures (e.g., widespread habitat degradation) that are already posing a significant threat to populations and species of butterflyfishes around the world (e.g., Pratchett et al., 2006; Wilson et al., Chapter 9). Butterflyfishes are foremost among coral reef fishes in their association with live corals, and often experience localised extinction following extensive coral loss (Wilson et al., 2006; Pratchett et al., 2008a). As coral reefs become increasingly degraded and coral cover continues to decline, many butterflyfishes may face a significantly increased risk of global extinction (Pratchett et al., 2008a, but see Lawton et al. 2011). The added pressure of direct harvesting may make some populations unviable, or significantly erode

A

B

Fig. 11.1 Wild harvesting of coral reef butterflyfishes is undertaken by artisanal fishers and marine aquarium collectors; A) *Chaetodon* butterflyfishes for sale at the food markets in Papete, Tahiti. Photo by Morgan Pratchett. B) *Heniochus* butterflyfishes in a public aquarium. Photo by Charles Delbeek.

Colour image of this figure appears in the colour plate section at the end of the book.

the capacity to withstand further environmental disturbances. Irrespective of environmental disturbances, butterflyfishes also tend to be long-lived (Yabuta and Berumen, Chapter 8; Delbeek, Chapter 12) and have low rates of population replenishment (Pratchett et al., 2008b; Pratchett et al., Chapter 13), making them highly vulnerable to overfishing.

THE GLOBAL MARINE AQUARIUM FISHERY

The global marine aquarium fishery is very extensive and rapidly expanding. The world-wide trade in ornamental reef fishes was estimated to be worth between US$24 million (Wood, 2001) and up to US$300 million (Wabnitz et al., 2003), with annual sales of between 14 million to 30 million fishes. The actual number of marine aquarium fishes caught on coral reefs is also likely to be well above the numbers actually reported and traded when taking account of mortality of fishes during capture and transport to international markets (Sadovy and Vincent, 2002). Marine aquarium fishes are currently harvested from coral reefs in more than 80 countries around the world, and largely represent species that would not otherwise be caught for food (Sadovy and Vincent, 2002). Effects of aquarium fisheries on major target species, as well as broader environmental impacts of current fishing practices, are largely unknown (Wood, 2001). However, it is clear that the industry is poorly regulated and some species, especially rare endemic species (e.g., Bangai cardinalfishes, *Apogon kauderni*), may face possible extinction in the wild as a direct consequence of high demand and excessive harvesting (Wood, 2001; Sadovy and Vincent, 2002). Detailed studies in specific locations (e.g., Philippines, Albaladejo and Corpuz 1982; Eritrea, Daw et al., 2001; Hawaii, Tissot and Hallacher, 1999, 2003) reveal that aquarium collectors are having significant and detectable effects on populations of major targeted species. Aside from over-harvesting of target species, the potential environmental and biological impacts of the marine aquarium industry also include, i) indirect effects associated with selective removal of target species, ii) incidental mortality of non-target species, iii) introductions of non-native fishes, and iv) damage to habitats inflicted during destructive fishing techniques (Andrews, 1990; Wood, 2001; Tissot and Hallacher, 2003).

A major limitation in assessing the global impact of the marine aquarium industry is an overwhelming lack of reliable data collected at all stages of the collection, trade and sale of marine aquarium fishes. At the point of capture, some countries impose significant regulations on marine aquarium fisheries (e.g., Australia), limiting the number of fishes that can be caught and restricting fishing activities to specific locations (Roelofs 2008). However, given that major target species do not otherwise have any significant commercial value, combined with the rapid emergence of the aquarium industry, there are poor records relating to the number of individuals of each species that are actually caught. For example, collectors are often not required to specify the actual species of butterflyfishes that are caught and only report total numbers for all butterflyfishes. Elsewhere around the world, catch reporting is entirely voluntary (e.g., Hawaii) which means it is impossible to establish the extent to which catches are under-reported (Tissot and Hallacher, 2003). These deficiencies in data collection prevent necessary analyses on current and projected catches for individual taxa, making it difficult to determine whether potentially vulnerable species are being caught and traded unsustainably (Bruckner, 2001). Such data are important, both for the industry to provide defensible data on the true extent of their impacts, and for regulators to assess whether there needs to be further limitations imposed.

Aquarium Catches of Coral Reef Butterflyfishes

In an attempt to gather a more complete picture of the trade in marine ornamental fishes, the Global Marine Aquarium Database (GMAD) was created to collate export and import data from 45 representative wholesale operations representing 30 exporting and 44 importing countries (Green, 2003). International trade data fail to account for fishes that die prior to export or are sold within their country of origin, but are likely to provide the best opportunity for objective and independent data on the scale of marine fish harvesting. Up until 2003, international trade of 27 million fishes per year, and >1,000 species, were recorded in the GMAD database (Green, 2003). While these numbers are substantial, it is also apparent that there are many omissions in the GMAD database (e.g., no data from the Red Sea, Green, 2003). Moreover, data on exports do not relate to import data for corresponding countries, which challenges the utility of these data. Wood (2001) for example, showed that reported imports of marine fishes into the USA were <65% of the reported exports to the USA. Consequently, the GMAD database has come to be viewed as a relative measure of international trade in marine aquarium fishes, demonstrating importance of different markets and major trade routes. Green (2003) summarised the database for marine fishes and revealed that 85% of exported fish came from the Western Pacific and 8% from the Caribbean, with the United States importing 68% of the fish traded globally, followed by the European Union and Southeast Asia at 11% each.

More reliable data on exports and imports of marine aquarium fishes might be expected to be available from federal departments responsible for monitoring international trade of live animals in developed countries (e.g., United States Fish and Wildlife Service), especially given the quarantine risks posed by introducing potential pest species and diseases.

In the United States, however, the United States Fish and Wildlife Service (USFWS) rarely report on specific numbers of individual species that are being imported, unless the species involved are covered by the *Convention on International Trade in Endangered Species* (CITES). There are also limited resources to track international imports at many major ports, which means a significant proportion of imports are not recorded. In compiling this chapter, we specifically requested import/export data on butterflyfishes from the USFWS, who acceded to our request by providing the number of distinct shipments of marine aquarium fishes into North America in each year from 2003–2008, but there were no records of the number of individuals of different species of fishes that were imported. Moreover, this data was conspicuous by the absence of any data from Los Angeles or Miami, which are major ports for international trade from the Pacific and Atlantic oceans, respectively. There were also no records of imports arriving from Fiji, which is a major source of marine aquarium fishes to North America. Wood (2001) reported that Fiji has one of the largest fisheries for aquarium fishes in the Pacific, and regularly exports in excess of 100,000 fishes each year, mostly to the USA.

Similar data was also sought from the Queensland Department of Agriculture, Fisheries and Forestry (DAFF), which maintains extensive records on all catches of aquarium fishes, as well as keeping track of catch-per-unit-effort as a means to assessing sustainability of this fishery (Roelofs, 2008). The only criticism of these data is that there is no information on the specific species that comprise these catches, at least for butterflyfishes. However, Queensland uses an alternative approach to managing aquarium fisheries based on a risk assessment framework. Using standard risk assessment methods applied in an ecological context provides an opportunity to develop management measures that mitigate identified risks in relatively data poor fisheries (Fletcher et al. 2002; Roelofs 2008; Roelofs & Silcock 2008). Roelofs (2008) applied this approach to the Queensland Marine Aquarium Fish Fishery (MAFF), assessing the ecological risk of harvesting on all collected fish species, including 45 species of butterflyfishes. The assessment identified that collected butterflyfish species were at low ecological risk from the fishery and did not require changes to the existing management regime for the Family. This ecological risk assessment process is subject to a review every three to five years with annual assessment of harvest numbers against sustainability triggers helping ensure harvest levels remain with sustainable limits over this period.

The Marine Aquarium Fish Fishery on the Great Barrier Reef is one of the most effectively managed aquarium fisheries in the world, and there is documented evidence that fisheries catches are currently well within sustainable limits (Roelofs, 2008). Catch data from DAFF reveal that a total of 126,795 individual butterflyfishes that were caught on the Great Barrier

Reef during the period 1998–2007. There was a consistent decline in total annual harvests through time, whereby the total number of butterflyfishes caught in 2006 was only 47% (10,204) of the number caught in 1998 (21,553). However, this decline in catches is largely reflective of a decline in effort, whereby the cumulative number of days spent fishing by all commercial aquarium collectors has declined from approximately 2,800 in 1998 to 1,600 in 2006 (Roelofs, 2008). In all, catch per unit effort (CPUE) has shown no change with time, and remains at approximately 130 fishes per day, for all target species (600 species) combined (Roelofs, 2008). Without detailed data on catch effort, declines in the catches of butterflyfishes at collection locations may have raised considerable alarm over the sustainability of this fishery. This is therefore, a perfect example of the need to compile data to support future management of aquarium fisheries throughout the world.

Importance of Butterflyfishes Relative to other Major Families

Despite data limitations it is clear that butterflyfishes are an important component of the international trade in marine aquarium fishes, consistently ranking among the top 3–6 families of coral reef fishes that are caught and sold by the global marine aquarium trade (e.g., Chan and Sadovy, 1998; Green, 2003; Bruckner, 2005; Roelofs, 2008; Fig. 11.2). Bruckner (2005)

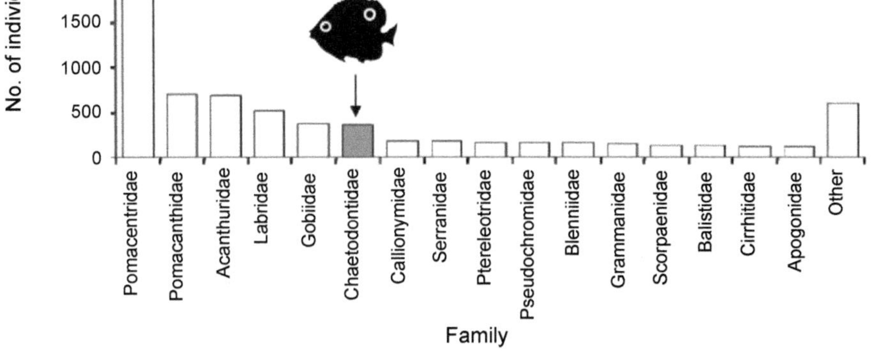

Fig. 11.2 Total number of individual fishes imported to all countries between 1988 and 2002, grouped by family. Only those families with a >1% contribution to the global aquarium trade are shown. Data are from GMAD (www.unep-wcmc.org/marine/GMAD/index.html).

reported that butterflyfishes are the third most heavily traded family of coral reef fishes, accounting for 11% of the total volume (number of fishes) and 10% of total value for the global marine aquarium trade, though these values appear to come directly from Pyle (1993), who estimated the commercial value of different aquarium fishes for only select Pacific countries (Wood, 2001). In the Global Marine Aquarium Database, butterflyfishes account for 4.3% (349,140/ 8,060,859 fishes) of all reported imports of marine aquarium fishes between 1988 and 2002 (www.unep-wcmc.org/marine/GMAD/ index.html), which places them sixth behind Pomacentridae (damselfishes), Pomacanthidae (angelfishes), Acanthuridae (surgeonfishes), Labridae (wrasses) and Gobiidae (gobies), in terms of total number of fishes traded (Fig. 11.2). However, the mean price (retail and wholesale) of butterflyfishes is much higher than for the gobies (e.g., Wood, 2001), so given approximately equal numbers of individuals being traded (Fig. 11.2), butterflyfishes probably rank fifth (and may be even higher) in terms of their economic contribution to international trade marine aquarium fishes.

The relative importance of butterflyfishes (versus other major families of coral reef fishes) in aquarium fisheries varies regionally. In Australia, for example, butterflyfishes accounted for only 6% (10,204/170,000) of aquarium fishes caught in 2006, and the number of damselfishes collected was 3–4 times the number of butterflyfishes (Roelofs, 2008). In Hawaii however, butterflyfishes are ranked 5th of the 11 most heavily targeted aquarium fishes, which together account for 90% of the total harvest for this region (Tissot and Hallacher, 2003). Elsewhere in the Pacific, catches of butterflyfishes are surpassed only by catches of damselfishes and angelfishes (Pyle, 1993). Within other collection locations, butterflyfishes are certainly among the most important marine aquarium fishes. Daw et al. (2001) compiled extensive data on marine tropical collection and exports from Eritrea from 1995 to 1997, and showed that the six species of butterflyfishes harvested from this location accounted for 25% of 103,855 fish exported.

Species Composition of Aquarium Collections

Reports vary as to how many species of butterflyfishes are actively traded; Green (2003) provided a very extensive list of butterflyfish species that have been caught and sold in the marine aquarium trade, but Wood (2001a) suggested that most of these species are not actively targeted and only about 21 butterflyfish species are commonly traded. Import data from GMAD (www.unep-wcmc.org/marine/GMAD/index.html) lists 72 species of butterflyfishes imported between 1988 and 2002, but the 38 most commonly traded species (listed in Table 11.1) accounted for >95% of all reported imports. The single most popular species of butterflyfish that is kept in captivity is probably the copper-banded or beaked coralfish,

Table 11.1 Apparent vulnerability to overfishing for the main species of butterflyfishes caught and traded for aquarium-use. Species are ranked based on total cumulative import records in the GMAD database, intended to provide a relative measure of species importance, rather than absolute estimates of catches.

Species	Total Imports	Diet	Min. pop^n doubling	Vulnerability	Longevity
Heniochus acuminatus	42,531	NC	<15 months	13.16	-
Chelmon rostratus	42,239	NC	<15 months	10.00	-
Forcipiger flavissimus	29,851	P	<15 months	11.30	18 years
Chaetodon melannotus	23,935	OC	4.5–14 years	33.00	20 years
Chaetodon auriga	23,159	FC	<15 months	11.94	-
Chaetodon xanthurus	20,773	FC	<15 months	10.00	-
Chaetodon lunula	17,231	FC	<15 months	10.00	9 years
Chaetodon collare	13,148	FC	<15 months	10.00	-
Chaetodon mertensii	12,000	FC	<15 months	10.00	-
Chaetodon pelewensis	10,594	FC	<15 months	10.00	-
Chaetodon punctatofasciatus	8,853	OC	<15 months	10.00	-
Chaetodon kleinii	7,346	FC	<15 months	10.00	-
Chaetodon multicinctus	6,972	OC	<15 months	10.00	-
Chaetodon vagabundus	6,804	FC	<15 months	11.94	-
Chaetodon quadrimaculatus	6,791	FC	<15 months	10.00	-
Chaetodon miliaris	5,957	P	<15 months	10.00	-
Chaetodon capistratus	5,280	FC	<15 months	10.00	-
Chaetodon falcula	4,418	NC	<15 months	10.00	-
Chaetodon ephippium	4,326	FC	<15 months	20.90	-
Chaetodon ulietensis	4,128	FC	<15 months	10.00	-
Chaetodon unimaculatus	3,520	OC	<15 months	10.00	-
Chaetodon fremblii	3,333	NC	<15 months	10.00	-
Chaetodon semilarvatus	2,929	OC	<15 months	11.94	-
Hemitaurichthys polylepis	2,919	NC	<15 months	10.00	-
Chaetodon fasciatus	2,456	FC	<15 months	11.30	-
Heniochus varius	2,370	FC	<15 months	10.00	-
Chaetodon citrinellus	2,147	FC	<15 months	10.00	10 years
Heniochus pleurotaenia	1,997	NC	<15 months	10.00	-
Chaetodon striatus	1,899	NC	<15 months	10.00	-
Chaetodon plebeius	1,712	OC	<15 months	10.00	-
Chaetodon lineolatus	1,709	NC	<15 months	20.90	-
Chaetodon rafflesii	1,505	FC	<15 months	10.00	-
Chaetodon flavirostris	1,479	FC	<15 months	10.00	-
Chaetodon sedentarius	1,421	NC	1.4–4.4 years	10.00	-
Chaetodon baronessa	1,318	FC	<15 months	10.00	-
Chaetodon paucifasciatus	1,302	NC	<15 months	10.00	-
Chaetodon ocellatus	1,296	NC	<15 months	10.00	-
Chaetodon rainfordi	1,243	OC	<15 months	10.00	-

Chelmon rostratus (Chan and Sadovy, 1998), though imports of the long-fin bannerfish, *Heniochus acuminiatus*, appear to exceed that of *Chelmon rostratus* (Fig. 11.3). These two species account for 24.4% of recorded imports of butterflyfishes, and their popularity probably comes from the fact that they are among the few coral reef butterflyfishes that do not generally eat live corals (Fig. 11.3).

Delbeek (Chapter 12) discusses the merits of keeping non-coral feeders over coral feeding butterflyfishes, and especially warns against keeping obligate coral feeders in aquaria (see also Allen et al., 1998; Michael, 2004). Importantly, obligate corallivores do not generally survive well, if at all, in captivity. One exception however, is the butterflyfish, *Chaetodon melannotus*, which is among the most popular species of butterflyfish from the genus *Chaetodon*. This butterflyfish is relatively small (maximum length <15 cm) and while it feeds almost exclusively on live corals (hard corals and soft corals) in the wild (Pratchett, 2005), *C. melannotus* will take a variety of foods offered in captivity, as does *C. punctatofasciatus* and *C. multcinctus* (Delbeek, Chapter 12). These species do, however, pose a significant threat to the survivorship of any corals or other colonial organisms placed with them in aquaria.

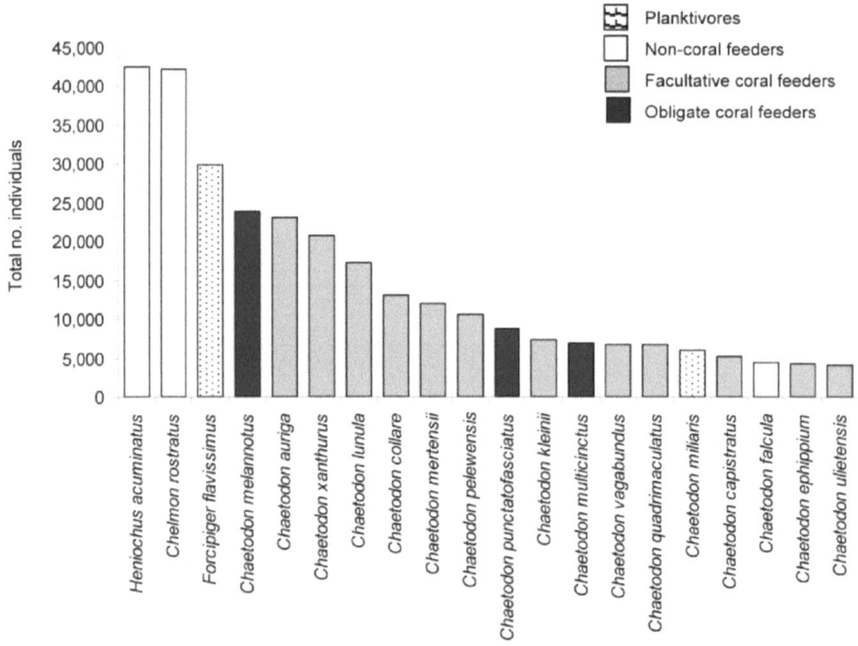

Fig. 11.3 Total number of individual fishes imported to all countries between 1988 and 2002, for the top-20 most frequently imported butterflyfishes. Data are from GMAD (www.unep-wcmc.org/marine/GMAD/index.html). Dietary habits of individual species were established based on data published in Cole and Pratchett—Chapter 5.

There is considerable consensus among fish ecologists, managers and aquarists, that corallivorous butterflyfishes and especially highly specialised species (e.g., *C. trifascialis*) are not suited to aquaria, mainly because they are difficult to sustain on readily available food sources. However, some of these species continue to be harvested, albeit in small numbers and only in certain locations. Wood (2001) complied data from trade reports from nine countries (Florida, Puerto Rico, Brazil, Eritrea, Saudi Arabia, Sri Lank, Maldives, Hawaii, and Cost Rica) and listed eight species of butterflyfishes that featured in the top-ten species of fishes exported for one or more these countries. Of these species, three are questionable aquarium fishes; *Chaetodon larvatus* is an obligate corallivore and does not survive very well, if at all, in captivity, *C. mesoleucos* has a highly variable success rate in aquaria and *C. semilarvatus* can do well if imported as a juvenile, but adults rarely survive (Michael, 2004). The question is then, what effect (if any) are localised collections of butterflyfishes having on the wild stocks of these species? Daw et al. (2001) surveyed densities of butterflyfishes at major collection sites and found that local populations of *C. larvatus*, *C. mesoleucos*, *C. semilarvatus*, and also *Heniochius intermedius* were significantly lower compared to nearby sites with limited aquarium collections. These data are not entirely convincing as they are based on a single survey and do not therefore show that actual depletion of these species associated with sustained and ongoing aquarium collections.

A recent review of ornamental fisheries in the Red Sea and Gulf of Eden (PERGSA, 2004) identified 5 species of butterflyfishes that are regularly harvested, but typically have very poor survivorship in captivity. These five species included four corallivorous species (*C. austriacus*, *C. melapterus*, *C. trifascialis*, and *C. larvatus*), but PERGSA (2004) also suggested that *Chaetodon auriga*, which is a very common aquarium specimen, is sometimes difficult to maintain. Importantly, field-based surveys did not detect any clear or consistent differences in the abundance of heavily harvested species when comparing mean densities of these butterflyfishes between sites where aquarium fishes are exploited versus unexploited sites (Fig. 11.4). The only species of butterflyfish that was found at significantly lower abundance in exploited sites versus unexploited sites was *C. paucifasciatus*, whereas some other species (*C. auriga*, and *C. trifascialis*) were actually more abundant at exploited sites (PERGSA, 2004). Given extreme variation in the abundance of butterflyfishes within and among reefs, especially among contrasting habitats (e.g., Pratchett and Berumen, 2008), the only effective test of harvesting impacts on target populations would be to implement monitoring programs that test for declines in the abundance of harvested species through time at exploited and unexploited sites.

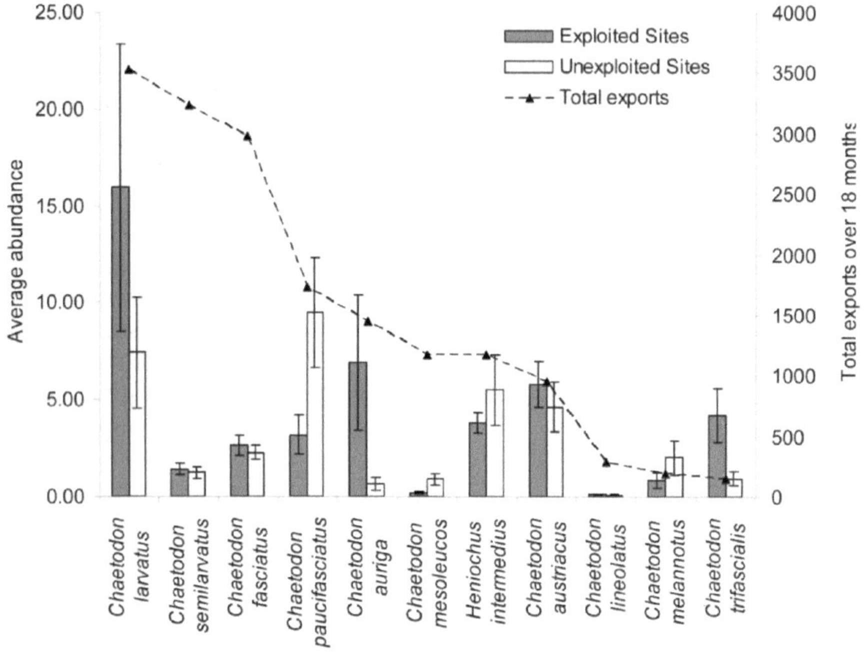

Fig. 11.4 Mean (±SE) abundance of butterflyfishes in the Red Sea and Gulf of Eden at sites were fishes are collected for the aquarium trade versus comparable unexploited sites. Species are ranked according to total exports of each species recorded during the period Nov 1990 to June 1991, and May 2001 to June 2002. Data were extracted from PERGSA (2004).

Improving Governance and Regulation of Marine Aquarium Fisheries

The global marine aquarium industry comprises numerous collectors in more than 80 countries, with many collectors operating in remote locations, making it very difficult to impose effective regulation and management at the point of collection. Some countries have effective ornamental management plans in place; others have poorly enforced regulations or none at all (Wood, 2001). Aquarium fisheries are characterised by being a high value-to-volume fishery, where target species are often very prone to over-fishing (Sadovy and Vincent, 2002). Importantly, rare fishes (due to natural rarity or limited supply) in the marine aquarium trade attract a premium, such that species may actually become increasingly valuable as they face increased fishing pressure due to excessive depletion of natural stocks. This is a stark contrast to most food-fish fisheries whereby it often becomes uneconomical to target fishes as they become increasingly rare. It is imperative therefore, that aquarium fisheries are well managed and effectively regulated

(Sadovy and Vincent, 2002). One definitive recommendation for improved management of the marine aquarium industry is to impose an embargo on certain butterflyfishes that are either not suited to aquariums or particularly vulnerable to over-harvesting. However, the aquarium industry is a luxury demand-driven industry and consumers can also have a profound impact on trade in individual species. This is where education of aquarium keepers comes into play as to which species are suitable and should be favoured. Even so, it is unclear what effect (if any) wild harvesting of these species is really having on natural populations.

The most significant limitation in understanding whether aquarium fisheries are in fact operating in a sustainable fashion is the lack of credible and ongoing data. Trade in marine ornamental fishes is a complicated matter and gathering meaningful and accurate data from all major exporting and importing regions is problematic, as has been extensively discussed by Bruckner (2001), Wood (2001), Balboa (2003), Larkin (2003) and Green (2003). Current data that is available (e.g., GMAD and specific import records) is plagued with major omissions and obvious errors. It is clear therefore, that independent research on aquarium fisheries within key countries should be considered a priority for the conservation of major target and potential sensitive species. In order to defend their activities, aquarium fishers may also need to become pro-active in providing necessary data to improve sustainability in fishing practices.

ARTISANAL FISHERIES

The term artisanal fisheries is used to label a wide variety of reef-based fishing operations and can be broadly defined as *"traditional fisheries involving fishing households (as opposed to commercial companies), using relatively small amount of capital and energy, relatively small fishing vessels (if any), making short fishing trips, close to shore, where fisheries catches are mainly for local consumption"* (FAO website). The distinction between commercial (also referred to as industrial fisheries, e.g., Hawkins and Roberts, 2002; or large-scale fisheries, e.g., McClanahan et al., 2008) and artisanal fisheries (including subsistence fishing) is important in terms of the relative selectivity of fishing practices. Whereas commercial fisheries tend to develop techniques that increase efficiency in capturing a narrow range of specific target species (Berkeset al., 2006), artisanal fisheries tend to develop small scale operations that use simple non-mechanised gears to capture a wide range of species (e.g., Hawkins and Roberts, 2002; Turner et al., 2007). Artisanal fisheries also differ from commercial fisheries in that much of the catch is for immediate personal consumption, is given away to friends or family, or sold at local markets (Kronen, 2004; Craig et al., 2008; Mangi and Roberts, 2006). For example in the Ofu-Olosega fishery in American Samoa, 89% of the catch is directly consumed by villagers and the remainder is shipped to other family members (Craig et al., 2008).

Artisanal fisheries are often considered to have much less impact compared to commercial fisheries, owing to their small scale and broad range of target species (Hawkins and Roberts, 2002), as well as long-standing traditional management practices that tend to limit over exploitation (e.g., McClanahan et al., 2006). However, burgeoning human populations in many tropical countries are placing extreme demands on coral reef resources (Hoegh-Guldberg et al., 2007; Bell et al., 2009), and there is clear evidence of significant fishing impacts even in locations where fishing is exclusively artisanal (Russ, 2002; Hawkins and Roberts, 2002). Munro (1996) estimated that artisanal fisheries throughout the tropics have an annual yield in excess of 6 million tonnes, which totally dwarfs current estimates of total annual yields from the widely condemned live food fish trade (50,000 tonnes, Sadovy and Vincent, 2002). In heavily fished areas, artisanal fisheries have undoubtedly caused changes in the structure of fish and benthic reef assemblages (e.g., Hawkins and Roberts, 2002), and contributed to regional extinctions of highly vulnerable reef fishes (e.g., Bellwood et al., 2002). Throughout much of the tropical western Pacific, coral reef fishes that are harvested mainly for local consumption, are often overexploited (Dalzell et al., 1996).

A variety of fishing methods and gear types are utilised in artisanal coral reef fisheries. These include hook and line, spear guns, hand spears, weirs, traps, gill nets, seine nets, drag nets, trawl nets, dynamite fishing, kite fishing, and gleaning (Jiddawi and Ohman, 2002; Cinner and McClanahan, 2006; Campbell and Pardede, 2006; Mangi and Roberts, 2006; Craig et al., 2008). Aside from destructive fishing methods, such as monofilament nets and blast fishing, certain fishing techniques are more likely to lead to undesirable changes in ecosystem structure due to excessive removal of fishes with critical ecological functions (e.g., herbivores) or species that are highly vulnerable to overfishing (McClanahan and Cinner, 2008; McClanahan et al., 2008). Species composition and size of fishes caught varies depending on the type of gear that is used (e.g., Jiddawi and Ohman, 2002; Mangi and Roberts, 2006; Turner et al., 2007). For example, a study in Kenya that compared the impacts of artisanal fishing gear on coral reefs found that beach seines have the lowest selectivity compared to five other gear types (McClanahan and Mangi, 2004) and catch a considerable numbers of juvenile fishes (Mangi and Roberts, 2006). As a consequence, several local villages have banned the use of beach seines and are seeing definite benefits in terms of fisheries yield (McClanahan et al., 2008).

Artisanal Catches of Coral Reef Butterflyfishes

For those artisanal coral reef fisheries for which specific data is available, fishes from the families Carangidae (jacks), Lutjanidae (snappers), Lethrinidae (emperors), Siganidae (rabbitfishes), Labridae (wrasses and parrotfishes), and Serranidae (groupers) are the most commonly caught species (Bellwood, 1988; Adams and Dalzell, 1994; Friedlander and Parrish, 1997; Mangi and Roberts, 2006). Butterflyfishes meanwhile, constitute only a very small percentage (<0.01%) of reported artisanal fisheries catches (e.g., Bellwood, 1988; Adams and Dalzell, 1994; Friedlander and Parrish, 1997; Mangi and Roberts, 2006; McClanahan and Cinner, 2008; Fig. 11.5) and are only vulnerable to exploitation in locations and fisheries that utilise relatively non-selective gear types (e.g., nets, traps and spears). In American Samoa, Adams and Dalzel (1996) showed that butterflyfishes accounted for 0.11% of the total catch (by volume) of reef fishes, whereas the major family caught (the Carangidae) accounted for 85% of the total catch (Fig. 11.5a). Similarly, in Papua New Guinea, butterflyfishes were caught by artisanal fishers using nets and spears, but only accounted for 0.12% of all individual fishes that were caught. Here, catches were numerically dominated by Acanthuridae and Lethrinidae, which accounted for 16.9% and 15.2% of catches, respectively (Fig. 11.5b).

Although butterflyfishes account for only a very small proportion of artisanal fisheries catches throughout tropical island nations, the limited data that is available suggests that there could be a significant number of individual butterflyfishes that are being harvested by artisanal fisheries. Even if only 0.001% of the 6 million tonnes of fishes harvested annually by artisanal fishers on coral reefs (Munro, 1996) comprise butterflyfishes, this could represent annual harvesting of in excess of 600,000 individual butterflyfishes, based on a mean weight of 100 g for most larger species of butterflyfishes (e.g., *C. auriga* and *C. vagabundus*). Opportunistic surveys of butterflyfishes available for sale at the fish markets in Tahiti (on 5 separate days over a 12 month period) revealed that there were in excess of 30 individuals per day (which would equate to 10,950 indviduals. yr^{-1}, Fig. 11.5a), though there were no butterflyfishes sold on 2 (out of the 5) days surveys were conducted (Pratchett, unpublished data).

Catches of butterflyfishes for food are expected to comprise mainly larger species and/or most abundant species (Fig. 11.5a). However, it is also true that any reasonable sized fishes (> 6 cm total length) caught in non-selective gear types (nets and traps) are rarely discarded (Mangi and Roberts, 2006), and so opportunistic catches of butterflyfishes could include the full range of coral reef species. During an extensive survey of fishing practices in Kenya, Mangi and Roberts (2006) recorded six butterflyfish species (*C. auriga*, *C. trifasciatus*, *C. falcula*, *C. melannotus*, *C. kleinii*, *C. fasciatus*)

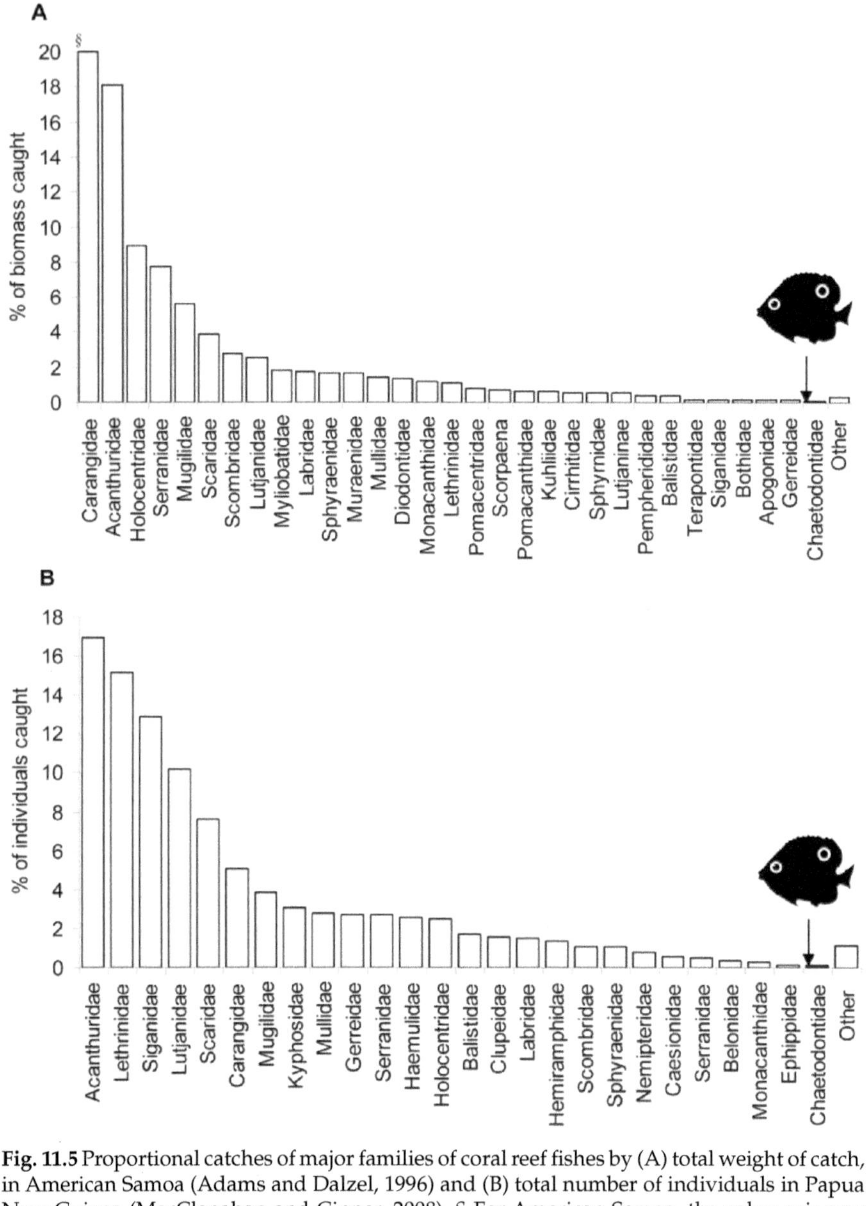

Fig. 11.5 Proportional catches of major families of coral reef fishes by (A) total weight of catch, in American Samoa (Adams and Dalzel, 1996) and (B) total number of individuals in Papua New Guinea (MacClanahan and Cinner, 2008). § For American Samoa, the value axis was constrained to 20% to help resolve small percentage values; Carangidae accounts for 85.74% of total catch by weight.

among a total of 195 species of fishes that were caught by artisanal fishes. However, *C. auriga* accounted for 72% (52/ 72) of individuals caught, which probably reflected it's abundance in shallow sandy habitats were beach seines were being used (Mangi and Roberts, 2006).

CHALLENGES TO THE MANAGEMENT OF SMALL-SCALE FISHERIES

One of the major challenges to the management of artisanal fisheries is that established fisheries management measures, such as spatial and temporal limits on catches and/or fishing effort, are difficult to implement and enforce when dealing with multitudes of individual fishers and small fishing operations operating largely independently. Throughout many developing tropical countries, fishermen tend to be loosely organised and there are very few points in the fishing operation where enforcement can be concentrated (Adams and Dalzell, 1994; McClanahan and Mangi, 2004). In addition, national and provincial governments often lack the necessary resources to monitor these fisheries and enforce regulations (Cinner and McClanahan, 2006; Zeller et al., 2006). However, considerable progress has been made towards addressing these management challenges. For example, co-management arrangements that involve local communities have had reasonable success in some artisanal fisheries (McClanahan et al., 2008), while the livelihoods approach advocated by Allison and Ellis (2001) provides a framework to manage artisanal fisheries within the wider context of a livelihood platform.

Within the broader scope of fisheries management there is a move towards ecosystem based fisheries management and integrated approaches that consider fisheries as integrated social-ecological systems that incorporate management, economic and ecosystem perspectives (Francis et al., 2007; McClanahan et al., 2008). Implementation and effectiveness of these ideals is however, still constrained by the limited availability of important information on fishing behaviour (including, spatial and temporal patterns of fishing, as well as the gear types), stock sizes, and stock responses to exploitation. Use of established fisheries management measures may be inappropriate in situations where traditional values and customs still shape participation in artisanal fisheries. For example, artisanal fishing activities in Tonga reflect the need to satisfy subsistence requirements and social obligations rather than trying to maximise economic benefit, as occurs in commercial fisheries (Kronen, 2004). In these cases, some biological and technical management tools may not achieve compliance because they lack social context necessary for acceptance (McClanahan et al., 2008).

For artisanal fisheries, the biggest challenge to effective management is the lack of data on harvest rates (Craig et al., 2008), and although catch data exists for some fisheries it often requires extensive fisheries-independent research effort to obtain this data (e.g., Zeller et al., 2006). However, perhaps one of the biggest challenges to the management of artisanal fisheries is reconciling a need to reduce catches in unsustainable fisheries with the dependence of many people on these fisheries for food or livelihoods. Artisanal fisheries are of fundamental importance for many tropical countries, providing food, income or livelihoods for up to 450 million people worldwide (McClanahan et al., 2008). In many areas, catches from artisanal fisheries constitute the major source of protein (Stobutzki et al., 2006; Craig et al., 2008; Bell et al., 2009), particularly in households where insufficient income is available to purchase alternative sources (Kuster et al., 2005). In addition to providing a crucial food source, fishing is often seen as an occupation of last resort when income from other sources fails, and fisherman are commonly characterised as the poorest of the poor (Béné, 2003). While the critical role that artisanal fisheries play in ensuring food security and providing livelihoods is acknowledged, there is also concern over the sustainability of some of these fisheries and calls for a reduction in catches (Stobutzki et al., 2006). Sustainable fisheries are essential for ensuring long-term food security, but rather than simply enforcing immediate reductions in catches of certain target species, reduced reliance on marine resources needs to be effected through redistribution of fisheries effort (e.g., targeting more near-shore pelagic resources, to reduce pressure on highly vulnerable reef-based species), and economic development aimed at providing diversified livelihoods. From the perspective of butterflyfishes, it is important to realise that the individual value of these fishes to the aquarium industry is often far greater than their food value. Where market chains permit, it would therefore, be preferable to develop appropriate and sustainable aquarium fisheries, rather than indiscriminately harvesting small, potentially high value, species for food.

SUSCEPTIBILITY OF BUTTERFLYFISHES TO OVERFISHING

Inherent vulnerability of fishes to overfishing depends largely on their life-history characteristics, whereby species with higher longevity and/or lower rates of population replenishment cannot sustain as higher fishing pressure compared to species with high population turnover (Cheung et al., 2005, 2007). Butterflyfishes are conspicuous among coral reef fishes in that populations appear to be maintained by very low rates of larval input. Pratchett et al. (2008b) showed that juvenile butterflyfishes (age <1 year) typically account for <25% of reef-wide densities of butterflyfishes, whereas for many other fishes populations are expected to be dominated by youngest cohorts. The only butterflyfishes for which juveniles outnumber adults were *C. aureofasciatus* and sometimes *C. rainfordi* (Pratchett et al., 2008b). A corollary of these findings is probably that butterflyfishes are very long-lived, whereby high adult abundance reflects that gradual accumulation of individuals across many different cohorts. To date, there has been very limited data published on the potential longevity of coral-reef butterflyfishes, but unpublished data from P. Nangle and G.P. Jones suggest that some butterflyfishes (e.g., *C. lineolatus*) may live up to 70 years. Berumen (2005) reports that several common small-bodied butterflyfishes consistently live more than 10 years, and may live considerably longer. Given low recruitment rates and assuming that they have considerable longevity (see *C. melannotus*, Table 11.1), butterflyfishes must be considered to be highly vulnerable to overfishing. However, this is not currently reflected in the species-specific vulnerability assessments provided on FishBase (Froese and Pauly, 2008), due largely to a lack of necessary data on population demographics (mainly, longevity), which is used to calculate vulnerability (Table 11.1).

CONCLUSIONS AND FUTURE DIRECTIONS

As a family, butterflyfishes receive relatively little fishing pressure (from either aquarium or artisanal fisheries), but are possibly much more susceptible to fishing pressure compared to many other coral reef fishes. Aside from their inherent vulnerability to over-fishing, many butterflyfishes (e.g., *C. bennetti*) are naturally very rare, and/ or are already experiencing severe reduction in population size due to widespread and ongoing degradation of coral reef ecosystems (Lawton et al., 2011). Moreover, perturbations to shallow coastal environments are increasing in both geographical scale and severity (e.g., coral bleaching, Hoegh-Guldberg, 1999), and simultaneous impacts across large (e.g., ocean) scales will greatly increase likelihood of species extinctions, even for widespread and previously abundant species (Lawton et al., 2011). Habitat degradation is undoubtedly the major threat to coral reef butterflyfishes, but direct harvesting compounds effects of coral loss. The number butterflyfishes that are being harvested around the world probably numbers in the millions each year, due to combined impacts of marine aquarium fisheries and artisanal fisheries. For the most part, harvesting is directed at larger generalist species that are less susceptible to habitat degradation, but all butterflyfishes are likely to be highly susceptible to over-harvesting due to their relatively low densities and low rates of population turnover.

The ecological consequences of localised extirpation of butterflyfishes from coral reefs is very unclear, though butterflyfishes do constitute the major group of corallivores and so loss of butterflyfishes may limit energetic transfer from corals to higher consumers (Cole et al. 2008). Of greatest concern, however, is the possible extinction of coral reef butterflyfishes. Given marked declines in the abundance of some species (e.g., *C. trifascialis*) throughout their geographic range (Lawton et al., 2011), increased effort is required to reduce any direct anthropogenic threats upon species that are already facing increased risk of extinction. One factor often correlated with high extinction risk and vulnerability is ecological specialisation (McKinney, 1997; Fisher and Owens, 2004; Colles et al., 2009), whereby highly specialised and obligate coral-feeding butterflyfishes probably warrant increased levels of protection (Lawton et al., 2011; Wilson et al., Chapter 9). This provides further imperative to reduce harvesting of specialised coral-feeding butterflyfishes, which do not perform well in aquaria anyway. It is reassuring to note however, that Lawton et al. (2011) demonstrated levels of high gene flow and connectivity among widely dispersed populations of coral-feeding butterflyfishes, which will counter increased extinction risk due to widespread habitat degradation.

REFERENCES

Adams, T.J. and P.J. Dalzell. 1994. Artisanal Fishing. Paper presented at the East-West Centre Workshop on Marine Biodiversity Issues in the Pacific Islands. University of Hawaii. pp. 19.

Adams, T.J. and P.J. Dalzell. 1996. Management of Pacific Island Inshore Fisheries. *3rd Australasian Fisheries Management Conference, Rottnest Island* pp. 225–237.

Albaladejo, V.D. and V.T. Corpuz. 1982. A market study of the aquarium fish industry of the Philippines: An assessment of the growth and the mechanics of the trade. *Proceedings of the 4th International Coral Reef Symposium, Manila* pp. 75–81.

Allen, G.R., R. Steene and M. Allen. 1998. *A guide to angelfishes and butterflyfishes*, Odyssey Publishing, Perth, Australia.

Allison, E.H. and F. Ellis. 2001. The livelihoods approach and management of small-scale fisheries. *Marine Policy* 25: 377–388.

Andrews, C. 1990. The ornamental fish trade and fish conservation. *Journal of Fish Biology* 37: 53–59.

Balboa, C.M. 2003. The consumption of marine ornamental fish in the United States: a description from US import data. *In: Marine Ornamental Species: Collection, Culture & Conservation*, J.C. Cato and C.L. Brown (eds.). Blackwell Publishing Company, Ames, USA pp. 65–76.

Bell, J.D., M. Kronen, A. Vunisea, W.J. Nash, G. Keeble, A. Demmke, S. Pontifex and S. Andréfouët. 2009. Planning the use of fish for food security in the Pacific. *Marine Policy* 33: 64–76.

Bellwood, D. 1988. Seasonal changes in the size and composition of the fish yield from reefs around Apo Island, central Philippines, with notes on methods of yield estimation. *Journal of Fish Biology* 32: 881–893.

Bellwood, D., P. Wainwright, C. Fulton and A. Hoey. 2002. Assembly rules and functional groups at global biogeographical scales. *Functional Ecology* 16: 557–562.

Bellwood, D.R., A.S. Hoey and J.H. Choat. 2003. Limited functional redundancy in high diversity systems: resilience and ecosystem function on coral reefs. *Ecology Letters* 6: 281–285.

Béné, C. 2003. Contribution of small-scale fisheries to rural livelihoods in a water multi-use context (with a particular emphasis on the role of fishing as "last resort activity" for the poor). FAO Fisheries Report. No. 735, Suppl., Rome, Italy pp. 20–49.

Berkes, F., T. Hughes, R. Steneck, J.A. Wilson, D. Bellwood, B. Crona, C. Folke, L. Gunderson, H. Leslie and J. Norberg. 2006. Globalization, roving bandits, and marine resources. *Science* 311: 1557–1558.

Berumen, M.L. 2005. The importance of juveniles in modelling growth: butterflyfish at Lizard Island. *Environmental Biology of Fishes* 72: 409–413.

Bruckner, A.W. 2001. Tracking the trade in ornamental coral reef organisms: the importance of CITES and its limitations. *Aquarium Sciences and Conservation* 3: 79–94.

Bruckner, A.W. 2005. The importance of the marine ornamental reef fish trade in the wider Caribbean. *Revista de Biologia Tropical* 53: 127–137.

Campbell, S. and S. Pardede. 2006. Reef fish structure and cascading effects in response to artisanal fishing pressure. *Fisheries Research* 79: 75–83.

Chan, T.T.C. and Y.J. Sadovy. 1998. Profile of the Marine Aquarium Fish Trade in Hong Kong. *Aquarium Sciences and Conservation* 2: 197–213.

Cheung, W.W.L., T.J. Pitcher and D. Pauly. 2005. A fuzzy logic expert system to estimate intrinsic extinction vulnerabilities of marine fishes to fishing. *Biological Conservation* 124: 97–111.

Cheung, W.W.L., R. Watson, T. Morato, T.J. Pitcher and D. Pauly. 2007. Intrinsic vulnerability in the global fish catch. *Marine Ecology Progress Series* 333: 1–12.

Cinner, J.E. and T.R. McClanahan. 2006. Socioeconomic factors that lead to overfishing in small-scale coral reef fisheries of Papua New Guinea. *Environmental Conservation* 33: 73–80.

Cole, A.J., M.S. Pratchett and G.P. Jones. 2008. Diversity and functional importance of coral-feeding fishes on tropical coral reefs. *Fish and Fisheries* 9: 286–307.

Colles, A., L.H. Liow and A. Prinzing. 2009. Are specialists at risk under environmental change? Neoecological, paleoecological and phylogenetic approaches. *Ecology Letters* 12: 849–863.

Craig, P., A. Green and F. Tuilagi. 2008. Subsistence harvest of coral reef resources in the outer islands of American Samoa: modern, historic and prehistoric catches. *Fisheries Research* 89: 230–240.

Dalzell, P.J., T.J.H. Adams and N.V.C. Polunin. 1996. Coastal fisheries in the Pacific Islands. *Oceanography and Marine Biology: An Annual Review* 34: 395–531.

Daw, T.M., G.C.C. Rogers, P. Mapson and J.E. Kynoch. 2001. Structure and management issues of the emerging ornamental fish trade in Eritrea. *Aquarium Sciences and Conservation* 3: 53–64.

Fisher, D.O. and I.P.F. Owens. 2004. The comparative method in conservation biology. *Trends in Ecology & Evolution* 19: 391–398.

Fletcher, W., J. Chesson, M. Fisher, K. Sainsbury, T. Hundloe, A. Smith and B. Whitworth. 2002. National ESD Reporting Framework for Australian Fisheries: The 'How To' Guide for Wild Capture Fisheries. FRDC Project 2000/145, Canberra, Australia.

Francis, R.C., M.A. Hixon, M.E. Clarke, S.A. Murawski and S. Ralston. 2007. Ten commandments for ecosystem-based fisheries scientists. *Fisheries* 32: 217–233.

Friedlander, A.M. and J.D. Parrish. 1997. Fisheries harvest and standing stock in a Hawaiian Bay. *Fisheries Research* 32: 33–50.

Froese, R. and D. Pauly, eds. 2008. *FishBase*. World Wide Web electronic publication: www.fishbase.org, version (02/2008).

Green, E. 2003. International trade in marine aquarium species: using the Global Marine Aquarium Database. *In: Marine Ornamental Species: Collection, Culture & Conservation*, J.C. Cato and C.L. Brown (eds.). Blackwell Publishing Company, Ames, USA pp. 29–48.

Grey, M., A.M. Blais and A.C.J. Vincent. 2005. Magnitude and trends of marine fish curio imports to the USA. *Oryx* 39: 413–420.

Hawkins, J.P. and C.M. Roberts. 2004. Effects of artisanal fishing on Caribbean coral reefs. *Conservation Biology* 18: 215–226.

Hoegh-Guldberg, O. 1999. Climate change, coral bleaching and the future of the world's coral reefs. *Marine and freshwater research* 50: 839–866.

Hoegh-Guldberg, O., P.J. Mumby, A.J. Hooten, R.S. Steneck, P. Greenfield, E. Gomez, C.D. Harvell, P.F. Sale, A.J. Edwards, K. Caldeira, N. Knowlton, C.M. Eakin, R. Iglesias-Prieto, N. Muthiga, R.H. Bradbury, A. Dubi and M.E. Hatziolos. 2007. Coral Reefs Under Rapid Climate Change and Ocean Acidification. *Science* 318: 1737–1742.

Hughes, T.P. 1994. Catastrophes, Phase Shifts, and Large-Scale Degradation of a Caribbean Coral Reef. *Science* 265: 1547–1551.

Hughes, T.P. 1996. Demographic approaches to community dynamics: a coral reef example. *Ecology* 77: 2256–2260.

Hughes, T.P., M.J. Rodrigues, D.R. Bellwood, D. Ceccarelli, O. Hoegh-Guldberg, L. McCook, N. Moltschaniwskyj, M.S. Pratchett, R.S. Steneck and B.L. Willis. 2007. Phase Shifts, Herbivory, and the Resilience of Coral Reefs to Climate Change. *Current Biology* 17: 360–365.

Jackson, J.B.C., M.X. Kirby, W.H. Berger, K.A. Bjorndal, L.W. Botsford, B.J. Bourque, R.H. Bradbury, R. Cooke, J. Erlandson, J.A. Estes, T.P. Hughes, S. Kidwell, C.B. Lange, H.S. Lenihan, J.M. Pandolfi, C.H. Peterson, R.S. Steneck, M.J. Tegner and R.R. Warner. 2001. Historical Overfishing and the Recent Collapse of Coastal Ecosystems. *Science* 293: 629–637.

Jiddawi, N. and M. Ohman. 2002. Marine fisheries in Tanzania. *Ambio* 31: 518.

Kronen, M. 2004. Fishing for fortunes?: A socio-economic assessment of Tonga's artisanal fisheries. *Fisheries Research* 70: 121–134.

Kuster, C., V. Vuki and L. Zann. 2005. Long-term trends in subsistence fishing patterns and coral reef fisheries yield from a remote Fijian island. *Fisheries Research* 76: 221–228.

Larkin, S.L. 2003. The US wholesale marine ornamental market: trade, landings, and market opinions. *In: Marine Ornamental Species: Collection, Culture & Conservation*, J.C. Cato and C.L. Brown (eds.). Blackwell Publishing Company, Ames, USA pp. 77–90.

Lawton, R.J., V. Messmer, M.S. Pratchett and L.K. Bay. 2011. High gene flow across large geographic scales reduces extinction risk for a highly specialised coral feeding butterflyfish. *Molecular Ecology* 20: 3584–3598.

Mangi, S.C. and C. Roberts. 2006. Quantifying the environmental impacts of artisanal fishing gear on Kenya's coral reef ecosystems. *Marine Pollution Bulletin* 52: 1646–1660.

McClanahan, T.R. and S.C. Mangi. 2004. Gear-based management of a tropical artisanal fishery based on species selectivity and capture size. *Fisheries Management and Ecology* 11: 51–60.

McClanahan, T.R. and J.E. Cinner. 2008. A framework for adaptive gear and ecosystem-based management in the artisanal coral reef fishery of Papua New Guinea. *Aquatic Conservation: Marine and Freshwater Ecosystems* 18: 493–507.

McClanahan, T.R., M.J. Marnane, J.E. Cinner and W.E. Kiene. 2006. A comparison of marine protected areas and alternative approaches to coral-reef management. *Current Biology* 16: 1408–1413.

McClanahan, T.R., C.C. Hicks and E.S. Darling. 2008. Malthusian overfishing and efforts to overcome it on Kenyan coral reefs. *Ecological Applications* 18: 1516–1529.

McKinney, M.L. 1997. Extinction vulnerability and selectivity: combining ecological and paleontological views. *Annual Review of Ecology Evolution and Systematics* 495–516.

Michael, S.W. 2004. *Angelfishes & Butterflyfishes*, TFH Publications, Neptune City, USA.

Morris, A.V., C.M. Roberts and J.P. Hawkins. 2000. The threatened status of groupers (Epinephelinae). *Biodiversity and Conservation* 9: 919–942.

Munro, G.R. 1996. Approaches to the economics of the management of high seas fishery resources: a summary. *The Canadian Journal of Economics* 29: 157–164.

Myers, R.A. and B. Worm. 2003. Rapid worldwide depletion of predatory fish communities. *Nature* 423: 280–283.

PERSGA/GEF. 2004. Current status of the ornamental fish trade in the Red Sea and Gulf of Aden with guidelines for a self-financing monitoring, control and surveillance programme and proposals for quotas. PERSGA, Jeddah, Kingdom of Saudi Arabia pp. 76.

Pratchett, M.S. 2005. Dietry overlap among coral-feeding butterflyfishes (Chaetodontidae) at Lizard Island, northern Great Barrier Reef. *Marine Biology* 148: 373–382.

Pratchett, M.S. and M.L. Berumen. 2008. Interspecific variation in distributions and diets of coral reef butterflyfishes (Teleostei: Chaetodontidae). *Journal of Fish Biology* 73: 1730–1747.

Pratchett, M.S., S.K. Wilson and A.H. Baird. 2006. Declines in the abundance of*Chaetodon* butterflyfishes following extensive coral depletion. *Journal of Fish Biology* 69: 1269–1280.

Pratchett, M.S., P.L. Munday, S.K. Wilson, N.A.J. Graham, J.E. Cinner, D.R. Bellwood, G.P. Jones, N.V.C. Polunin and T.R. Mcclanahan. 2008a. Effects of climate-induced coral bleaching on coral-reef fishes—Ecological and economic consequences. *Oceanography and Marine Biology: An Annual Review* 46: 251–296.

Pratchett, M.S., M.L. Berumen, M.J. Marnane, J.V. Eagle and D.J. Pratchett. 2008b. Habitat associations of juvenile versus adult butterflyfishes. *Coral Reefs* 27(3): 541–551.

Pyle, R.L. 1993. Marine aquarium fish. FFA Report 92/55. Pacific Islands Forum Fisheries Agency, Honiara, Solomon Islands pp. 135–176.

Roberts, C.M. 1995. Rapid Build-up of Fish Biomass in a Caribbean Marine Reserve. *Conservation Biology* 9: 815–826.

Roelofs, A. 2008. Annual status report 2007: Marine Aquarium Fish Fishery. Department of Primary Industries & Fisheries, Brisbane, Australia pp. 12.

Roelofs, A. and R. Silcock. 2008. A sustainability assessment of marine fish species collected in the Queensland marine aquarium trade, Department of Primary Industries & Fisheries, Brisbane.

Russ, G.R. 1991. Coral Reef Fisheries: Effects and Yields. In *The Ecology of Fishes on Coral Reefs*, P.F. Sale (ed.) Academic Press, Inc, San Diego, USA pp. 601–636.

Russ, G.R. 2002. Yet another review of marine reserves as reef fishery management tools. In *Coral reef fishes: dynamics and diversity in a complex ecosystem*, P.F. Sale (ed.). Academic Press, New York, USA pp. 421–444.

Sadovy, Y.J. and A.C.J. Vincent. 2002. Ecological issues and the trades in live reef fishes. In *Coral reef fishes: dynamics and diversity in a complex ecosystem*, P.F. Sale (ed.). Academic Press, New York, USA pp. 391–340.

Stobutzki, I.C., G.T. Silvestre and L.R. Garces. 2006. Key issues in coastal fisheries in South and Southeast Asia, outcomes of a regional initiative. *Fisheries Research* 78: 109–118.

Tissot, B.N. and L.E. Hallacher. 1999. Impacts of aquarium collectors on coral reef fishes in Kona, Hawaii. Department of Land and Natural Resources, Division of Aquatic Resources, Honolulu, USA.

Tissot, B.N. and L.E. Hallacher. 2003. Effects of aquarium collectors on coral reef fishes in Kona, Hawaii. *Conservation Biology* 17: 1759–1768.

Turner, R., A. Cakacaka, N.A.J. Graham, N.V.C. Polunin, M. Pratchett, S. Stead and S. Wilson. 2007. Declining reliance on marine resources in remote South Pacific societies: ecological versus socio-economic drivers. *Coral Reefs* 26: 997–1008.

Wabnitz, C., M. Taylor, E. Green and T. Razak. 2003. *From Ocean to Aquarium: The global trade in marine ornamental species*, UNEP/Earthprint, Cambridge, UK.

Wilson, S.K., N.A.J. Graham, M.S. Pratchett, G.P. Jones and N.V.C. Polunin. 2006. Multiple disturbances and the global degradation of coral reefs: are reef fishes at risk or resilient? *Global Change Biology* 12: 2220–2234.

Wood, E.M. 2001. Collection of coral reef fish for aquaria: global trade, conservation issues and management strategies. Marine Conservation Society, Ross-on-Wye, UK pp. 80.

Zeller, D., S. Booth, P. Craig and D. Pauly. 2006. Reconstruction of coral reef fisheries catches in American Samoa, 1950–2002. *Coral Reefs* 25: 144–152.

CHAPTER 12

Captive Care and Breeding of Coral Reef Butterflyfishes

J. Charles Delbeek

INTRODUCTION

With brilliant colours, intriguing patterns, and an elegant shape, it is not surprising that butterfishes (family Chaetodontidae) are highly prized by marine aquarists, and one of the most important families of coral reef fishes in the marine aquarium industry (e.g., Wood, 2001). Estimates of the number of butterflyfishes caught for the aquarium industry vary greatly (Pyle, 1993; Wood, 2001; Green, 2003; Lawton et al., Chapter 11), but the Global Marine Aquarium Database (www.unep-wcmc.org/marine/GMAD/index.html) reports that 347,487 butterflyfishes, from 72 different species, were imported (mostly to north America) as part of the global trade in marine aquarium fishes between 1988 and 2002. These numbers almost certainly under-represent the true extent of harvesting of butterflyfishes for the marine aquarium industry, but it is clear that butterflyfishes (along with the angelfishes, damselfishes, wrasses, and gobies) are the top five most heavily traded families of coral reef fishes (Lawton et al., Chapter 11).

With the exception of rare angelfishes, rarely collected butterflyfishes are amongst the most highly desired coral reef fishes and can command prices of several hundred dollars in certain markets. In an extensive survey of the Hong Kong marine aquarium fish trade, Chan and Sadovy (1998) recorded only 3 individual fishes priced at more than HK$1,000 where the mean retail price was HK$60 per fish, and two (out of 3) of these were butterflyfishes. Many of the most highly prized butterflyfishes are not necessarily rare in

California Academy of Sciences, Steinhart Aquarium, San Francisco, California, 94118, USA.
E-mail: CDelbeek@calacademy.org

the wild, but are largely inaccessible due to their specific habitat or limited geographic range. For example, some species (e.g., *Chaetodon burgessi*, *C. mitratus*, and *Prognathodes* spp.) are only found in deeper water (>35m depth), requiring the use of specialised diving equipment to collect. Others (e.g., *Chaetodon smithi*, *C. litus* and *C. sanctaehelenae*) are only found in remote locations (Allen et al., 1998), which are rarely visited by collectors. Therefore, these species are rarely offered for sale and when they are, they command a high price.

Maintaining butterflyfishes in captivity is not any more challenging than most other coral reefs fishes such as angelfishes (Pomacanthidae), but there are many species of butterflyfishes that are generally unsuitable for captivity, mostly because of their highly specialised dietary and habitat requirements. Most notably, there are many butterflyfishes that feed almost exclusively on live corals in the wild (Hourigan et al., 1988; Roberts and Ormond, 1992; Pratchett, 2005; Cole and Pratchett, Chapter 5) and do not readily accept alternative food sources in captivity (Fenner, 1998; Michael, 2004). Moreover, coral feeding butterflyfishes can consume up to 50g of coral tissue per day (Cole et al., 2011), and will quickly deplete scleractinian corals placed within the same aquaria. The mariculture of stony corals such as *Acropora*, has increased dramatically in the last decade and many of these corals can be easily grown and asexually propagated by hobbyists, making it feasible for some hobbyists to actually grow enough live coral to serve as food for corallivorous butterflyfishes. However, it is often easier to choose butterflyfishes that do not normally eat scleractinian corals, which will leave display coral alone, especially if they are provided with sufficient alternative prey (Plate 12.1).There are many butterflyfishes that do not require live corals and are much hardier aquarium species (e.g., *Chaetodon auriga*, *C. collare*, *C. lunula*, *Chelmon rostratus*, *Forcipiger flavissimus* and *Heniochus diphreutes*; Plate 12.2). Fortunately, it is these species that represent the major component of the trade in butterflyfishes (Global Marine Database). However, there is ongoing harvesting of some inappropriate species. With limited regulation of aquarium fisheries in most parts of the world and considerable challenges facing ongoing management of the international marine aquarium industry (Lawton et al., Chapter 11) it is largely up to individual consumers to ensure that they do not support the wild harvesting and international trade of inappropriate species. Fortunately, these species do not appear to represent a major component of the trade in butterflyfishes (Global Marine Database).

The purpose of this chapter is to combine information on the biology of butterflyfishes in their natural environment with contemporary knowledge of marine aquarium systems, to provide important information relevant to selecting and keeping butterflyfishes in aquaria. It is not the intention of this chapter to provide a complete introduction on marine aquarium keeping;

Plate 12.1 *Chaetodon ulientensis* in an aquarium full of scleractinian corals at the Waikiki Aquarium. This butterflyfish is not an obligate coral feeder, but like many *Chaetodon* butterflyfishes will normally eat small quantities of coral (Pratchett, 2005). If however, there is plenty of alternative prey, or there is a large quantity of coral, then any effects from feeding on display corals should be minimal. This specimen has been on display since 1986 and would only pick at soft corals and anemones. Photograph by author.

Colour image of this figure appears in the colour plate section at the end of the book.

Plate 12.2 *Heniochus diphreutes* are easy to keep in aquaria, but do require large tanks. This exhibit is at the Monterey Bay Aquarium. They normally feed in mid-water (on plankton) and will readily take most foods offered in aquaria. Unlike some *Heniochus* sp. they rarely feed on benthic prey, and will leave coral alone. Photograph by author.

Colour image of this figure appears in the colour plate section at the end of the book.

however, it is prudent to briefly cover a few of the basics with specific reference to keeping butterflyfishes. For further information, there are a myriad of excellent books available on marine aquarium keeping (e.g., Allen et al., 1998; Michael, 2004), not to mention the plethora of online forums dedicated to keeping of butterflyfishes and other specific marine fishes (e.g., www.meerwasserforum.com, www.recif-france.com, www.reefcentral.com, www.reefs.org, www.zeewaterforum.nl). Allen et al. (1998), Fenner (1998) and Michael (2004) also provide a species by species description for the majority of butterflyfishes available in the aquarium trade and these texts are highly recommended for comprehensive information on the captive care of butterflyfishes.

AQUARIUMS SYSTEMS FOR MAINTAINING BUTTERFLYFISHES

When butterflyfishes were first made widely available for the aquarium trade (in the 1950s), they deservedly had the reputation of being very delicate and difficult to maintain in captivity (Wood, 2001). It was not until the mid-1970s in Europe and the late 1980s in North America, that success with several "difficult" marine ornamental species, including several different butterflyfishes, dramatically increased. This change coincided with fundamental changes in the general approach to keeping and maintaining marine aquaria (Delbeek and Sprung, 2005). Prior to this, marine aquaria were best categorised as sterile glass boxes with dead bleached coral skeletons as decoration and equally pale looking fish. The resurgence of marine aquaria can be directly attributed to the advent of the use of "live" rock as the primary decoration in the aquarium (Delbeek and Sprung, 2005). These new systems became known as "minireefs" and had an emphasis on the use of "live" rock, aerobic biological filtration, bright lighting, macroalgae, foam fractionation, and a collection of invertebrates and fishes that would not harm soft or stony corals. The emphasis on living biota, however, meant that benthic feeding fishes such as angelfishes and butterflyfishes became less popular, because they would invariably feed on prized corals, sponges and anemones (e.g., Allen et al., 1998). In the last decade, however, aquarists have been experimenting with keeping a range of different fishes in reef aquaria and are discovering that some (but not all) butterflyfishes can be safely kept along side live corals and other sessile invertebrates (Delbeek and Sprung, 2005; Delbeek, 2008; Wandell, 2010). There are now a wide range of butterflyfishes that can be maintained in reef aquaria with relative ease by following a few simple guidelines.

Filtration

Most species of butterflyfishes come from relatively shallow (typically <30 meters) coral reef environments (Allen et al., 1998), which are characterised by brightly lit, nutrient poor and highly oxygenated waters. For this reason, butterflyfishes are extremely sensitive to slight changes in environmental factors such as pH, temperature, salinity, dissolved oxygen, dissolved inorganic nitrogen and dissolved organics, all of which must be kept as close as possible to natural levels (Michael, 2004). With the advent of modern filtration and lighting techniques, duplicating this environment in a closed system has become relatively easy to accomplish, following Fosså and Nilsen (1996); Tullock (1997); Adey and Loveland (1998); Fenner (1998); Paletta (1999); Borneman (2001), and/or Delbeek and Sprung (2005).

Biological filtration (aerobic and anaerobic) can be easily provided through the use of "live" rock. "Live" rock is formed from dead coral skeletons and consolidated calcium carbonate based sediments. These become covered with bacteria, sponges and other sessile invertebrates both on the exterior and interior of the rock. As such, they comprise small, complete "ecosystems" able to completely process the nitrogenous compounds produced by fishes and invertebrates in closed systems. "Live" rock contains both aerobic and anaerobic zones that allow for the complete cycling of nitrogen from nitrogenous waste to nitrogen gas by nitrifying and denitrifying bacteria (Carlson, 1986).

If the use of "live" rock is not feasible, the use of trickle filters or fluidised bed filters can provide sufficient aerobic filtration. These filters allowed for the extremely efficient conversion of ammonia to nitrate within the filter by increasing the efficiency of oxygen diffusion through a thin film of water to the nitrifying bacteria growing on biological media, with a resulting increase in oxygenation of the water. However, trickle filters have become less popular, and aquariums using only "live" rock for biological filtration are the norm today. In addition to the use of "live" rock, the use of separate chambers outside of the aquarium designed to house macroalgae as well as thick sediments of sand or mud has also become popular in the last five years. These "refugia" have resulted in improved water quality by attaining natural reef levels of nitrate and phosphate, as well as providing a predator-free habitat for micro-crustaceans such as copepods, mysis shrimp, and polychaetes, whose larvae contribute to the food chain within the aquarium. The use of a foam fractionator is also highly recommended as it removes fine particulates, inorganic phosphate, and dissolved organics; some models can also increase dissolved oxygen levels. A foam fractionator is the perfect adjunct to a biological filter as they remove many of the nitrogenous compounds that would otherwise be metabolised by heterotrophic bacteria into ammonia and eventually nitrate, hence indirectly lowering nitrate levels.

Ozone gas (O_3) can also be used to control bacterial levels in closed systems. In the 1970s, ozone was first introduced as a tool to sterilise aquarium water when used in conjunction with a foam fractionator; this was especially common in Europe where it was believed that the only way to keep delicate fishes such as butterflyfishes was in systems that were as sterile as possible. Today, ozone continues to be used in public aquaria to control bacterial levels in large systems, especially those that contain marine mammals. However, in most fish and coral systems ozone is only used in very small amounts to control the accumulation of yellowing compounds in the water of closed systems, and to increase the efficiency of foam fractionators, not as a means to sterilise the water (Delbeek and Sprung, 2005).

The use of granular activated carbon is another method by which yellowing compounds as well as a host of organic compounds can be removed from the water. Activated carbon removes a wide variety of organic molecules by trapping them within the carbon pores (absorption) or by chemically bonding them (adsorption). Adsorption relies on the fact that many organic molecules are polar in nature. Therefore, when a polar molecule comes close to a polar surface such as GAC, it becomes attached, effectively removing the molecule from solution. Used in marine aquaria, activated carbon reduces the organic load thereby reducing the amount of material available for conversion by bacteria into nitrogenous waste, but must be replenished frequently as it becomes saturated with dissolved compounds (Delbeek and Sprung, 2005).

System Sizes

Given their rapid growth and relatively large maximum size (Allen, 1985) and generally large feeding territories in the wild (>100m^2; Berumen and Pratchett, 2007), it should not come as a great surprise that butterflyfishes tend to do best in captivity in large systems. Michael (2004) recommends aquaria no smaller than 280 L for most species, and no smaller than 380 L for larger species such as *C. lineolatus*. This may be one of the reasons that longevity of captive butterflyfishes is much higher in large public aquaria which can typically provide systems of several thousand litres capacity. Butterflyfishes that are placed in an aquarium system that is too small are generally smaller, have slower growth and do not live as long. While is it true that fish will often not grow much larger than their environment, this is not natural and such fish will often exhibit gross abnormalities as their growth becomes stunted (J.C. Delbeek, pers. obs.). For example, at the Waikiki Aquarium, juvenile butterflyfishes were frequently collected that were only a few months from settlement on to the reef. Most of these fish readily took to artificial diets but unless moved into larger exhibits as they grew, their growth became stunted. When moving a fish from a small system to a large system, one will invariably witness a rapid increase in size within a few months, even when fish have been confined to a small system for a year or more (J.C. Delbeek, pers. obs.). Tank size can also alter the types of interactions that butterflyfish exhibit to each other and tank mates. For example, in small aquaria with a high density of butterflyfish, dominance hierarchies are more likely to be formed, while in large tanks with lower densities, territories are more likely to become established (Michael, 2004).

Most problems associated with housing butterflyfishes in aquaria that are too small are directly attributed to unexpected growth and adult size of these fishes. Unfortunately, many marine ornamental fishes are imported as juveniles (Wood, 2001), and this is also true for butterflyfish. There are a variety of reasons for this such as greater abundance, ease in shipping (small bags + less water = lower shipping costs and more fish per box) and more attractive colours and patterns (less so for butterflyfishes). In general, smaller specimens of some butterflyfish species are more likely to adapt to captive diets than adults, and this is especially true for those species with more specialised diets. Therefore, these fish often do much better in captivity and stand a much better chance of surviving in the long term than adults. Unfortunately, this often leads to juvenile fish being purchased and placed in aquaria that might be a suitable size initially, but will eventually prove to be too small as the fish grows.

Butterflyfishes should be housed in aquaria that are as large as possible and ideally should maximise surface area over volume. This translates into systems that are wide and long, as opposed to tall and narrow. Doing so not only provides for a greater surface area for providing a rock work structure, but also maximises foraging and swimming area, and provides for a greater surface area to volume ratio. The latter is important as it maximises atmospheric gas exchange and helps to maintain carbon dioxide and oxygen levels at equilibrium with the atmosphere and stabilizes pH.

Aquascaping

In their natural environment, most butterflyfishes remain in very close proximity to benthic substrates and will periodically shelter beneath corals and overhangs. Moreover, most butterflyfishes have specific and sheltered locations, often outside of their usual feeding territories, where they overnight (Yabuta, 1997). As such, it is important to provide captive butterflyfishes with several distinct hiding places. The use of "live" rock or other materials to build a varied aquascape that creates numerous small caves, crevices and overhangs is ideal for butterflyfish to feel comfortable. Even those species that form mid-water schools (e.g., *Heniochus diphreutes, Chaetodon miliaris, Hemitaurichthys* spp.) must be provided with enough shelter to allow each fish to have a secure place to spend the night. Failure to do so eventually results in stressed fish that will stop eating and waste away. A varied and complex habitat also mitigates competitive exclusive and aggression amongst congeners. Although it is important to provide plenty of hiding places, the tank should not be completely filled with rock work and should have adequate open areas to allow for sufficient swimming space; another reason to have as large an exhibit as possible (Michael, 2004).

Aquaria today use a wide variety of depths and materials on the bottom, with some having a simple bare bottom with no substratum while others use a coral sand bed of several centimetres thickness. In the case of these deep sand beds, it is believed that they also operate to reduce nitrogenous wastes as well as providing for a population of infauna that contributes planktonic larvae to the water column. Conversely, the bottom of the aquarium can be covered with a thin layer of crushed coral, or coral sand, which has a more aesthetic than biological function. The merits of both approaches are discussed in Delbeek and Sprung (2005).

Lighting

Butterflyfishes that come from waters less than 30 m can tolerate a wide range of light intensity and will do equally well under fluorescent LED, or metal halide lighting. If however, the fish exhibit night-time colouration during the day, then light levels should be increased. Otherwise light levels have not been shown to be of critical importance in the maintenance of butterflyfish. The only exception to this may be deep water species such as *Prognathodes* spp. and some *Chaetodon* spp. (e.g., *C. tinkeri*, *C. burgessi*, *C. declevis* and *C. flavocoronatus*) that tend to live at depths greater than 30 m and experience relatively low light levels in their normal habitat. However, the Waikiki Aquarium has kept several *C. tinkeri* over the years in exhibits ranging from very little light to a specimen that was kept in an exhibit at the entrance to the Aquarium that received daylight and was also brightly lit with six T5 fluorescent lamps, without any noticeable differences in behaviour (J.C. Delbeek, pers. obs.).

Water Chemistry

As has been mentioned, most butterflyfish come from shallow waters characterised by low dissolved organic and inorganic nitrogen and phosphorous, as well as high dissolved oxygen levels. The levels of these compounds should follow values considered acceptable in closed marine aquarium systems namely: ammonia < 0.01 mg/L, nitrite < 0.01 mg/L, nitrate < 40 ppm, dissolved oxygen 7–10 mg/L, dissolved organic phosphate < 0.01 mg/L and dissolved inorganic phospate < 0.05 mg/L. Similarly, salinity values should be close to natural seawater (32–35 ppt, ~ 40 ppt in the Red Sea), as should pH (8.0–8.3). Temperatures can range from 24–28°C but can be as cool as 16–18°C for deep water or temperate species.

Water Motion

One aspect of marine aquaria that is often overlooked is water motion. Water motion allows for efficient gas exchange between the tank water and the atmosphere. Strong water movement also prevents particulates from settling on the bottom and clogging gravel and sand beds. By utilising strong water movement, detritus is kept suspended in the water column and allows the filtration system to remove this material. Water motion may also aid fish by simulating water currents and surge on reefs keeping the fish active. These factors all have an impact on water quality, which is of utmost importance when keeping butterflyfishes.

DISEASE

Like many tropical marine fishes, butterflyfishes are very susceptible to disease. Fenner (1998) likens butterflyfish to the proverbial "canary in a coal mine" in that whenever water quality begins to deteriorate or the fish become stressed by extrinsic factors such as fluctuating temperature, butterflyfishes will be amongst the first fishes to show signs of distress. They are known hosts for parasites such as sp. *ciliates* (e.g., *Amyloodinium, Brooklynella, Cryptocaryon*), as well as monogentic flukes, digeneans, nematodes, copepods, and acanthocephalans (Michael, 2004). Most external parasites can be dealt with using copper sulfate, quinine hydrochloride, chloroquine diphosphate, formalin, freshwater dips or hyposalinity treatments (see Fenner, 1998 and Michael, 2004 for methods). Internal parasites can be treated with Praziquantal treatment using a 10 mg/L bath for 3 hours (Michael, 2004). Since many of these treatments are not safe for use with invertebrates, treating sick fish in live coral tanks becomes extremely problematic. For this reason, quarantining fish in an isolated system should be standard practice for not only fish destined for live coral systems but for any system. This allows for careful examination of the fish for disease and to get it eating prepared foods. If disease is found, then the fish can be treated appropriately.

Butterflyfishes are also highly susceptible and can quickly succumb to parasitic diseases such as the ciliated protozoan *Cryptocaryon irritans*, also known as saltwater "itch". The use of an ultraviolet (UV) steriliser can help to curtail this parasite, however, the dosages required to affect a protozoan must be at least 200,000 microwatt-sec/m^2, which is a significant dosage requiring a large unit (Escobal, 2000). The greatest utility for a UV unit is in the control of bacterial levels in fish holding systems since bacteria require much lower dosages to be eliminated (4000–26,000 microwatt-sec/m^2) (Escobal, 2000).

NATURAL VERSUS CAPTIVE DIETS

One the greatest challenges faced by those who wish to keep butterflyfishes is providing them with adequate food. Fishes within the family Chaetodontidae can generally be classified into three main groupings: i) obligate corallivores, ii) facultative corallivores, and iii) non-coral feeders which include benthic feeders that tend to feed on small motile invertebrates and zooplanktivores (e.g., Harmelin-Vivien and Bouchon-Navaro, 1983; Sano, 1989; Pratchett, 2005; Cole and Pratchett, Chapter 5). All but the first group are generally suitable for aquaria, but it is not always obvious which species belong in which dietary categories. Extensive field-based research on the natural diets of butterflyfishes is revealing that some butterflyfishes target different prey in different parts of their geographic range. For example, Pratchett (2005) reports that *Chaetodon auriga* from Australia's Great Barrier Reef do not generally feed on corals (see also Allen, 1985), whereas this species appears to feed extensively on corals elsewhere throughout its range (e.g., Indian Ocean, Graham, 2007; central Pacific, Harmelin-Vivien and Bouchon-Navaro, 1983). In general, non-coral feeding butterflyfishes are most suited and easiest to keep in aquaria. Some coral feeding butterflyfishes (obligate and facultative coral-feeders) can be maintained on alternative diets, but still pose a risk to corals and other sessile invertebrates kept in the same aquaria.

Corallivores

Coral feeding butterflyfish species are almost always more difficult to keep in captivity compared to non-corallivorous congeners. Importantly, some species are extremely specific and only feed on a single coral genus (e.g., *Chaetodon trifascialis* = *Acropora* spp.) while others feed on a wider variety (e.g., *C. lunulatus* = *Montipora*/*Porites*/*Pocillipora*/*Pavona*) (e.g., Pratchett et al., 2004; Pratchett, 2005). Moreover, some species feed on actual coral polyps (e.g., *C. baronessa*) while others feed mainly on coral mucus (e.g., *C. ornatissimus*). Other species feed primarily on soft coral polyps as opposed to stony coral polyps (e.g., *C. melannotus*) (Michael, 2004). With the advent of live coral aquariums and the ease with which several species of coral can be asexually propagated and grown, there are anecdotal reports that some advanced aquarium hobbyists have been able to keep obligate corallivores in captivity by growing and providing live coral as food. It is important to note however, that adult specimens of obligate hard coral feeders require access to upwards of 2 m^2 of live coral; otherwise they will rapidly deplete available corals. Condé (1986) describes attempts at maintaining *C. meyeri* at the Nancy Aquarium, France. A juvenile specimen (60 mm) received in 1979 was kept in a tank with the stony corals *Goniopora* spp., *Plerogrya sinuosa* and *Trachyphyllia* spp. It immediately began to feed on the *Trachyphyllia* but after three weeks it was eating chopped mussels and *Trachyphyllia* equally. Condé (1986) also reported that it would browse on "small anemones", presumably *Aiptasia*, growing on the walls of the exhibit and ate a complete raw mussel every day.

In other cases, it is more a matter of scale. For example, the Waikiki Aquarium has a 26.5 m^3 outdoor exhibit that has several large colonies of *Montipora capitata*; this exhibit has been in operation since 1986. In August 2006, a juvenile *C. ornatissimus* was added to the exhibit and as of April 2008, it had tripled in size but has never been observed to feed on any offered foods; clearly it is subsisting on the live coral heads in the exhibit. There are several similar reports of obligate butterflyfish species flourishing in large outdoor exhibits in Hawaii where they can feed on the glass anemone (*Aiptasia* spp.), live coral and crustaceans (D. Chai, K. Izumi, pers. comms.). Therefore, it is possible to maintain obligate corallivores on a live coral diet, provided the exhibit is large enough to accommodate enough live coral or when propagated corals can be provided as needed.

Another promising development for the captive care of corallivorous butterflyfishes is the collection of late pre-settlement larval fishes using floating light traps or crest nets. Pioneered by a group of French researchers in Polynesia in the late 1990s, these techniques provided relatively high catches of larvae for many species that are highly valued within the aquarium industry (including butterflyfishes) and it was shown that these larval butterflyfishes readily accepted and could be reared on prepared diets, and maintained on these diets in aquaria for several years thereafter (Dufour, 2002). Butterflyfishes normally considered impossible to keep such as *Chaetodon lunulatus*, *C. speculum* and *C. trifascialis* were collected in light traps, reared on live newly hatched *Artemia* and later kept in captivity in public aquaria in France on prepared foods (V. Dufour, pers. comm.; G. LeCallion, pers. comm.; Delbeek, 2006a, b). However, if these fish were then introduced into an aquarium with the appropriate genus of live coral, they did resume their corallivorous habits (G. LeCallion, pers. comm.), so they would not be safe for a hobbyist's live coral tank. The drawback to this technique is that the method for collecting the post-larval fishes is currently not selective, therefore it is not possible to provide a reliable supply of specific species; something that must be attained in order for this method to become economically viable in the aquarium trade.

Non-corallivores

Non-corallivorous butterflyfishes often have highly varied diets in the wild, which may be why they are more accepting of artificial food preparations in captivity. Species in the genera *Chelmon*, *Chelmonops*, *Forcipiger* and *Prognathodes* for example, have elongated snouts adapted for feeding on a suite of small discrete food items, including crustaceans and polychaetes (Allen, 1985), though some will also pick at the tube feet of echinoderms (J.C. Delbeek, pers. obs.) and these fishes adapt well to captive diets. Mid-water schooling species such as *C. sanctaehelenae*, *C. miliaris*, *C. smithi*, *Hemitaurichthys* spp. and several *Heniochus* spp. are zooplanktivores and also do very well in captivity (J.C. Delbeek, pers. obs.; Michael, 2004). Allen et al. (1998), Fenner (1998) and Michael (2004) present tables and indicators that rate butterflyfish on their suitability for captivity based on their dietary needs, suitability for reef aquariums housing live corals and other invertebrates, and survivorship in captivity. For the most part, non-corallivorous butterflyfishes rank more highly than corallivorous species in terms of their ease of keeping in aquaria (e.g., Allen et al., 1998), but caution should be exercised when comparing these ratings as the criteria used are highly subjective and suitability can be based only on keeping fish in aquaria based on personal experience (Fenner, 1998), some give no indication as to what the recommendations for suitability are based on (Allen et al., 1998) while others rank them on their ability to be kept in live coral displays as well as in fish only exhibits (Michael, 2004).

With many butterflyfishes, especially those considered picky eaters or difficult to get to feed in captivity, it is best to collect juveniles as opposed to larger, adult fishes. Species such as the Hawaiian endemics *C. fremblii* and *C. multicinctus* will accept prepared foods much more readily when juvenile (less than 2–3 months post settlement) then when older (J.C. Delbeek, pers. obs.). Even hardy butterflyfish species can be difficult to get to feed upon first arrival. The use of live foods such as adult brine shrimp (*Artemia*), mysis shrimp (*Mysis bahia*) and worms such as tubificid (*Tubifex*) and black worms (*Lumbriculus*) can usually get a reluctant butterflyfish to feed (Fenner, 1998). For some species, placing an opened clam shell on the bottom of the tank with the meat exposed can also entice some of the more reluctant species to feed.

Once feeding, a varied diet is best for butterflyfishes, as it is with the care of many marine tropical fishes. The use of dried prepared foods such as flakes and pellets, frozen prepared foods (e.g., mysis, copepods, brine shrimp), and finely chopped whole marine foods (e.g., squid, shrimp, fish) will provide a well-balanced diet for the more catholic species.

CAPTIVE BREEDING VERSUS WILD HARVESTING

As for most species of coral reef fishes, demand for butterflyfishes in the international aquarium industry is currently met entirely through wild harvesting, which may pose a significant threat to populations that are already threatened by extensive habitat degradation (Lawton et al., Chapter 11). There have been some moves towards harvesting younger (e.g., early pre-settlement fishes) rather than removing established individuals from the reef (Dufour, 2002), but if large-scale harvesting of pre-settlement larvae was implemented then this might impact natural populations by reducing larval input that is critical to their maintenance and recovery. Really what is needed to sustain the extensive and rapidly growing demand for coral reef fishes is to breed these fishes in captivity, but this is proving very difficult for some fish families, such as butterflyfishes. The greatest obstacle in rearing butterflyfishes, like many coral reef fishes, is maintaining the larvae during the extensive period (up to 2 months) when they would normally be in open ocean pelagic environments. For example, very little is known about what butterflyfishes eat during their extensive larval stage (e.g., Leis, 1989). However, even before considering the appropriate diets and care of larvae, it is necessary to induce spawning and/or facilitate fertilisation of gametes.

While there have been numerous papers on spawning by butterflyfishes in the wild (e.g., Yabuta and Kawashima, 1996), spawning of butterflyfishes is rarely been observed in captivity. However, it is not unknown for butterflyfishes to spawn in aquaria; Suzuki et al. (1980) described the spawning behaviour and larval development of *C. nippon* in an aquarium. Several individuals were kept in a 3.6 m^3 aquarium exhibit along with other coastal Japanese fishes. One to two days before spawning, the abdomen of the female began to expand dramatically (presumably due to hydration of the eggs). Spawning occurred between a single female and several males in the early evening from 18:00–22:00 with eggs hatching after 28 hours at a water temperature of 22.3–23.7°C. Courtship occurred between a single male and female, which would circle each other and then rise to the water surface with the male nudging the female upwards at which point sperm and egg release occurred. Immediately after spawning, other males in the aquarium would then rush in and also release sperm, a behaviour that may be the result of having so many conspecifics (24 at the beginning of the study in August 1974, and 12 at the end in October 1975) in a small space. Feeding was attempted using the rotifer, *Branchionus plicatilis*, oyster larvae (*Crassostrea gigas*), and sea urchin larvae (*Temnopleurus reevesi*). Only the oyster larvae were found in the stomachs of larval fish yet all butterflyfish larvae died by day eight.

Hioki (1997) lists four species of butterflyfishes that spawned at the Tokai University Aquarium. *Forcipiger flavissimus* and *C. lunula* released eggs that did not hatch, while those released by *C. modestus* and *C. nippon* hatched but never fed. Tanaka et al. (2001) described the courtship behaviour and development of eggs and larvae for the deep-water butterflyfish, *C. modestus* at the Tokai University aquarium in Tokai, Japan. Courtship between multiple males and a single female occurred nightly for 18 months between 18:00–22:00 at a temperature that varied between 18.6 and 26.5°C. Eggs and sperm were broadcast into the water column; the eggs were 0.75–0.80 mm in diameter. After hatching, development of the larvae was described over eight days after which they died, presumably due to exhaustion of the egg yolk and the subsequent lack of a suitable first food item.

Madden and May (1977) reported on attempts to induce spawning in five species of Hawaiian marine tropical fish via injection of Human Chorionic Gonadotropin (HCG), two of which were butterflyfish, *C. miliaris* and *F. flavissimus*. Viable eggs were obtained from the latter species only; however, the larvae did not survive past day 6, again most likely a sign that an acceptable first food was not obtainable. The fertility of the three clutches of eggs produced using this method appeared to depend on the timing of the injections in relation to ovarian development (Ogawa and Brown, 2001).

Even if spawning does not occur, this does not prohibit captive breeding of butterflyfishes. For example, aquaculture of many fishes involves "stripping" of eggs from adult fish and then mixing egg and sperm *ex situ* and hatching the resulting fertilised eggs. This has been attempted with *F. flavissimus* but the larvae did not survive past a few days (S. Kraul, pers. comm.). At present, there do not appear to be any records of successful rearing of butterflyfishes in captivity, though there are unconfirmed accounts of successful rearing of very small cohorts of *Chelmon rostratus*, where two adult individuals spawned naturally in a large dedicated aquarium and several larvae were successfully raised on a diet of mixed plankton (M. Edler, pers. comm.). In 2011, a consortium of North American public aquaria in partnership with the University of Florida's Tropical Aquaculture Lab (TAL) and the American Zoo and Aquarium Association (AZA) formed the Rising Tide initiative. Between them, these institutions house large populations of spawning marine tropical fish, by collecting and shipping eggs and larvae from these events overnight to Florida, the expertise and facilities there could be used to attempt to rear many novel species. Recently, the staff at TAL were able to rear larval *Heniochus diphreutes* to 37 days post hatching from eggs provided by the AZAs Columbus Zoo and Aquarium (for more information in this exciting program see http://risingtideconservation.blogspot.com).

As mentioned previously, the collection and rearing of post-settlement butterflyfish has been accomplished and does offer a means to acquire juvenile fish that readily feed on prepared foods. It is estimated that up to 90% of fish larvae (post-larval stage) do not survive the first week of settlement on a coral reef (Doherty et al., 2004). This then represents a large amount of fish that could be harvested before settlement, with less environmental impact compared to direct harvesting of post-settlement fishes. Lead by Vincent Dufour in French Polynesia, a company was established in 1999 in France with a station on the island of Moorea in Tahiti. The collected post-larvae were then reared in shore-based facilities and exported to Europe for sale to the aquarium trade; some of which were sent to the Nancy Aquarium for trials (e.g., butterflyfishes, surgeonfishes, damselfishes, and blennies) or sold to locals for grow-out and consumption (e.g., groupers and snappers) (Dufour, 2002). In the first year of operation, 25,000 tank-reared fish were exported to France. By selecting only post-larval fish for rearing, there is no need to be concerned with developing first foods since they are already large enough to feed on newly hatched *Artemia* naupili or even microencapsulated feeds. This also allows for the rearing of obligate corallivores on artificial diets as mentioned previously.

LONGEVITIY OF BUTTERFLYFISHES IN CAPTIVITY

Given the very limited data on population demographics of butterflyfishes in the wild (e.g., Berumen, 2005), much of what is known about the longevity of butterflyfishes species come from the records of public aquariums, however, very little of this information is available in published form. Some notable exceptions are Condé (1986), Frakes (1993), and Delbeek and Randall (2010). Condé (1986) describes not only the systems used to maintain fish at the aquarium, but also presents data on how many years each fish had been on display at the aquarium. Condé reported that the maximum length of time in captivity for *Forcipiger flavissimus* was 18 years; however, the fish was still alive at the time of publication. The maximum age for a *Chaetodon* species was *C. rafflesi* at 14.5 years, which was also still alive at the time of publication in 1986. Frakes (1993) reported that a *C. ephippium* he first saw at the aquarium in 1986 was still present on a subsequent visit in 1992, marking 20 years in captivity. Delbeek and Randall (2010) reported on several fish at the Waikiki Aquarium that have been on exhibit for over 25 years including a *C. ulietensis* that has been in the collection since 1988. While longevity in captivity is not always a good indicator of natural lifespan, it does point out that many marine ornamental species have the potential to live for decades.

There are several factors that play a role in longevity in fishes, but diet, disease and environmental stress are the most critical. With proper diet and low stress, fish have a tendency to be more disease resistant. Stress can be the result of poor water quality such as low dissolved oxygen levels, fluctuating temperature or salinity, improper light levels (too bright or too dim), lack of suitable hiding places (i.e., poor aquascape) and aggression from conspecifics or other species in the aquarium. While some butterflyfish species form schools or mated pairs, others tend to be solitary. Furthermore, although some species may school in nature or form pairs, this does not mean that this behaviour will be exhibited in the aquarium. All too often, aquaria are much too small and too many fish are placed in close proximity to behave naturally. Again, references such as Allen et al. (1998), Fenner (1998) and Michael (2004) should be consulted for species specific information on species compatibility and behaviour in aquaria.

CONCLUSIONS

Butterflyfishes are considered one of the most delicate of coral reef fish species to maintain in captivity. However, with technological advances in aquarium keeping and improved shipping and supply lines, combined with increased understanding of their basic biology, it is relatively easy to keep many species of butterflyfishes in aquaria. There remain, however, species that are not suitable for the average aquarium keeper, primarily those that are obligate corallivores. Advances in coral culture and the appearance of large live coral exhibits at public aquaria within the last decade, allow for the display of some of these species due to the sheer volume of these displays and the amount of live coral they can contain. Until an artificial diet can be developed that can provide the proper amount and type of nutrition for these species, they should not be attempted by any but the most experienced of aquarists. There are so many hardy species of butterflyfish available that the demand for corallivorous species will most likely remain a niche market for several years to come. Finally, with advances in marine fish breeding and rearing resulting in success with various difficult species (e.g., *Centropyge* spp., see Baensch, 2002, *Synchiropus splendidus* and *Oxymonacanthus longirostris* to name just a few, see also Wittenrich, 2007), it is not inconceivable that cultured butterflyfish, bred and reared completely in captivity that will feed on prepared diets, will become a viable option within the next decade. Until then aquarists must remain conscious of potential environmental impacts of ongoing wild harvesting and carefully select butterflyfishes that they know have been collected with minimal environmental impacts and choose species that they are capable of maintaining.

REFERENCES

Adey, W.H. and K. Loveland. 2007. *Dynamic aquaria: building living ecosystems*, Academic Press, San Diego, USA.

Allen, G.R. 1985. *Butterfly and angelfishes of the world*. Vol. 2, Mergus Publishers, Melle, West Germany.

Allen, G.R., R. Steene and M. Allen. 1998. *A guide to angelfishes and butterflyfishes*, Odyssey Publishing, Perth, Australia.

Baensch, F. 2002. The culture and larval development of three pygmy angelfish species. *Freshwater and Marine Aquarium Magazine* 25: 4–12.

Berumen, M.L. 2005. The importance of juveniles in modelling growth: butterflyfish at Lizard Island. *Environmental Biology of Fishes* 72: 409–413.

Berumen, M.L. and M.S. Pratchett. 2008. Trade-offs associated with dietary specialization in corallivorous butterflyfishes (Chaetodontidae: *Chaetodon*). *Behavioral Ecology and Sociobiology* 62: 989–994.

Borneman, E.H. 2001. *Aquarium corals: selection, husbandry, and natural history*, T.F.H. Publications, Neptune City, USA.

Carlson, B.A. 1986. Aquarium systems for living corals. *International Zoo Yearbook* 26: 1–9.

Chan, T.T.C. and Y.J. Sadovy. 1998. Profile of the Marine Aquarium Fish Trade in Hong Kong. *Aquarium Sciences and Conservation* 2: 197–213.

Cole, A.J., R.J. Lawton, M.S. Pratchett and S.K. Wilson. 2011. Chronic coral consumption by butterflyfishes. *Coral Reefs* 30: 85–93.

Condé, B. 1986. Longevity of marine tropicals at the Nancy Aquarium. *SeaScope* 3 Summer 1986.

Delbeek, J.C. 2006a. Advancements in marine fish aquaculture. *Marine Fish and Reef 2006 Annual* 8: 111–116.

Delbeek, J.C. 2006b. Kewken van zeewater vissen. *Het Zee-Aquarium* 56: 36–41 (in Dutch).

Delbeek, J.C. 2008. Taming butterflyfish. *Marine Fish and Reef USA Annual* 2009: 52–60.

Delbeek, J.C. and J. Sprung. 2005. *The Reef Aquarium: Science, Art, and Technology*, Ricordea Publishing, Coconut Grove, USA.

Delbeek, J.C. and J.E. Randall. 2010. Aquarium fish longevity. *Reefs Magazine Online Fall 2010*. http://www.reefsmagazine.com/showthread.php?t=82591

Doherty, P.J., V. Dufour, R. Galzin, M.A. Hixon, M.G. Meekan and S. Planes. 2004. High mortality during settlement is a population bottleneck for a tropical surgeonfish. *Ecology* 85: 2422–2428.

Dufour, V. 2002. Reef fish post-larvae collection and rearing programme for the aquarium market. *SPC Live Reef Fish Information Bulletin* 10: 31–32.

Escobal, P.R. 2000. *Aquatic systems engineering: devices and how they function*. 2 ed., Dimension Engineering Press, Oxnard, USA.

Fenner, R.M. 1996. *Conscientious Marine Aquarist*, Microcosm Ltd., Shelburne, USA.

Fosså, S.A. and A.J. Nilsen. 1996. *The Modern Coral Reef Aquarium, Volume 1*, Birgit Schemttkamp Verlag, Bornheim, Germany.

Frakes, T.A. 1993. The Nancy Aquarium—revisited. *SeaScope* 10 Spring 1993.

Graham, N.A.J. 2007. Ecological versatility and the decline of coral feeding fishes following climate driven coral mortality. *Marine Biology* 153: 119–127.

Green, E. 2003. International trade in marine aquarium species: using the Global Marine Aquarium Database. In *Marine Ornamental Species: Collection, Culture & Conservation*, J.C. Cato and C.L. Brown (eds.). Blackwell Publishing Company, Ames, USA pp. 29–48.

Harmelin-Vivien, M. and Y. Bouchon-Navaro. 1983. Feeding diets and significance of coral feeding among chaetodontid fishes in Moorea (French Polynesia). *Coral Reefs* 2: 119–127.

Hioki, S. 1997. Spawning and larval rearing of marine fishes at the Marine Science Museum. *Proceedings of the 4th Annual Marine Aquarium Congress, Tokyo* pp. 361–366.

Hourigan, T.F., T. Tricas and E.S. Reese. 1988. Coral reef fishes as indicators of environmental stress in coral reefs. In *Marine Organisms as Indicators*, D.F. Soule and G.S. Kleppel (eds.). Springer Verlag, New York, USA pp. 107–135.

Leis, J.M. 1989. Larval biology of butterflyfishes (Pisces, Chaetodontidae): what do we really know? *Environmental Biology of Fishes* 25: 87–100.

Madden, W.D. and R.C. May. 1977. Ornamental fish culture project. Final Report, Marine Aquarium Council Task Order No. 136. The Oceanic Institute and the Hawaii Institute of Marine Biology pp. 23.

Michael, S.W. 2004. *Angelfishes & Butterflyfishes*, TFH Publications, Neptune City, USA.

Ogawa, T. and C.L. Brown. 2001. Ornamental reef fish aquaculture and collection in Hawaii. *Aquarium Sciences and Conservation* 3: 151–169.

Paletta, M.S. 1999. *The New Marine Aquarium: Step-by-Step Setup and Stocking Guide*, Microcosm Ltd., Shelburne, USA.

Pratchett, M.S. 2005. Dietry overlap among coral-feeding butterflyfishes (Chaetodontidae) at Lizard Island, northern Great Barrier Reef. *Marine Biology* 148: 373–382.

Pratchett, M.S., S.K. Wilson, M.L. Berumen and M.I. McCormick. 2004. Sublethal effects of coral bleaching on an obligate coral feeding butterflyfish. *Coral Reefs* 23: 352–356.

Pyle, R.L. 1993. Marine aquarium fish. FFA Report 92/55. Pacific Islands Forum Fisheries Agency, Honiara, Solomon Islands. pp. 135–176.

Roberts, C.M. and R.F.G. Ormond. 1992. Butterflyfish social behaviour, with special reference to the incidence of territoriality: a review. *Environmental Biology of Fishes* 34: 79–93.

Sano, M. 1989. Feeding habits of Japanese butterfyfishes (Chaetodontidae). *Environmental Biology of Fishes* 25: 195–203.

Suzuki, K., Y. Tanaka and S. Hioki. 1980. Spawning behavior, eggs, and larvae of the butterflyfish, *Chaetodon nippon*, in an aquarium. *Japanese Journal of Ichthyology* 26: 334–341.

Tanaka, Y., S. Hioki and K. Suzuki. 2001. Spawning behavior, eggs, and larvae of the butterflyfish, *Chaetodon modestus*, in an aquarium. *Journal of the Faculty of Marine Science and Technology-Tokai University* 51: 89–100.

Tullock, J.H. 1997. *Natural reef aquariums: simplified approaches to creating living saltwater microcosms*, Microcosm Ltd., Shelburne, USA.

Wandell, M. 2010. *Reef safe butterflyfishes? Reefs Magazine Online Summer 2010.* http://www.reefsmagazine.com/forum/reefs-magazine/78104-reef-safe-butterflyfishes-matt-wandell.html

Wittenrich, M.L. 2007. *The Complete Illustrated Breeder's Guide to Marine Aquarium Fishes.* T.F.H. Publications, Neptune City, NJ, USA.

Wood, E.M. 2001. Collection of coral reef fish for aquaria: global trade, conservation issues and management strategies. Marine Conservation Society, Ross-on-Wye, UK pp. 80.

Yabuta, S. 1997. Spawning migrations in the monogamous butterflyfish, *Chaetodon trifasciatus. Ichthyological Research* 44: 177–182.

Yabuta, S. and M. Kawashima. 1997. Spawning behavior and haremic mating system in the corallivorous butterflyfish, *Chaetodon trifascialis*, at Kuroshima Island, Okinawa. *Ichthyological Research* 44: 183–188.

CHAPTER 13

Butterflyfishes as a Model Group for Reef Fish Ecology: Important and Emerging Research Topics

Morgan S. Pratchett,[1,a,*] *Karen M. Chong-Seng,*[1,b] *David A. Feary,*[2] *Andrew S. Hoey,*[1,c] *Christopher J. Fulton,*[3] *Jessica P. Nowicki,*[1,d] *Adam K. Dewan,*[4] *Stefan P.W. Walker*[1,e] and *Michael L. Berumen*[5]

INTRODUCTION

In his preface to a special issue (and dedicated workshop) on the biology of butterflyfishes, Motta (1989) suggested that butterflyfishes have received disproportionate scientific attention compared to other common and conspicuous families of coral reef fishes. In support of this assertion, the number of scientific publications that consider butterflyfishes (ISI Web of Knowledge; 382 publications since 1927) is far greater than for many other families of nominal reef fishes (e.g., angelfishes, surgeonfishes, and rabbitfishes); the only families that have been more intensively studied are the Pomacentridae (damselfishes), Serranidae (groupers) and Labridae (parrotfishes and wrasses), which probably reflects their high diversity, commercial and functional importance, respectively. In this respect, research on butterflyfishes has contributed greatly to general understanding of the

---

Authors' affiliations given at the end of the chapter.

biology and ecology of coral reef fishes (e.g., Almany et al., 2007; Lawton et al., 2011). The fields of research in which butterflyfishes have played essential roles also extends well beyond the notion that butterflyfishes are "indicators" of overall reef health, which is the usual justification for studying this relatively unique group of reef fishes (e.g., Öhman et al., 1998; Bozec et al., 2005; Shokri et al., 2005).

The extent to which butterflyfishes are featured in scientific studies of coral reef fishes probably reflects their amenability to a broad range of different research topics and approaches. Added to that, butterflyfishes are extremely conspicuous, and while rarely abundant, they do represent a significant and consistent component of the ichthyofauna on coral reefs. From the perspective of feeding ecology (Pratchett, Chapter 6), obligate coral feeding butterflyfishes are unique in that it is possible to compare proportional use of different prey resources (in this case coral species) within and among species of butterflyfish from *in situ* feeding observations, and relate the relative use of different prey resources to their proportional availability in the local environment. This circumvents many of the usual complexities associated with ascertaining biases in prey detection and capture (e.g., Ferry-Graham et al., 2002), but is also proving to be very useful in testing many of the well-established (and as yet untested) theories related to ecological specialisation (Pratchett, Chapter 6; Wilson et al., Chapter 9). Moreover, different butterflyfishes have very distinctive colouration and body shape, which is strongly reflective of phylogenetic affinities, revealing links between ecology and evolution (e.g., Montanari et al., 2012). The strong dependence of butterflyfishes on corals, and their associated vulnerability to widespread coral depletion (e.g., De'ath et al., 2012), has also generated considerable research interest in this family (e.g., Lawton et al., 2011). Most notably, coral-feeding butterflyfishes may be among the first species of reef fishes that go extinct due to global climate change, or alternatively may provide important insights into the mechanisms that prevent global extinctions despite increasing incidence of large-scale disturbances (Lawton et al., 2011).

Given extensive and ongoing research on butterflyfishes (especially coral reef species) consideration needs to be given to whether butterflyfishes are broadly reflective of coral reef fishes. Given their strong dependence on corals and reef habitats, butterflyfishes (especially, *Chaetodon* species) are widely regarded as the epitome of coral reef fishes, but can this group of fishes yield insights into the general ecology and evolution of reef fishes? In terms of demography, butterflyfishes exhibit a general acanthuroid growth pattern common to many reef fishes, with rapid growth to asymptotic size within 1–2 years after settlement (Berumen, 2005). Their maximum size (Munday and Jones, 1998) and longevity (Berumen et al., 2012b) are also intermediate and broadly reflective of the major families associated with coral reefs (Gobiidae, Blenniidae, Pomacentridae, Pomacanthidae, Siganidae, Acanthuridae, Lutjanidae and Serranidae). Like most reef fishes, butterflyfishes are gonochoric, fertilisation is external, and there is a distinct dispersive (planktonic) larval phase (Choat and Bellwood, 1991). However, it is unknown whether butterflyfishes are capable of changing sex. Many reef fishes change sex, with protogynous hermaphrodism being the normal pattern of reproductive ontogeny (Warner, 1984), which maximises individual fitness in cases where large males can monopolise spawning opportunities. Some butterflyfishes (e.g., *Chaetodon trifascialis*) have haremic mating systems, where there would be fitness advantages associated with protogyny, but butterflyfishes characteristically exhibit strong pair bonding, considered to be evidence of monogamous mating (Whiteman and Côté, 2004). Monogamous fishes are known to change sex, but only under specific conditions of aggressive dominance between mates (e.g., Fricke and Fricke, 1977) and limited mating opportunities (Munday, 2002). As such, butterflyfishes provide an important test case for theories relating to the evolution of sex change.

What really sets butterflyfishes apart from most other families of coral reef fishes is their strong association with benthic habitats, and especially coral-dominated habitats (e.g., Harmelin-Vivien and Bouchon-Navaro, 1983; Halford et al., 2004; Pratchett et al., 2006). However, research on the diet and habitat requirements of coral reef fishes is increasingly showing that a very broad range of fishes rely on corals for food (Cole et al., 2008), or habitat (Coker et al., 2012). Jones et al. (2004) suggested that up to 75% of reef fishes rely on live corals, mostly at settlement. Accordingly, many species (>60%) of fishes exhibit significant declines in abundance following extensive coral depletion (Sano et al., 1987; Jones et al., 2004; Pratchett et al., 2011). It is clear however, that highly specialised coral-feeding butterflyfishes are among the first and worst affected (Wilson et al., 2006; Pratchett et al., 2008; Wilson et al., Chapter 9). As such, butterflyfishes provide an important test case for understanding broad-scale and long-term effects of coral loss and reef degradation on coral reef fishes (Pratchett et al., 2009, 2011). Moreover, responses of butterflyfishes to coral loss vary spatially, temporally and taxonomically (e.g., Pratchett et al., 2006). Ongoing research is therefore necessary to explain this response diversity, which is considered to be critically important for ecological stability and ecosystem resilience (Elmqvist et al., 2003; Bellwood et al., 2004).

Aside from established topics of research (e.g., feeding preferences, and responses to reef degradation), there are also several emerging areas of research in which butterflyfishes are expected to represent key study groups. This is either because butterflyfishes provide extreme examples, to be considered in comparative studies with other perciform fishes (e.g., Fulton, 2007) or in themselves represent a well-defined and monophyletic group of fishes that exhibit considerable diversity in their biology and ecology (e.g., Berumen et al., 2012b). The most important of these topics (to be discussed in turn) are swimming ecology and biomechanics, larval biology and connectivity, climate-induced range shifts, demographic consequences of feeding ecology, and proximate causes of different socialities. This is by no means a comprehensive review of recent and pioneering research involving butterflyfishes, but we present these emerging topics with the hope of stimulating even more research on this family.

SWIMMING ECOLOGY AND BIOMECHANICS

A long-held paradigm is that deep-bodied fishes (e.g., butterflyfishes) should incur higher drag than their fusiform-shaped counterparts, which in turn should exert increased locomotor costs, and/ or limit deep-bodied fish to slower sustained speeds (Lighthill, 1969; Webb, 1984; Tokić and Yue, 2012). However, butterflyfish seem to be breaking all the rules in this regard, having some of the deepest body profiles (i.e., highly laterally compressed) of any demersal fish, but the capacity to sustain exceptional speeds for their size (Fulton, 2007). Indeed, recent biomechanical analyses placed 13 butterflyfishes within the context of 92 reef fish species (from 17 families) to reveal that body shape is far less important for swimming speed performance than previously assumed (Fulton, 2010; Walker et al., submitted). Delving further into how these chaetodontiform swimmers achieve such performance should provide crucial insights into the consequences of novel extremes for fish functional morphology and biology.

Butterflyfishes are also playing a central role in advancing our understanding of behavioural ecology and ecophysiology. For instance, recent field-based examinations found exceptional levels of flexibility in the behavioural response of chaetodontid fishes to a rapidly changing hydrodynamic environment (Heatwole and Fulton, 2013). Despite notions that reef fishes may need to shelter when wave climates intensify (Johansen et al., 2008), butterflyfishes have shown that adaptive swimming postures and behaviours can enable reef fish to remain stable and continue active foraging under turbulent conditions. Questions about the underlying energetic cost of living in wave-swept environments (Fulton, 2010) have also prompted research into whether shifts in the foraging ecology and demography of coral reef butterflyfishes across wave energy gradients has an energetic basis (C.J. Fulton and C. Cvitanovic, unpublished data). Given the prey- and habitat-specificity of many butterflyfishes (Pratchett, 2005; Cole and Pratchett, Chapter 5; Wilson et al., Chapter 9), they provide ideal model species to develop some of the first quantitative eco-energetic models for actively foraging coral reef fishes. Combining the fields of demography, physiology and ecology, this field of research should provide key insights into how reef fish populations and communities vary spatially in terms of energy acquisition and demands, which has important ramifications for predicting vulnerability to climate change.

LARVAL BIOLOGY AND CONNECTIVITY

The early life-history of coral reef fishes, encompassing the numerous developmental and ecologically differentiated life stages from fertilisation to recruitment, represents the most critical phase in their life-cycle (e.g., Leis and McCormick, 2002). Like most marine fishes, the life cycle of butterflyfishes is characterised by having a pelagic larval stage, during which recently fertilised larvae may be dispersed vast distances from their natal habitats. These larvae must then locate appropriate settlement habitat within a specified time (generally <1 month) to facilitate individual transitions to the more sedentary reef-associated stage of their life-cycle (e.g., Sponaugle and Cowen, 1997). For most reef fishes (e.g., butterflyfishes, wrasses), the pelagic stage starts prior to fertilisation, as adult fishes release their unfertilised gametes directly into open water. Fertilised eggs then develop into larvae within the pelagic environment and gradually attain both sensory abilities and physical capabilities necessary to settle within coral reef habitats. While fertilisation success by pelagic-spawning coral reef fishes is often very high (up to 90%, e.g., Petersen et al., 2001), very high rates of mortality in the pelagic stage and during settlement mean that very few of the fertilised eggs will successfully colonise coral reef habitats and complete the transformation to juvenile fishes. Clearly therefore, the early life-history represents a significant bottleneck in the lifecycle of coral reef fishes.

For coral reef butterflyfishes, larval densities (both in the plankton and during settlement) are typically very low (Leis, 1989) perhaps more so than for any other major family of coral reef fishes. Larval butterflyfishes are very conspicuous as they are relatively large compared to most other larval reef fishes and most species have a specialised larval form (tholichthys; Burgess, 1978) distinguished by a heavily armoured head-shield (Fig. 13.1). Moreover, larval butterflyfishes largely reflect the distinct colouration of adult butterflyfishes, making them easy to identify to species. Despite this, there are few records of larval butterflyfishes in collections of larval fishes (e.g., Leis, 1989), and even where they are reported, butterflyfishes consistently represent only a very small fraction (typically, <0.1%) of larval fishes caught. However, it is unknown whether the common techniques used to capture larvae (e.g., light traps and crest nets) consistently under-sample butterflyfishes, or whether the limited representation of butterflyfishes in larval samples is truly reflective of their limited abundance in pelagic and near-reef waters (Choat et al., 1993).

Fig. 13.1 Pre-settlement larvae (tholichthys) of *Chaetodon plebeius*, showing the conspicuous head shield, which is characteristic of larval butterflyfishes. Photo by C. Wen.

Colour image of this figure appears in the colour plate section at the end of the book.

Despite their apparent rarity in larval samples, butterflyfishes have featured prominently in studies on the larval ecology of coral reef fishes. For example, Almany et al. (2007) selected a species of butterflyfish (*Chaetodon vagabundus*) as a model species to examine the role of self-recruitment in structuring pelagically spawning reef fish communities. In addition, butterflyfishes have featured prominently in studies examining larval swimming behaviour. For example, Stobutzki and Bellwood (1997) compared the swimming speed and duration of 56 larval butterflyfishes (encompassing 10 different species) with a range of other coral reef fish larvae. While the swimming ability (maximum distance swum) was high relative to the body size of larval butterflyfishes, there were several other families (Acanthuridae, Lethrinidae and Lutjanidae) that were capable of much longer periods of sustained swimming (Stobutzki and Bellwood, 1997).

The potential dispersal of larval fishes may depend upon the planktonic larval duration (PLD), which varies greatly within and among families of coral reef fishes (Bay et al., 2006). Published estimates of the PLD recorded for coral reef butterflyfishes range from 20 to 56 days (Table 13.1). While there is an apparent relationship between PLD and geographic range of coral-reef butterflyfishes (Fig. 13.2), larval fishes may often settle on the same reef where they were originally spawned (self-recruitment; Jones et al., 1999). Almany et al. (2007) showed that self-recruitment is possible, if

Table 13.1 Estimates of planktonic larval duration (PLD) for coral reef butterflyfishes.

Species	**PLD (days)**	**Reference**
Chaetodon aureofasciatus	20–43	Leis and Carson-Ewart, 2003
Chaetodon auriga	40–53	Leis, 1989
	47	Almany, 2004
Chaetodon baronessa	38	Leis, 1989
Chaetodon collare	43	Leis, 1989
Chaetodon flavirostris	45	Almany, 2004
Chaetodon kleinii	56	Leis, 1989
Chaetodon lineolatus	54	Leis, 1989
Chaetodon lunulatus	39.7 (4.8 SE)	Soeparno et al., 2012
Chaetodon miliaris	32–36	Leis, 1989
Chaetodon plebeius	21–56	Leis and Carson-Ewart, 2003
	21–56	Leis, 1989
	39	Brothers et al., 1983
	45.2	Fowler 1989
Chaetodon rainfordi	20–43	Leis, 1989
	27.8	Fowler, 1989
	35	Brothers et al., 1983
Chaetodon speculum	26	Leis, 1989
	28.9 (4.0 SE)	Soeparno et al., 2012
Chaetodon trifascialis	30–34	Leis, 1989
	39.7 (4.8 SE)	Soeparno et al., 2012
Chaetodon unimaculatus	35	Leis, 1989
Chaetodon vagabundus	38	Almany et al., 2007
	42.3 (1.6 SE)	Soeparno et al., 2012
Chelmon rostratus	26–28	Brothers et al., 1983
	23.2	Fowler, 1989
Forcipiger flavissimus	41–57	Brothers and Thresher, 1985
Heniochus diphreutes	30–34	Brothers and Thresher, 1985

not common, for *C. vagabundus*, despite having a PLD in excess of 30 days. Berumen et al. (2012a) continued this work with *C. vagabundus*, finding that self-recruitment appears to be persistent through time. This same study expanded spatially to include documentation of successful dispersal of larvae to reefs over 30 km from their natal reef. Such dispersal away from natal reefs is important in maintaining genetic exchange among meta-populations, colonising new habitats, or recolonising areas following local extirpations (Planes et al., 2009; Berumen et al., 2012a).

For coral reef butterflyfishes that do settle on their natal reef, it remains unclear how far the larvae actually disperse before they start swimming

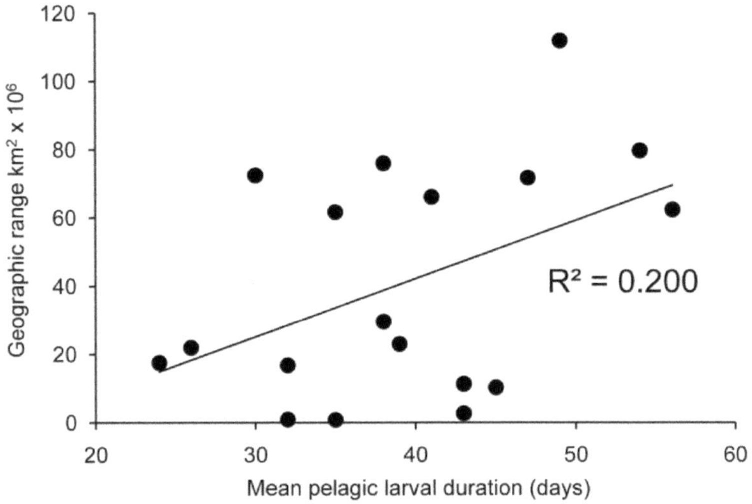

Fig. 13.2 Relationship between the maximum planktonic larval duration (PLD) and geographic range for 18 species of coral reef butterflyfishes. Geographic range was estimated based on the extent of occurrence, following Jones et al. (2002).

home. Coral reef fishes begin their pelagic phase as weak ineffective swimmers, which are largely at the mercy of localised hydrodynamic regimes, but gradually develop into very strong effective swimmers that have considerable control over where they settle (Leis and McCormick, 2002). One possibility is that coral reef butterflyfishes use their extensive swimming abilities and acute sensory abilities to minimise the distance travelled during the pelagic larval stage, e.g., by holding their position within on-coming currents adjacent to their natal reef (Leis et al., 2007). However, exact mechanisms that allow larval butterflyfishes to remain in close proximity to and/ or detect their natal reefs remain unknown (Almany et al., 2007).

Limited insights into the feeding ecology of larval reef fishes have been obtained by studying gut contents of larval fishes caught using plankton nets (Sampey et al., 2007). Sampey et al. (2007) examined gut contents for 591 larval fishes from 50 different families, including two individual butterflyfishes. While most larval fishes feed extensively on copepods, which are the most common zooplankton, butterflyfishes appeared to specialise on chaetognaths (Sampey et al., 2007). As butterflyfishes are likely to continue featuring in many larval ecology studies, further work is required to understand similarities and differences in the larval biology of butterflyfishes versus other coral reef fishes.

RANGE SHIFTS

Historically, the abundance and diversity of butterflyfish populations have been quantified against the structure and diversity of the underlying benthic characteristics (Findley and Findley, 2001; Öhman et al., 1998; Bozec et al., 2005). The majority of this research has focused on the importance of particular benthic characteristics (i.e., live coral cover and/or particular coral species/growth forms) in structuring butterflyfish communities (Öhman et al., 1998; Pratchett et al., 2004). More recently, this work has been utilised to identify how butterflyfish, and tropical reef fish communities as a whole, may respond to changes in habitat associated with climate-induced degradation (e.g., coral bleaching, outbreaks of crown-of-thorns starfish, disease, storms). While the loss of live coral and subsequent collapse of structural complexity are having dramatic effects on the structure and function of coral reef fish communities (Pratchett et al., 2008; Graham et al., 2009; Wilson et al., Chapter 9), there is increasing awareness that changes in physical oceanographic factors, in parallel with changes in benthic habitat characteristics, will be important in structuring tropical fish communities. Most notably, rising sea surface temperatures are projected to affect the life histories and performance of individual fishes (i.e., growth, mortality, and reproduction), leading to shifts in the size and composition of communities, and ultimately the structure and functioning of entire ecosystems (e.g., Holbrook et al., 1997; Portner and Knust, 2007; Dulvy et al., 2008; Munday et al., 2008; Last et al., 2011). While much of the evidence for the effects of increasing sea surface temperature on fishes is derived from temperate systems (e.g., Perry et al., 2005; Neuheimer et al., 2011), there has been an increasing interest in examining the demographic and physiological effects of climate change on tropical fishes, including butterflyfishes (Nilsson et al., 2009; Johansen and Jones, 2011). Collectively, this research is showing that predicted increases in seawater temperatures may lead to elevated growth rates, earlier maturation, reduced longevity, reduced aerobic performance, and poleward shifts in population distributions.

Projected changes in ocean climate have led to suggestions that subtropical (i.e., marginal) coral reefs may provide a 'habitat refuge' from the impacts of climate change for coral reef fish communities. This may be especially relevant for butterflyfish communities, as they exhibit considerable overlap in community structure between tropical and subtropical reef systems, and have relatively long PLDs giving them considerable dispersal potential. Research to date has focused on how the structure and function of butterflyfish populations differ between tropical and marginal reef systems, and has shown that while considerable overlap in species presence occurs between these systems, marginal reefs typically have lower butterflyfish abundance and diversity, with obligate corallivores under-represented on marginal reefs (Feary et al., 2010; Pratchett et al., 2013). Although differences in butterflyfish community structure between marginal and tropical reefs have been predominantly associated with differences in the cover and composition of coral communities, the potential for butterflyfish populations to be susceptible to extremes in oceanographic factors (i.e., temperature, salinity) has also been argued (Feary et al., 2010). Subtropical reefs, while providing a potential habitat refuge, can experience greater seasonal variation in oceanographic factors than their lower-latitude counterparts (Riegl et al., 2011; Feary et al., 2012). In particular, seasonal variations in seawater temperature of up to 12°C are relatively common on high latitude reefs (i.e., Japan 16–28°C; Lord Howe Island: 17–27°C; Gulf of Oman: 22–32°C), and exceed 20°C on the marginal reefs of the Persian Gulf (15–36°C: Feary et al., 2012; Riegl and Purkis, 2012a). There is a growing field of research focusing on butterflyfishes that examines how such extremes in oceanographic variables affect the resource acquisition (i.e., feeding rates, resource use and competition) (e.g., Pratchett et al., 2013) and energy demands and physiology (e.g., metabolic rates and swimming abilities) of butterflyfishes on these marginal reefs (D.A. Feary, unpublished data). This work will not only provide a mechanistic basis for the factors that are shaping butterflyfish communities on these marginal reefs, but will provide important insights into how butterflyfishes and other reef fish families may acclimate or adapt to environmental extremes (Feary et al., 2012). In particular, the marginal reefs of the Persian Gulf provide a unique system to investigate the potential ability of fishes to adapt to temperatures that are forecast for tropical reefs in the next 50–100 years (Riegl et al., 2011; Riegl and Purkis, 2012b).

Climate change is resulting in the extension of tropical fishes from both tropical and subtropical regions into temperate environments (Booth et al., 2007; Hirata et al., 2011; Soeparno et al., 2012; Feary et al., 2013). Despite this, there is still little understanding of the species-specific traits that may constrain or promote successful establishment of populations in temperate regions (Feary et al., 2013). There is increasing work illustrating that a distinct subset of butterflyfishes may potentially be showing substantial shifts in their latitudinal distribution (Booth et al., 2011; Feary et al., 2013). Work within south-eastern Australia (Booth et al., 2007; Figueira and Booth 2010; Feary et al., 2013), Western Australia (Hutchins and Pearce, 1994), Japan (Hirata et al., 2011; Soeparno et al., 2012) and the eastern US (McBride and Able, 1988; McBride, 1996) have all shown that butterflyfishes can form a numerically dominant component of seasonally abundant tropical fish communities within temperate environments. As such, there is now increasing impetus in examining butterflyfish species to determine the importance of larval characteristics (e.g., size at settlement, planktonic larval duration, pelagic swimming ability) that may augment or moderate latitudinal distributional shifts (Feary et al., 2013). Such work is forming a vital foundation for understanding the demographic factors within tropical fish species that may be important in constraining or enhancing latitudinal shifts in species populations.

Although seasonal incursions of tropical fishes into temperate reef environments are increasing in frequency and magnitude, there is still little information on the demographic consequences of such shifts. Current research focusing on tropical butterflyfish species, is examining how post-settlement demographic population parameters (e.g., growth rate) may enhance individual, and thus population, success within temperate environments (D.A. Feary, unpublished data; Y. Nakamura, unpublished data). This work, albeit in its infancy, is showing that species-specific phenotypic plasticity in post-settlement development may be vital in species recruitment and survival within temperate environments. The role of other post-settlement processes in structuring tropical reef fish communities, including competition for suitable resources, species-specific differences in reproduction, and differential effects of temperate predation, are important areas of future research for understanding how butterflyfish communities may persist in temperate environments (Feary et al., 2013). Such work will provide researchers with the potential suite of mechanisms that may enhance settlement, recruitment and long-term survival of tropical reef fishes within temperate habitats.

DEMOGRAPHIC CONSEQUENCES OF FEEDING ECOLOGY

Demographic and life history characters of coral reef fishes can vary both within and among species (e.g., Robertson et al., 2005), and this is an important consideration with respect to species conservation and management. Life history and demographic parameters are likely to vary with the density of conspecifics or other competitors (Gust et al., 2002), predation risk (Stallings, 2008), and resource availability (Berumen and Pratchett, 2008), but these factors may be difficult to control (or even measure) for most species when conducting large-scale comparisons (e.g., Crossman et al., 2005). The effects of resources on demographics of coral reef fishes in particular has been poorly understood, primarily due to the problem of quantifying and qualifying the resources available for most fishes at scales relevant to individuals and subpopulations. Ecological variation among coral reef butterflyfishes may therefore provide an important opportunity to better understand the effects of dietary composition and ecological specialisation on species-specific life histories.

Foraging theory has long recognised that animals should select the resources that are most beneficial or more specifically, those resources which maximise individual fitness (e.g., Pyke, 1984). In most cases, it is very difficult to demonstrate the long-term consequences of foraging choices on life history or demographic traits, let alone lifetime reproductive success. However, recent advances in genetic parentage analysis, used to assess self-recruitment (as described above), allows matching individual juvenile fishes to their parents from a pool of potential parents (Planes et al., 2009). In principle, this approach could be used to assess the relative fitness of different individuals and relate this back to specific biological traits, such as dietary composition, habitat or territory size.

Despite the extensive scientific interest in butterflyfishes, there is a notable paucity of data on key demographic rates (e.g., growth) and life history traits for this family. The best resource for maximum size data may be identification guides that are specific to the family Chaetodontidae (e.g., Allen et al., 1998; Kuiter, 2002). Most age-based studies for butterflyfishes have been conducted on a Hawaiian endemic species, *Chaetodon miliaris* (Ralston, 1976a). Other life history traits for *C. miliaris* have also been extensively studied (e.g., Ralston, 1976b, 1981; Stimson, 1990; DeMartini and Friedlander, 2004). Stimson (1990) showed that recruitment rates for *C. miliaris* were inversely related to adult densities, suggesting that there is intra-specific competition among ontogenetic stages and that butterflyfish populations are resource rather than recruitment limited (see also Harmelin-Vivien, 1989). In Australia, Fowler (1988, 1989) investigated the growth of three butterflyfishes from the lagoon of One Tree Island on the Great Barrier Reef, but none of these fish were more than a year old. These studies provided some of the earliest estimates of planktonic larval duration, as well revealing very rapid early post-settlement growth across three different species (*Chaetodon rainfordi*, *Chaetodon plebeius* and *Chelmon rostratus*).

Butterflyfishes display a typical acanthuroid growth form (*sensu* Choat and Robertson, 2002) with size and age decoupled after an initial rapid period of growth. This means that size is not necessarily indicative of age beyond the juvenile stage. Berumen (2005) established size-at-age curves for four species from Lizard Island, Great Barrier Reef, and found that the fishes achieved 92% of their maximum size within the first two years of their life. Despite their relatively small size, many of these individuals were found to be living more than a decade (Berumen et al., 2012b). Zekeria et al. (2006) provided the first size-at-age information for a Red Sea species, *Chaetodon larvatus*. This study likewise found rapid initial growth followed by near-asymptotic growth rate after the juvenile phase; several individuals were found to be more than a decade old and one individual was determined to be 14 years old.

Berumen et al. (2012b) presented the first comparative study of butterflyfish life history characters, sampling four species (*Chaetodon citrinellus, C. lunulatus, C. melannotus*, and *C. trifascialis*) in both the southern and northern Great Barrier Reef. Only one of the four species (*C. melannotus*) conformed to predictions that individuals from higher latitudes will grow faster than conspecifics at lower latitudes (see Kingsolver and Huey, 2008). Also *C. melannotus* was the only species that exhibited greater longevity at the higher latitude location, which has been shown for other reef fishes (e.g., Robertson et al., 2005; Trip et al., 2008). For butterflyfishes, trophic resource use is likely to vary greatly, and disentangling the role that food resources play (vs. other environmental factors) in shaping life history characters remains a major challenge (Berumen et al., 2012b). In this respect, butterflyfishes, particularly corallivores with conspicuous and quantifiable food resources, offer an ideal system in which these life history theories may be further tested.

Increasing effects of global climate change present a challenge for managers in the near future. For reef fishes in particular, climate change poses complex problems. As poikilotherms, reef fishes are likely to be impacted by new temperature ranges and alteration in the seasonal patterns they are exposed to. Increased intensity of storms may create even further patchiness in the success of recruitment events. In addition to direct effects on the fishes themselves, climate change will simultaneously alter the very reef structure that these fishes rely on for shelter, habitat, and (particularly in the case of butterflyfishes) food. Understanding the factors that influence life histories or demographic characters of reef fish populations will better help us to prepare for and manage the impacts of climate change on reef fish populations.

NEUROBIOLOGY AND SOCIALITY

Coral reef fishes exhibit among the most diverse forms of sociality of all vertebrates. When categorised by intraspecific behavioural interactions (Barlow, 1974), at least eight distinct social systems are present in coral reef fishes (Sale, 1978). These social systems include solitary living, pairs, feeding and resting aggregations, schools, migratory spawning aggregations, territorial leks, and hierarchical groups (Sale, 1978). When overlaid on a detailed phylogeny, hypotheses relevant to the proximate and ultimate mechanisms of sociality can be proposed. Coral reef fish families in which closely related species exhibit diverse social systems, such as the butterflyfishes (Yabuta and Berumen, Chapter 8), are powerful tools to test these hypotheses.

There is substantial support for hypotheses that link environmental conditions, specifically spatio-temporal resource availability and predictability, to specific reef fish socialites (Allen, 1972; Wilson, 1975; Hourigan, 1989; van Rooji, 1996; Overholtzer and Motta, 1999; Thompson et al., 2007; Gold et al., 2011). For example, food distribution is highly correlated to mobility and group size with the butterflyfishes (Hourigan, 1989). Because coral availability can be limited, but temporally and spatially stable, it can be defended from other conspecifics. Thus, site attachment to, and defence of, coral feeding grounds is a common feature of corallivorous species (Hourigan, 1989). Competition for feeding territory and mates favours the formation of cooperative territorial pair bonds in which males predominantly defend territories, thereby allowing inhabiting females to maximise foraging efficiency—an action that is likely to favour both female and male reproductive output (Schoener, 1971; Hourigan, 1989). Comparatively, territory and resource defence appears to be less advantageous for planktivorous butterflyfishes, due to the high spatio-temporal patchiness of plankton availability. Rather, planktivorous butterflyfishes tend to form large feeding aggregations and display little intraspecific aggression (e.g., Ralston, 1981); such social traits are thought to be adaptive due to the high potential for predation during plankton consumption in the water column (Hourigan, 1989).

In contrast to our understanding of ultimate explanations, proximate, neurobiological explanations for the maintenance and evolution of reef fish sociality are severely lacking. There are several reasons for this. Firstly, behaviourally relevant neural circuits that have been identified in genetically tractable species (e.g., mice) cannot be critically examined in teleost reef fishes. This is because homologies between the forebrains of tetrapods and teleosts are yet to be fully verified, due to developmental differences (Braford, 1995, 2009; Northcutt, 1995). Secondly, even if homologies between tetrapods and teleosts were verified, detailed neuroanatomical descriptions are rare for teleosts, and are virtually absent in coral reef teleosts. Thirdly, in many reef fish families, thoroughly described social systems are available for few species (Yabuta and Berumen, Chapter 8), thereby impeding examination of closely related species that differ in sociality, within the context of phylogenetic history.

The diversity of social systems present in closely related vertebrate species indicates that the evolution and maintenance of these behaviours is regulated by a mechanism that can be modulated over short time periods. Thus, it is unlikely that neural circuits of behaviour are 'rewired' to elicit species-specific behaviour. Instead, neuromodulators may act on conserved neural circuits to differentially regulate social behaviour. Evidence from research in tetrapods suggests that the nonapeptides isotocin (IT) or oxytocin (mammalian homolog) and arginine vasotocin (AVT) or vasopressin (mammalian homolog), may play a central role in modulating various pro-social and antagonistic behaviours, perhaps in a species-specific manner

(Goodson and Bass, 2001). For example, these nonapeptides promote social recognition (Dantzer et al., 1988), preference for species-typical group size and familiar conspecifics (Goodson et al., 2009), the formation and maintenance of social bonds (Young et al., 2011), group member guarding (Young et al., 2011), cooperation within and competition between groups (De Dreu, 2012), and aggression (Bosch and Neuman, 2012).

As these nonapeptides are highly conserved across vertebrates (Archer and Chauvet, 1995), they represent a promising proximate mechanism for the regulation of social behaviour in reef fishes. Preliminary investigations have demonstrated the role of IT/AVT in inhibiting conspecific social approach (Thompson and Walton, 2004) and interspecific cooperation (Soares et al., 2012) as well as promoting shoal preference (Braida et al., 2012), the formation of mate affiliation and non-mate aggression (Oldfield and Hofmann, 2011), territorial aggression (Santangelo and Bass, 2006; Kleszczyńska et al., 2012), parental care (O'Connell et al., 2012), and intragroup submission (Reddon et al., 2012). In butterflyfishes, the density of AVT immunoreactive varicosities (potential AVT release sites) within the ventral nucleus of the ventral telencephalon was strongly related to sociality (pair bonding vs. shoaling and harems) among 7 species of butterflyfishes (Dewan et al., 2011).

Despite some recent advances, many fundamental questions about the neurobiology of reef fish social behaviour and sociality remain to be addressed. In particular, further research is needed to clarify the socio-functional roles of isotocin in reef fishes, and the extent to which they relate to those of mesotocin in birds, and oxytocin in mammals. Furthermore, the brain regions that are modulated by AVT/IT in order to elicit species-specific behaviour need to be elucidated. Butterflyfishes are an excellent model system for testing proximate, neurobiological hypotheses for the evolution, divergence, and maintenance of species-specific sociality. Within this family, species exhibit diverse social groupings (e.g., solitary, pairs, trios, or shoals), mating systems (e.g., monogamous, polygamous, haremic), and conspecific behaviour (e.g., territorial/highly aggressive, nonterritorial/nonaggressive) (Hourigan, 1989; Ludwig, 1984; Ralston, 1981; Walsh, 1987; Yabuta and Berumen, Chapter 8). Furthermore, the protracted spawning season of many species (Tricas and Hiramoto, 1989; Ralston, 1981) and the lack of parental care yield adult social behaviours that are temporally stable. These diverse but consistent social behaviours can be mapped onto a robust molecular and morphological phylogeny (Fessler and Westneat, 2007; Smith et al., 2003). By using a comparative approach, hypotheses relevant to the proximate mechanisms of behaviour can be tested whilst controlling for evolutionary history. For example, Dewan et al. (2011) analysed the density and distribution of AVT across 7 butterflyfish species within 4 phylogenetic clades. This study provided evidence that the density of AVT release sites was a strong predictor of butterflyfish social behaviour and may function in a similar way across vertebrates.

Butterflyfishes are also useful for examining neurobiological mechanisms of intraspecific social behaviour. Species within this family exhibit highly characterised behavioural displays in response to conspecifics and other food competitors (Hourigan, 1989; Strang, 2005; Yabuta, 1999, 2000, 2002) that can be easily observed in wild populations and are maintained in laboratory settings (A.K. Dewan, personal observation). Butterflyfishes also communicate with visual, acoustic and hydrodynamic signals, and these forms of communication are also maintained in the laboratory setting (Tricas et al., 2006; Boyle and Tricas, 2010). This rich, and more importantly, well characterised social behaviour provides a strong foundation for studies focused on proximate mechanisms of behavioural modulation. For example, the frequency of individual visual displays of aggression was recently correlated to features of the AVT system in a territorial butterflyfish (Dewan and Tricas, 2011).

While the behavioural and ecological literature has provided a strong foundation for the utility of a butterflyfish model system for reef fish socio-neurobiological examination, several key aspects need to be included. Firstly, evolutionary analyses using butterflyfishes are dependent upon a well-established phylogenetic outgroup. Unfortunately, the closest extant relatives of butterflyfishes, the scats (Scatophagidae) and angelfishes (Pomacanthidae) or more distantly related kyphosids (Kyphosidae) and microcanthids (Microcanthinae) have not received much research. Additional behavioural, ecological, and neuroanatomical studies on these fishes are necessary. Secondly, the ability to uncover neural circuits of behaviour and test for homologous brain mechanisms across species is dependent upon well-characterised neuroanatomy. Ongoing neuroanatomical descriptions will allow for manipulation of behavioural circuits with neuroanatomical precision. Moreover, as the brain organisation of butterflyfishes is relatively well conserved (Dewan et al., 2011), these neuroanatomical descriptions should be applicable to all members of this speciose family. When combined with previously established behavioural and ecological knowledge, addressing these knowledge gaps will establish butterflyfishes as the most powerful model system for analysing proximate, neurobiological mechanisms for the evolution, divergence, and maintenance of reef fish sociology.

CONCLUSION AND FUTURE DIRECTIONS

Although butterflyfishes are among the best-studied family of coral reef fishes, there remain considerable gaps in our knowledge of their biology and ecology. Most notably, the demography and reproductive biology of all butterflyfishes has been very poorly studied. There is also a strong taxonomic bias in all previous research, whereby there are entire genera (especially, non-reefal and deep-water clades) for which almost nothing is known. Technological advances in recent years have facilitated increased research and a much greater appreciation of mesophotic reefs, which is likely to lead to significant advances and profound changes in established knowledge of the abundance, distribution, habitat associations and feeding ecology of butterflyfishes from deep water habitats (e.g., Bridge et al., 2011, 2012). This is particularly important for butterflyfishes (e.g., *Prognathodes*) that are generally only found in deep water (>30 m depth), but it is also likely that this research will extend known geographical ranges and diversity in the ecology of species that to date, have only been studied in shallow water reef habitats.

There are also many new and emerging topics of research in which butterflyfishes will play a key role. In particular, butterflyfishes have been and will remain at the centre of research on responses of coral reef fishes to ongoing climate change (e.g., Pratchett et al., 2008). Climate change is widely regarded as the most significant threat to the future structure and functioning of coral reef ecosystems, largely due to projected changes in the composition and abundance of habitat-forming corals (Hoegh-Guldberg et al., 2007). These changes in habitat structure will clearly affect coral-dependent and reef-associated fishes (e.g., Berumen and Pratchett, 2008), but the specific responses to both local and large-scale coral loss are poorly understood (Crosby et al., Chapter 10). At localised scales, behavioural responses to resource depletion by individual butterflyfishes (including, shifts in dietary composition and changes in the intensity of inter-specific competition) need to be considered, while broad-scale species level responses (e.g., range shifts, larval dispersal and meta-population structure) are important for establishing risk of extinction. Importantly, significant research on butterflyfishes (as captured in the current monograph) has contributed greatly to understanding of reef fish ecology, but there is clear and definite need for much more research, and butterflyfishes are a very rewarding study group of coral reef fishes.

REFERENCES

Acher, R. and J. Chauvet. 1995. The neurohypophysial endocrine regulatory cascade: precursors, mediators, receptors, and effectors. *Frontiers in Neuroendocrinology* 16: 237–289.

Allen, G.R. 1972. *The Anemonefishes: Their classification and biology*, T.F.H. Publications, Neptune, USA.

Allen, G.R., R. Steene and M. Allen. 1998. *A guide to angelfishes and butterflyfishes*, Odyssey Publishing, Perth, Australia.

Almany, G.R., M.L. Berumen, S.R. Thorrold, S. Planes and G.P. Jones. 2007. Local replenishment of coral reef fish populations in a marine reserve. *Science* 316: 742–744.

Barlow, G.W. 1974. Contrasts in Social Behavior between Central American Cichlid Fishes and Coral-reef Surgeon Fishes. *American Zoologist* 14: 9–34.

Bay, L.K., K. Buechler, M. Gagliano and M.J. Caley. 2006. Intraspecific variation in the pelagic larval duration of tropical reef fishes. *Journal of Fish Biology* 68: 1206–1214.

Bellwood, D.R., L. van Herwerden and N. Konow. 2004. Evolution and biogeography of marine angelfishes (Pisces: Pomacanthidae). *Molecular Phylogenetics and Evolution* 33: 140–155.

Berumen, M.L. 2005. The importance of juveniles in modelling growth: butterflyfish at Lizard Island. *Environmental Biology of Fishes* 72: 409–413.

Berumen, M.L. and M.S. Pratchett. 2008. Trade-offs associated with dietary specialization in corallivorous butterflyfishes (Chaetodontidae: *Chaetodon*). *Behavioral Ecology and Sociobiology* 62: 989–994.

Berumen, M.L., H.J. Walsh, N. Raventos, S. Planes, G.P. Jones, V. Starczak and S.R. Thorrold. 2010. Otolith geochemistry does not reflect dispersal history of clownfish larvae. *Coral Reefs* 29: 883–891.

Berumen, M.L., G.R. Almany, S. Planes, G.P. Jones, P. Saenz-Agudelo and S.R. Thorrold. 2012a. Persistence of self-recruitment and patterns of larval connectivity in a marine protected area network. *Ecology and Evolution* 2: 444–452.

Berumen, M.L., E.D.L. Trip, M.S. Pratchett and J.H. Choat. 2012b. Differences in demographic traits of four butterflyfish species between two reefs of the Great Barrier Reef separated by 1,200 km. *Coral Reefs* 1–9.

Booth, D.J., W.F. Figueira, M.A. Gregson, L. Brown and G. Beretta. 2007. Occurrence of tropical fishes in temperate southeastern Australia: Role of the East Australian Current. *Estuarine, Coastal and Shelf Science* 72: 102–114.

Booth, D.J., N. Bond and P. Macreadie. 2011. Detecting range shifts among Australian fishes in response to climate change. *Marine and Freshwater Research* 62: 1027–1042.

Bosch, O.J. and I.D. Neumann. 2012. Both oxytocin and vasopressin are mediators of maternal care and aggression in rodents: From central release to sites of action. *Hormones and Behavior* 61: 293–303.

Boyle, K.S. and T.C. Tricas. 2010. Pulse sound generation, anterior swim bladder buckling and associated muscle activity in the pyramid butterflyfish, *Hemitaurichthys polylepis*. *The Journal of experimental biology* 213: 3881–3893.

Bozec, Y.M., S. Doledec and M. Kulbicki. 2005. An analysis of fish habitat associations on disturbed coral reefs: chaetodontid fishes in New Caledonia. *Journal of Fish Biology* 66: 966–982.

Braford, Jr., M.R. 1995. Comparative aspects of forebrain organization in the ray-finned fishes: touchstones or not? *Brain, Behavior and Evolution* 46: 259–274.

Braford, Jr., M.R. 2009. Stalking the everted telencephalon: comparisons of forebrain organization in basal ray-finned fishes and teleosts. *Brain, Behavior and Evolution* 74: 56–76.

Braida, D., A. Donzelli, R. Martucci, V. Capurro, M. Busnelli, B. Chini and M. Sala. 2012. Neurohypophyseal hormones manipulation modulate social and anxiety-related behavior in zebrafish. *Psychopharmacology* 220: 319–330.

Bridge, T.C.L., T.J. Done, R.J. Beaman, A. Friedman, S.B. Williams, O. Pizarro and J.M. Webster. 2011. Topography, substratum and benthic macrofaunal relationships on a tropical mesophotic shelf margin, central Great Barrier Reef, Australia. *Coral Reefs* 30: 143–153.

Bridge, T.C.L., K.E. Fabricius, P. Bongaerts, C.C. Wallace, P.R. Muir, T.J. Done and J.M. Webster. 2012. Diversity of Scleractinia and Octocorallia in the mesophotic zone of the Great Barrier Reef, Australia. *Coral Reefs* 31: 179–189.

Brothers, E.B. and R.E. Thresher. 1985. Pelagic duration, dispersal, and the distribution of Indo-Pacific coral reef fishes. *The Ecology of Coral Reefs* 3: 53–69.

Brothers, E.B., D.M.B. Williams and P.F. Sale. 1983. Length of larval life in twelve families of fishes at "One Tree Lagoon", Great Barrier Reef, Australia. *Marine Biology* 76: 319–324.

Burgess, W.E. 1978. *Butterflyfishes of the world*, T.F.H. Publications, Neptune City, USA.

Choat, J.H. and D.R. Bellwood. 1991. Reef fishes: their history and evolution. In *The Ecology of Fishes on Coral Reefs*, P.F. Sale (ed.). Academic Press, Inc., San Diego, USA pp. 39–66.

Choat, J.H. and D.R. Roberston. 2002. Age-Based Studies. *In: Coral reef fishes: dynamics and diversity in a complex ecosystem*, P.F. Sale (ed.). Academic Press, San Diego, USA pp. 57–80.

Choat, J.H., P.J. Doherty, B.A. Kerrigan and J.M. Leis. 1993. A comparison of towed nets, purse seine, and light-aggregation devices for sampling larvae and pelagic juveniles of coral reef fishes. *Fishery Bulletin* 91: 195–209.

Coker, D.J., N.A.J. Graham and M.S. Pratchett. 2012. Interactive effects of live coral and structural complexity on the recruitment of reef fishes. *Coral Reefs* 31: 919–927.

Cole, A.J., M.S. Pratchett and G.P. Jones. 2008. Diversity and functional importance of coral-feeding fishes on tropical coral reefs. *Fish and Fisheries* 9: 286–307.

Crossman, D.J., J.H. Choat and K.D. Clements. 2005. Nutritional ecology of nominally herbivorous fishes on coral reefs. *Marine Ecology Progress Series* 296: 129–142.

Dantzer, R., G.F. Koob and M. Le Moal. 1988. Septal vasopressin modulates social memory in male rats. *Brain Research* 457: 143–147.

De Dreu, C.K.W. 2012. Oxytocin modulates cooperation within and competition between groups: An integrative review and research agenda. *Hormones and Behavior* 61: 419–428.

De-ath, G., K.E. Fabricius, H. Sweatman and M. Puotinen. 2012. The 27-year decline of coral cover on the Great Barrier Reef and its causes. *Proceedings of the National Academy of Sciences* 109: 17995–17999.

DeMartini, E.E. and A.M. Friedlander. 2004. Spatial patterns of endemism in shallow-water reef fish populations of the Northwestern Hawaiian Islands. *Marine Ecology Progress Series* 271: 281–296.

Dewan, A.K. and T.C. Tricas. 2011. Arginine vasotocin neuronal phenotypes and their relationship to aggressive behavior in the territorial monogamous multiband butterflyfish, *Chaetodon multicinctus*. *Brain Research* 1401: 74–84.

Dewan, A.K., M.L. Ramey and T.C. Tricas. 2011. Arginine vasotocin neuronal phenotypes, telencephalic fiber varicosities, and social behavior in butterflyfishes (Chaetodontidae): potential similarities to birds and mammals. *Hormones and Behavior* 59: 56–66.

Dulvy, N.K., S.I. Rogers, S. Jennings, V. Stelzenmüller, S.R. Dye and H.R. Skjoldal. 2008. Climate change and deepening of the North Sea fish assemblage: a biotic indicator of warming seas. *Journal of Applied Ecology* 45: 1029–1039.

Elmqvist, T., C. Folke, M. Nyström, G. Peterson, J. Bengtsson, B. Walker and J. Norberg. 2003. Response diversity, ecosystem change, and resilience. *Frontiers in Ecology and the Environment* 1: 488–494.

Feary, D.A., J.A. Burt, A.G. Bauman, P. Usseglio, P.F. Sale and G.H. Cavalcante. 2010. Fish communities on the world's warmest reefs: what can they tell us about the effects of climate change in the future? *Journal of Fish Biology* 77: 1931–1947.

Feary, D.A., J.A. Burt, G.H. Cavalcante and A.G. Bauman. 2012. Extreme Physical Factors and the Structure of Gulf Fish and Reef Communities. In *Coral Reefs of the Gulf*, B.M. Riegl and S.J. Purkis (eds.). Springer Netherlands pp. 163–170.

Feary, D.A., W.F. Figueira, M.S. Pratchett, E. Emslie, A.J. Fowler, O. De Luiz and Y. Nakamura. 2013. Latitudinal shifts in coral reef fishes: why some species do and other don't. *Fish and Fisheries*. DOI: 10.1111/faf.12036.

Ferry-Graham, L.A., D.I. Bolnick and P.C. Wainwright. 2002. Using functional morphology to examine the ecology and evolution of specialization. *Integrative and Comparative Biology* 42: 265–277.

Fessler, J.L. and M.W. Westneat. 2007. Molecular phylogenetics of the butterflyfishes (Chaetodontidae): Taxonomy and biogeography of a global coral reef fish family. *Molecular Phylogenetics and Evolution* 45: 50–68.

Figueira, W.F. and D.J. Booth. 2009. Increasing ocean temperatures allow tropical fishes to survive overwinter in temperate waters. *Global Change Biology* 16: 506–516.

Findley, J.S. and M.T. Findley. 2001. Global, regional, and local patterns in species richness and abundance of butterflyfishes. *Ecological Monographs* 71: 69–91.
Fowler, A.J. 1988. Aspects of the population ecology of three species of chaetodonts at One Tree Reef, southern Great Barrier Reef. PhD dissertation, University of Sydney, Sydney, Australia.
Fowler, A.J. 1989. Description, interpretation and use of the microstructure of otoliths from juvenile butterflyfishes (family Chaetodontidae). *Marine Biology* 102: 167–181.
Fricke, H. and S. Fricke. 1977. Monogamy and sex change by aggressive dominance in coral reef fish. *Nature* 266: 830–832.
Fulton, C.J. 2007. Swimming speed performance in coral reef fishes: field validations reveal distinct functional groups. *Coral Reefs* 26: 217–228.
Fulton, C.J. 2010. The role of swimming in reef fish ecology. *In: Fish swimming: an etho-ecological perspective*, P. Domenici and B.G. Kapoor (eds.). Science Publishers, Enfield pp. 374–406.
Gold, J.R., G. Voelker and M.A. Renshaw. 2011. Phylogenetic relationships of tropical western Atlantic snappers in subfamily Lutjaninae (Lutjanidae: Perciformes) inferred from mitochondrial DNA sequences. *Biological Journal of the Linnean Society* 102: 915–929.
Goodson, J.L. and A.H. Bass. 2001. Social behavior functions and related anatomical characteristics of vasotocin/vasopressin systems in vertebrates. *Brain Research Reviews* 35: 246–265.
Goodson, J.L., S.E. Schrock, J.D. Klatt, D. Kabelik and M.A. Kingsbury. 2009. Mesotocin and nonapeptide receptors promote estrildid flocking behavior. *Science* 325: 862–866.
Graham, N.A.J., S.K. Wilson, M.S. Pratchett, N.V.C. Polunin and M.D. Spalding. 2009. Coral mortality versus structural collapse as drivers of corallivorous butterflyfish decline. *Biodiversity and Conservation* 18: 3325–3336.
Gust, N., J.H. Choat and J. Ackerman. 2002. Demographic plasticity in tropical reef fishes. *Marine Biology* 140: 1039–1051.
Halford, A., A.J. Cheal, D. Ryan and D.M. Williams. 2004. Resilience to Large-Scale Disturbance in Coral and Fish Assemblages on the Great Barrier Reef. *Ecology* 85: 1892–1905.
Harmelin-Vivien, M.L. 1989. Implications of feeding specialization on the recruitment processes and community structure of butterflyfishes. *Environmental Biology of Fishes* 25(1): 101–110.
Harmelin-Vivien, M.L. and Y. Bouchon-Navaro. 1983. Feeding diets and significance of coral feeding among chaetodontid fishes in Moorea (French Polynesia). *Coral Reefs* 2: 119–127.
Heatwole, S.J. and C.J. Fulton. 2013. Behavioural flexibility in coral reef fishes responding to a rapidly changing environment. *Marine Biology* 160: 677–689.
Hirata, T., S. Oguri, S. Hirata, H. Fukami, Y. Nakamura and K. Yamaoka. 2011. Seasonal changes in fish assemblages in an area of hermatypic corals in Yokonami, Tosa Bay, Japan. *Japanese Journal of Ichthyology* 58: 49–64.
Hoegh-Guldberg, O., P.J. Mumby, A.J. Hooten, R.S. Steneck, P. Greenfield, E. Gomez, C.D. Harvell, P.F. Sale, A.J. Edwards, K. Caldeira, N. Knowlton, C.M. Eakin, R. Iglesias-Prieto, N. Muthiga, R.H. Bradbury, A. Dubi and M.E. Hatziolos. 2007. Coral Reefs Under Rapid Climate Change and Ocean Acidification. *Science* 318: 1737–1742.
Holbrook, S.J., R.J. Schmitt and J.S. Stephens Jr. 1997. Changes in an assemblage of temperate reef fishes associated with a climate shift. *Ecological Applications* 7: 1299–1310.
Hourigan, T.F. 1989. Environmental determinants of butterflyfish social systems. *Environmental Biology of Fishes* 25: 61–78.
Hutchins, J. and A. Pearce. 1994. Influence of the Leeuwin Current on recruitment of tropical reef fishes at Rottnest Island, Western Australia. *Bulletin of Marine Science* 54: 245–255.
Johansen, J. and G. Jones. 2011. Increasing ocean temperature reduces the metabolic performance and swimming ability of coral reef damselfishes. *Global Change Biology* 17: 2971–2979.

Johansen, J.L., D.R. Bellwood and C.J. Fulton. 2008. Do coral reef fishes exploit flow refuges in high-flow habitats. *Marine Ecology Progress Series* 360: 219–226.

Jones, G.P., M.J. Milicich, M.J. Emslie and C. Lunow. 1999. Self-recruitment in a coral reef fish population. *Nature* 402: 802–804.

Jones, G.P., M.I. McCormick, M. Srinivasan and J.V. Eagle. 2004. Coral decline threatens fish biodiversity in marine reserves. *Proceedings of the National Academy of Sciences of the United States of America* 101: 8251–8253.

Jones, G.P., M.J. Caley and P.L. Munday. 2002. Rarity in coral reef fish communities. In *Coral reef fishes: dynamics and diversity in a complex ecosystem*, P.F. Sale (ed.) Academic Press, San Diego, USA pp. 81–101.

Kingsolver, J.G. and R.B. Huey. 2008. Size, temperature, and fitness: three rules. *Evolutionary Ecology Research* 10: 251–268.

Kleszczyńska, A., E. Sokołowska and E. Kulczykowska. 2012. Variation in brain arginine vasotocin (AVT) and isotocin (IT) levels with reproductive stage and social status in males of three-spined stickleback (*Gasterosteus aculeatus*). *General and Comparative Endocrinology* 175: 290.

Kuiter, R.H. 2002. *Butterflyfishes, Bannerfishes and their Relatives. A comprehensive Guide to Chaetodontidae and Microcanthidae*. The Marine Fish Families Series, TMC Publishing, Chorleywood, UK.

Last, P.R., W.T. White, D.C. Gledhill, A.J. Hobday, R. Brown, G.J. Edgar and G. Pecl. 2011. Long-term shifts in abundance and distribution of a temperate fish fauna: a response to climate change and fishing practices. *Global Ecology and Biogeography* 20: 58–72.

Lawton, R.J., V. Messmer, M.S. Pratchett and L.K. Bay. 2011. High gene flow across large geographic scales reduces extinction risk for a highly specialised coral feeding butterflyfish. *Molecular Ecology* 20: 3584–3598.

Leis, J.M. 1989. Larval biology of butterflyfishes (Pisces, Chaetodontidae): what do we really know? *Environmental Biology of Fishes* 25: 87–100.

Leis, J.M. and M.I. McCormick. 2002. *The Biology, Behavior and Ecology of the Pelagic, Larval Stage of Coral Reef Fishes*. Edited by P.F. Sale. Coral reef fishes: dynamics and diversity in a complex ecosystem, Academic Press, San Diego, USA.

Leis, J.M. and B.M. Carson-Ewart. 2003. Orientation of pelagic larvae of coral-reef fishes in the ocean. *Marine Ecology Progress Series* 252: 239–253.

Leis, J.M., K.J. Wright and R.N. Johnson. 2007. Behaviour that influences dispersal and connectivity in the small, young larvae of a reef fish. *Marine Biology* 153: 103–117.

Lighthill, M. 1969. Hydromechanics of aquatic animal propulsion. *Annual Review of Fluid Mechanics* 1: 413–446.

Ludwig, G.M. 1984. Constrasts in morphology and life history among Hawaiian populations of two longnose butterflyfishes, *Forcipiger longirostris* and *F. flavissimus*: a possible case of character displacement. PhD dissertation, University of Hawaii, Honolulu, USA.

McBride, R. 1996. On the Rarity of Banded Butterflyfish in the Mid Atlantic. *Underwater Naturalist* 23: 18–20.

McBride, R.S. and K.W. Able. 1998. Ecology and fate of butterflyfishes, *Chaetodon* spp., in the temperate, western North Atlantic. *Bulletin of Marine Science* 63: 401–416.

Montanari, S.R., L. van Herwerden, M.S. Pratchett, J.P.A. Hobbs and A. Fugedi. 2012. Reef fish hybridization: lessons learnt from butterflyfishes (genus Chaetodon). *Ecology and Evolution* 2: 310–328.

Motta, P.J. 1989. Preface. *Environmental Biology of Fishes* 25: 7–8.

Munday, P.L. 2002. Bi-directional sex change: testing the growth-rate advantage model. *Behavioral Ecology and Sociobiology* 52: 247–254.

Munday, P.L. and G.P. Jones. 1998. The ecological implications of small body size among coral-reef fishes. *Oceanography and Marine Biology: An Annual Review* 36: 373–411.

Munday, P.L., G.P. Jones, M.S. Pratchett and A.J. Williams. 2008. Climate change and the future for coral reef fishes. *Fish and Fisheries* 9: 261–285.

Neuheimer, A., R. Thresher, J. Lyle and J. Semmens. 2011. Tolerance limit for fish growth exceeded by warming waters. *Nature Climate Change* 1: 110–113.

Nilsson, G.E., N. Crawley, I.G. Lunde and P.L. Munday. 2009. Elevated temperature reduces the respiratory scope of coral reef fishes. *Global Change Biology* 15: 1405–1412.

Northcutt, R.G. 1995. The Forebrain of Gnathostomes: In Search of a Morphotype. *Brain, Behavior and Evolution* 46: 275–288.

O'Connell, L.A., B.J. Matthews and H.A. Hofmann. 2012. Isotocin regulates paternal care in a monogamous cichlid fish. *Hormones and Behavior* 61: 725–733.

Öhman, M.C., A. Rajasuriya and S. Svensson. 1998. The Use of Butterflyfishes (Chaetodontidae) as Bio-indicators of Habitat Structure and Human Disturbance. *Ambio* 27: 708–716.

Oldfield, R.G. and H.A. Hofmann. 2011. Neuropeptide regulation of social behavior in a monogamous cichlid fish. *Physiology & Behavior* 102: 296–303.

Overholtzer, K.L. and P.J. Motta. 1999. Comparative resource use by juvenile parrotfishes in the Florida Keys. *Marine Ecology Progress Series* 177: 177–187.

Perry, A.L., P.J. Low, J.R. Ellis and J.D. Reynolds. 2005. Climate change and distribution shifts in marine fishes. *Science* 308: 1912–1915.

Petersen, C.W., R.R. Warner, D.Y. Shapiro and A. Marconato. 2001. Components of fertilization success in the bluehead wrasse, Thalassoma bifasciatum. *Behavioral Ecology* 12: 237–245.

Planes, S., G.P. Jones and S.R. Thorrold. 2009. Larval dispersal connects fish populations in a network of marine protected areas. *Proceedings of the National Academy of Sciences* 106: 5693–5697.

Pörtner, H.O. and R. Knust. 2007. Climate change affects marine fishes through the oxygen limitation of thermal tolerance. *Science* 315: 95–97.

Pratchett, M.S. 2005. Dietry overlap among coral-feeding butterflyfishes (Chaetodontidae) at Lizard Island, northern Great Barrier Reef. *Marine Biology* 148: 373–382.

Pratchett, M.S., S.K. Wilson, M.L. Berumen and M.I. McCormick. 2004. Sublethal effects of coral bleaching on an obligate coral feeding butterflyfish. *Coral Reefs* 23: 352–356.

Pratchett, M.S., S.K. Wilson and A.H. Baird. 2006. Declines in the abundance of *Chaetodon* butterflyfishes following extensive coral depletion. *Journal of Fish Biology* 69: 1269–1280.

Pratchett, M.S., P.L. Munday, S.K. Wilson, N.A.J. Graham, J.E. Cinner, D.R. Bellwood, G.P. Jones, N.V.C. Polunin and T.R. Mcclanahan. 2008. Effects of climate-induced coral bleaching on coral-reef fishes—Ecological and economic consequences. *Oceanography and Marine Biology: An Annual Review* 46: 251–296.

Pratchett, M.S., S.K. Wilson, N.A.J. Graham, P.L. Munday, G.P. Jones and N.V.C. Polunin. 2009. Multi-scale temporal effects of climate-induced coral bleaching on motile reef organisms. In *Coral Bleaching: patterns and processes, causes and consequences*, M. van Oppen and J. Lough (eds.). Springer, Berlin, Germany pp. 139–158.

Pratchett, M.S., A.S. Hoey, S.K. Wilson, V. Messmer and N.A.J. Graham. 2011. Changes in Biodiversity and Functioning of Reef Fish Assemblages following Coral Bleaching and Coral Loss. *Diversity* 3: 424–452.

Pratchett, M.S., A.S. Hoey, D.A. Feary, A.G. Bauman, J.A. Burt and B.M. Riegl. 2013. Functional composition of *Chaetodon* butterflyfishes at a peripheral and extreme coral reef location, the Persian Gulf. *Marine Pollution Bulletin* DOI:10.1016/j.marpolbul.2012.10.013.

Pyke, G.H. 1984. Optimal foraging theory: a critical review. *Annual Review of Ecology and Systematics* 1984: 523–575.

Ralston, S. 1976a. Age determination of a tropical reef butterflyfish utilizing daily growth rings of otoliths. *Fisheries Bulletin* 74: 990–994.

Ralston, S. 1976b. Anomalous Growth and Reproductive Patterns in Populations of *Chaetodon miliaris* (Pisces, Chaetodontidae) from Kaneohe Bay, Oahu, Hawaiian Islands. *Pacific Science* 30: 395–403.

Ralston, S. 1981. Aspects of the reproductive biology and feeding ecology of *Chaetodon miliaris*, a Hawaiian endemic butterflyfish. *Environmental Biology of Fishes* 6: 167–176.

Reddon, A.R., C.M. O'Connor, S.E. Marsh-Rollo and S. Balshine. 2012. Effects of isotocin on social responses in a cooperatively breeding fish. *Animal Behaviour* 84: 753–760.

Riegl, B.M. and S.J. Purkis. 2012a. Coral Reefs of the Gulf: Adaptation to Climatic Extremes in the World's Hottest Sea. In *Coral Reefs of the Gulf*, B.M. Riegl and S.J. Purkis (eds.). Springer Netherlands pp. 1–4.

Riegl, B.M. and S.J. Purkis. 2012b. Dynamics of Gulf Coral Communities: Observations and Models from the World's Hottest Coral Sea. In *Coral Reefs of the Gulf*, B.M. Riegl and S.J. Purkis (eds.). Springer Netherlands pp. 71–93.

Riegl, B.M., S.J. Purkis, A.S. Al-Cibahy, M.A. Abdel-Moati and O. Hoegh-Guldberg. 2011. Present limits to heat-adaptability in corals and population-level responses to climate extremes. *PLoS ONE* 6: e24802.

Robertson, D.R., J.L. Ackerman, J.H. Choat, J.M. Posada and J. Pitt. 2005. Ocean surgeonfish *Acanthurus bahianus*. I. The geography of demography. *Marine Ecology Progress Series* 295: 229–244.

Sale, P.F. 1978. Reef fishes and other vertebrates: a comparison of social structures. *In: Contrasts in Behavior: Adaptations in the Aquatic and Terrestrial Environments*, E.S. Reese and F.J. Lighter (eds.). Wiley, New York, USA, pp.

Sampey, A., A. McKinnon, M. Meekan and M. McCormick. 2007. Glimpse into guts: overview of the feeding of larvae of tropical shorefishes. *Marine Ecology Progress Series* 339: 243–257.

Sano, M., M. Shimizu and Y. Nose. 1987. Long-term effects of destruction of hermatypic corals by *Acanthaster planci* infestation on reef fish communities at Iriomote Island, Japan. *Marine Ecology Progress Series* 37: 191–199.

Santangelo, N. and A.H. Bass. 2006. New insights into neuropeptide modulation of aggression: field studies of arginine vasotocin in a territorial tropical damselfish. *Proceedings of the Royal Society B: Biological Sciences* 273: 3085–3092.

Schoener, T.W. 1971. Theory of feeding strategies. *Annual Review of Ecology Evolution and Systematics* 2: 369-404.

Shokri, M.R., S.M.R. Fatemi and M.P. Crosby. 2005. The status of butterflyfishes (Chaetodontidae) in the northern Persian Gulf, IR Iran. *Aquatic Conservation: Marine and Freshwater Ecosystems* 15: S91–S99.

Smith, W., J.F. Webb and S.D. Blum. 2005. The evolution of the laterophysic connection with a revised phylogeny and taxonomy of butterflyfishes (Teleostei: Chaetodontidae). *Cladistics* 19: 287–306.

Soares, M.C., R. Bshary, R. Mendonça, A.S. Grutter and R.F. Oliveira. 2012. Arginine Vasotocin Regulation of Interspecific Cooperative Behaviour in a Cleaner Fish. *PLoS ONE* 7: e39583.

Soeparno, Y. Nakamura, T. Shibuno and K. Yamaoka. 2012. Relationship between pelagic larval duration and abundance of tropical fishes on temperate coasts of Japan. *Journal of Fish Biology* 80: 346–357.

Sponaugle, S. and R.K. Cowen. 1997. Early life history traits and recruitment patterns of Caribbean wrasses (Labridae). *Ecological Monographs* 67: 177–202.

Stallings, C.D. 2008. Indirect effects of an exploited predator on recruitment of coral-reef fishes. *Ecology* 89: 2090–2095.

Stimson, J.S. 1990. Density dependent recruitment in the reef fish *Chaetodon miliaris*. *Environmental Biology of Fishes* 29: 1–13.

Stobutzki, I.C. and D.R. Bellwood. 1997. Sustained swimming abilities of the late pelagic stages of coral reef fishes. *Marine Ecology Progress Series* 149: 35–41.

Strang, D.A. 2005. The social evolution of the butterflyfish, *Chaetodon multicinctus*: Monogamy, sexual selection, juvenile behavior, and territorial aggression. PhD dissertation, University of Hawaii, Manoa, Hawaii.

Thompson, R.R. and J.C. Walton. 2004. Peptide effects on social behavior: effects of vasotocin and isotocin on social approach behavior in male goldfish (*Carassius auratus*). *Behavioral Neuroscience* 118: 620–626.

Thompson, V.J., P.L. Munday and G.P. Jones. 2007. Habitat patch size and mating system as determinants of social group size in coral-dwelling fishes. *Coral Reefs* 26: 165–174.
Tokić, G. and D.K.P. Yue. 2012. Optimal shape and motion of undulatory swimming organisms. *Proceedings of the Royal Society B: Biological Sciences* 279: 3065–3074.
Tricas, T.C. and J.T. Hiramoto. 1989. Sexual differentiation, gonad development, and spawning seasonality of the Hawaiian butterflyfish, *Chaetodon multicinctus*. *Environmental Biology of Fishes* 25: 111–124.
Tricas, T.C., S.M. Kajiura and R.K. Kosaki. 2006. Acoustic communication in territorial butterflyfish: test of the sound production hypothesis. *Journal of Experimental Biology* 209: 4994–5004.
Trip, E.L., J.H. Choat, D.T. Wilson and D.R. Robertson. 2008. Inter-oceanic analysis of demographic variation in a widely distributed Indo-Pacific coral reef fish. *Marine Ecology Progress Series* 373: 97–109.
van Rooij, J.M., F.J. Kroon and J.J. Videler. 1996. The social and mating system of the herbivorous reef fish *Sparisoma viride*: one-male versus multi-male groups. *Environmental Biology of Fishes* 47: 353–378.
Walker, J.A., M.E. Alfaro and C.J. Fulton. submitted. Fineness ration, streamlined drag, and swimming performance in coral reef fishes. *Journal of Morphology*.
Walsh, W.J. 1987. Patterns of recruitment and spawning in Hawaiian reef fishes. *Environmental Biology of Fishes* 18: 257–276.
Warner, R.R. 1984. Mating behavior and hermaphroditism in coral reef fishes. *American Scientist* 72: 128–136.
Webb, P.W. 1984. Body form, locomotion and foraging in aquatic vertebrates. *American Zoologist* 24: 107–120.
Whiteman, E. and I. Côté. 2007. Monogamy in marine fishes. *Biological Reviews* 79: 351–375.
Wilson, E.O. 1975. *Sociobiology: the new synthesis*, Belknap Press of Harvard University Press, Cambridge, USA.
Wilson, S.K., N.A.J. Graham, M.S. Pratchett, G.P. Jones and N.V.C. Polunin. 2006. Multiple disturbances and the global degradation of coral reefs: are reef fishes at risk or resilient? *Global Change Biology* 12: 2220–2234.
Yabuta, S. 1999. Behavioral rules and tail-up display in extra-and intra-pair interactions of the butterflyfish, *Chaetodon lunulatus*. *Journal of Ethology* 17: 79–86.
Yabuta, S. 2000. Behaviors in agonistic interaction of the butterflyfish (*Chaetodon lunulatus*). *Journal of Ethology* 18: 11–15.
Yabuta, S. 2002. Uncertainty in partner recognition and the tail-up display in a monogamous butterflyfish. *Animal Behaviour* 63: 165–173.
Young, K.A., K.L. Gobrogge, Y. Liu and Z. Wang. 2011. The neurobiology of pair bonding: insights from a socially monogamous rodent. *Frontiers in Neuroendocrinology* 32: 53–69.
Zekeria, Z.A., S. Weertman, B. Samuel, T. Kale-Ab and J.J. Videler. 2006. Growth of *Chaetodon larvatus* (Chaetodontidae: Pisces) in the southern Red Sea. *Marine Biology* 148: 1113–1122.

[1]ARC Centre of Excellence for Coral Reef Studies, James Cook University, Townsville, QLD 4811, Australia.
[a]E-mail: morgan.pratchett@jcu.edu.au
[b]E-mail: karen.chongseng@my.jcu.edu.au
[c]E-mail: andrew.hoey1@jcu.edu.au
[d]E-mail: jessica.nowicki@my.jcu.edu.au
[e]E-mail: stefan.walker@jcu.edu.au
[2]School of the Environment, University of Technology, Sydney, PO Box 123, Broadway NSW 2007, Australia.
E-mail: david.feary@uts.edu.au
[3]ARC Centre of Excellence for Coral Reef Studies, Research School of Biology, The Australian National University, Building 116 Daley Road, Canberra ACT 0200, Australia.
E-mail: christopher.fulton@anu.edu.au
[4]Department of Neurobiology, Northwestern University, 2205 Tech Drive Hogan 2-160 Evanston, IL 60208, USA.
E-mail: a-dewan@northwestern.edu
[5]Red Sea Research Center, King Abdullah University of Science and Technology, Thuwal, 23955, Saudi Arabia.
E-mail: michael.berumen@kaust.edu.sa
*Corresponding author

Index

T

U

V

W

Colour Plate Section

Chapter 2

Fig. 2.5 Prey-capture diversity in the Chaetodontidae. **(A)** Facultative cleaning: *Heniochus acuminatus* cleaning Giant Sunfish at Nusa Lembogan, Indonesia; **(B)** Facultative planktivory: *Hemitaurichthys polylepis* plankton-feeding off Osprey Reef, Australia; **(C)** Invertivory: *Chelmonops curiosus* feeding on attached invertebrates at Rapid Bay Jetty, South Australia; **(D)** Soft-coral feeders: *Chaetodon capistratus* feeding on soft coral in the Caribbean. **(E)** Obligate hard-coral feeders: *Chaetodon trifascialis* feeding on scleractinian coral polyps (*Acropora hyacinthus*) in Moorea, French Polynesia [All photos have been released from copyright by the authors and editors].

Chapter 3

Fig. 3.1 Hybrid butterflyfishes. **(A)** *C. milliaris* x *C. tinkeri*, **(B)** *C. multicinctus* x *C. quadrimaculatus*, **(C)** *C. ephippium* x *C. semion*, **(D)** *C. auriga* x *C. lunula* (All photos: G. Allen).

Fig. 3.2 Heterospecific pairs of butterflyfishes; **(A)** *C. trifasciatus* and *C. lunulatus* at Christmas Island (Indian Ocean), **(B)** *C. decussatus* and *C. vagabundus* at Ashmore Reef, Western Australia. More than 10 heterospecific pairs of these species were observed at each of the respective locations. Hybrid *C. trifasciatus* x *C. lunulatus* have also been observed at Christmas Island (Photos: JP Hobbs).

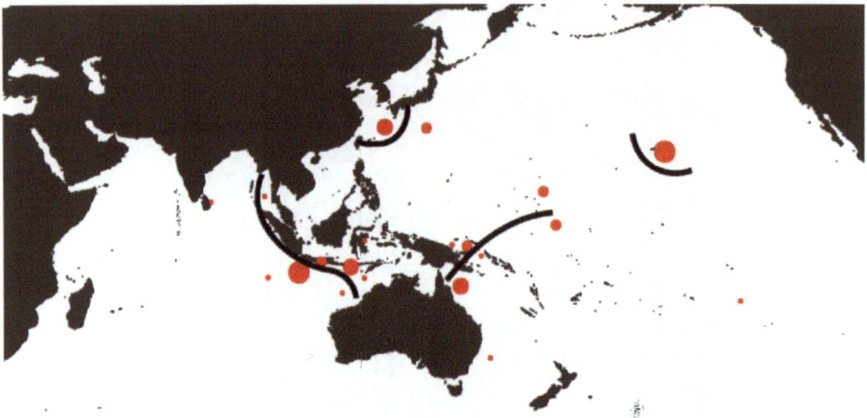

Fig. 3.3 The geographic distribution of reported locations of butterflyfish hybrids and heterospecific pairs. Some hybrids or heterospecific pairs have been observed at multiple locations and for these cases each reported location is represented by a dot (with the caveat that it is 300 km from another reported location). The size of the circle is representative of the number of species pairs that have formed heterospecific pairs or produced hybrids at a reported location, with the largest circles symbolising four cases. Not all instances from Table 3.1 are presented because locations were not always reported. The positions of four recognised biogeographic borders are illustrated with black lines (Blum, 1989; Bellwood and Wainwright, 2002; Hsu et al., 2007).

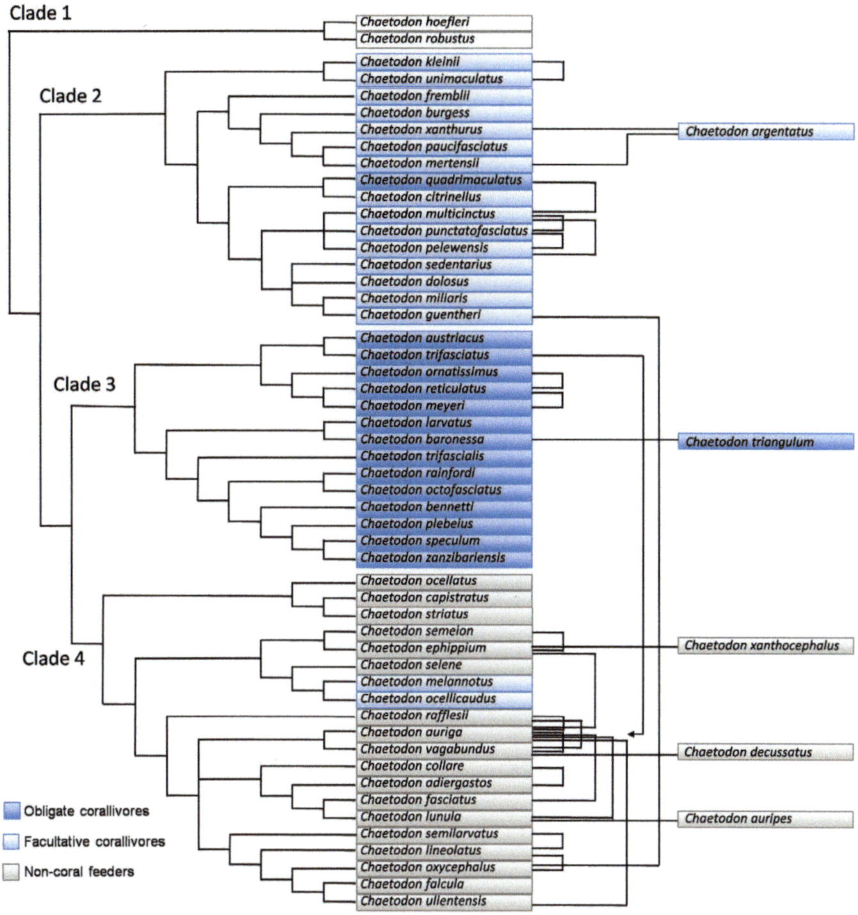

Fig. 3.4 Phylogenetic relationships among pairs of butterflyfishes known to hybridise. The phylogeny presented is based on Fessler and Westneat (2007), indicating 4 major clades that broadly correspond with major trophic guilds (indicated by colours). Hybridisations involving species within the current phylogeny are shown by linkages between species on the right side of the figure.

Chapter 4

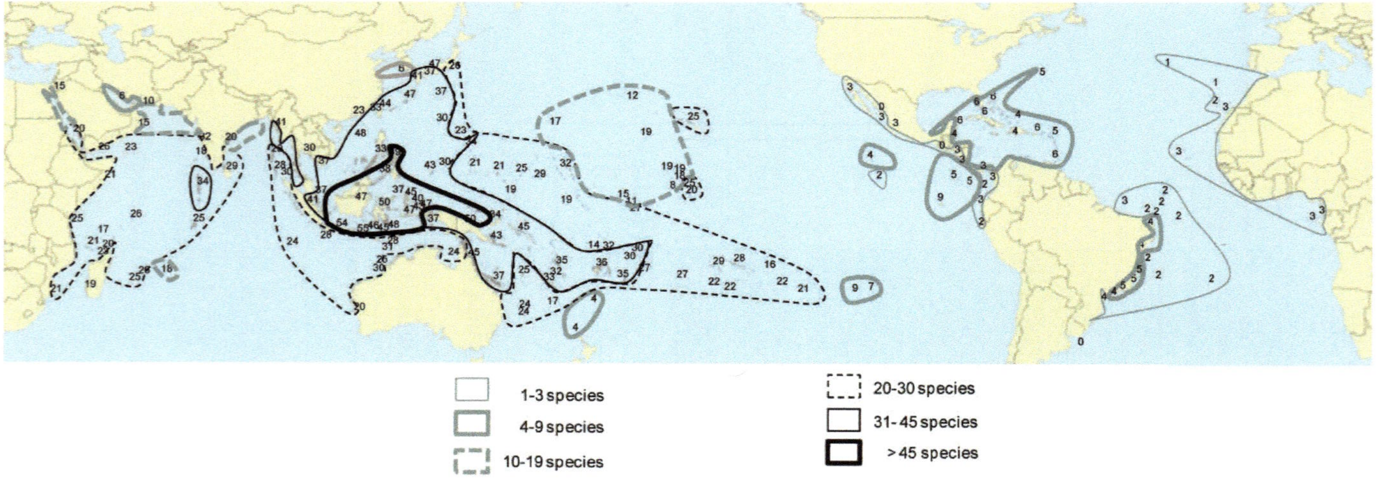

Fig. 4.1 World wide map of the alpha diversity distribution of butterflyfishes.

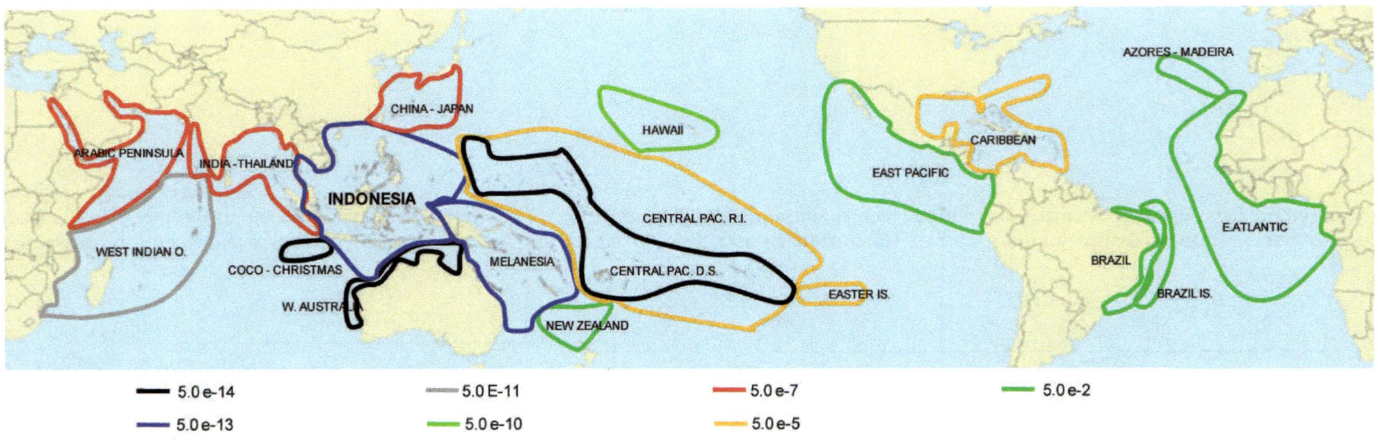

Fig. 4.2 Map of the 19 biogeographic regions determined by stepwise clustering of 175 butterflyfish species lists around the tropics. The lines indicate Ward's minimum variance distance levels.

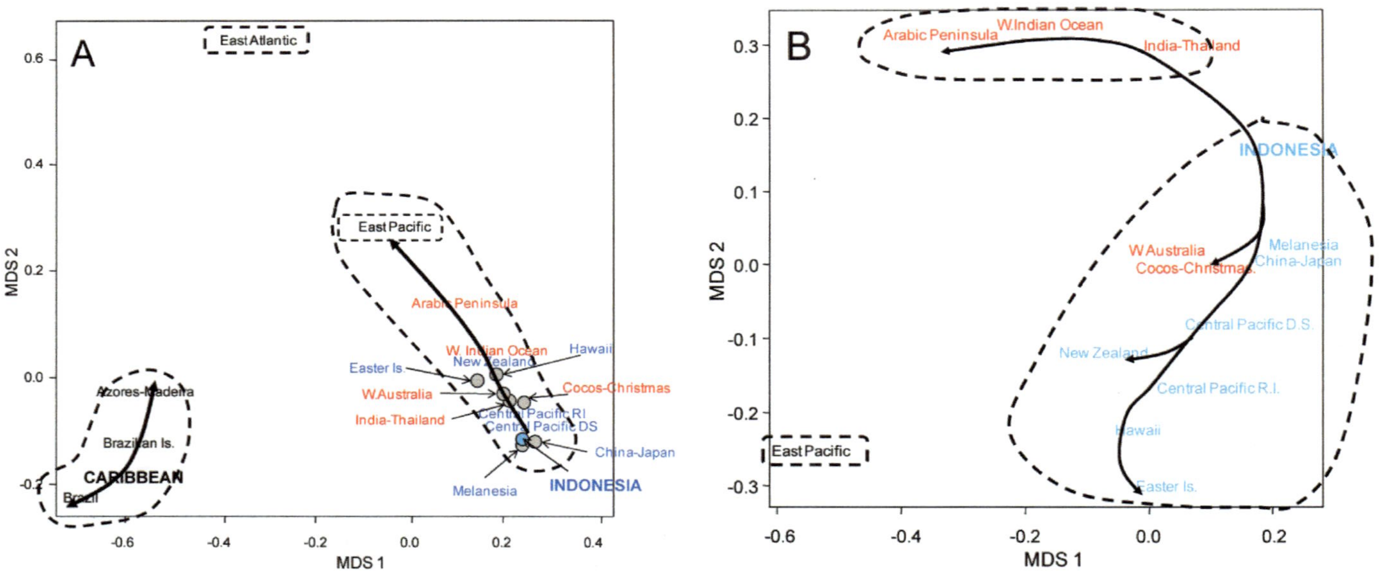

Fig. 4.4 Multidimensional scaling **(A)** all 19 biogeographic regions, **(B)** only Indo-Pacific regions. Regions from the Indian Ocean are in red; regions from the west and central Pacific are in blue. MDS1 and MDS2 are axes of the multidimensional scaling. The arrows indicate the shortest pathways from one region to another.

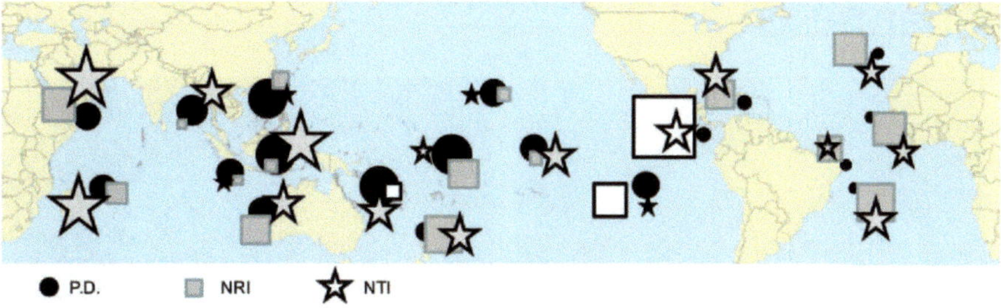

Fig. 4.14 Regional values of PD, NRI and NTI for Chaetodontidae across the world. PD: phylogenetic distance, radius of circle is proportional to value of PD. NRI: Net Relatedness Index, size of square is proportional to NRI value, grey represents >0 and white <0 values. NTI: Nearest Taxon Index, size of star is proportional to NTI value, black represents values <–0.3; +0.3>, grey represents >0.3 and white <–0.3 values.

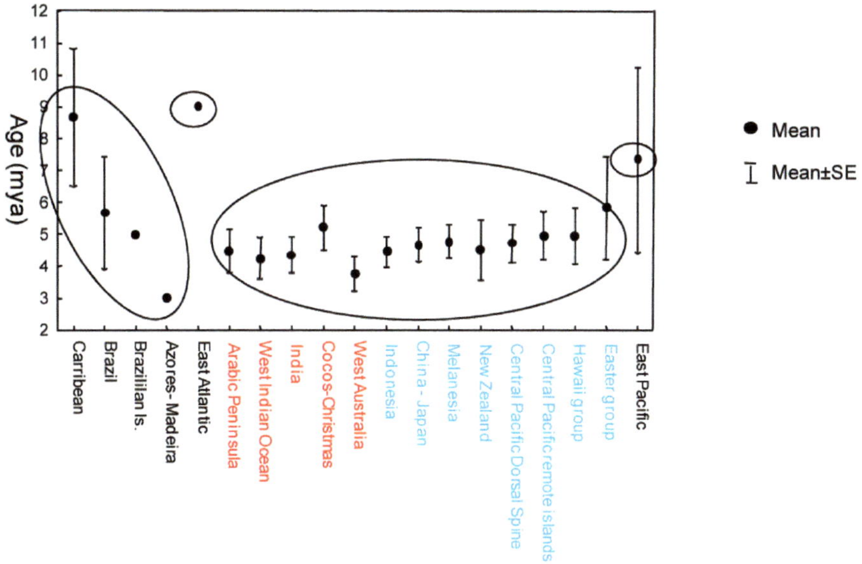

Fig. 4.15 Average age (mya) of species in each of the 19 biogeographic regions. Regions are ordered according to longitude. Indian Ocean regions are in red, west Pacific regions are in blue.

Chapter 6

Plate 6.1 Closely related *Chaetodon* species with strong similarities in their appearance are also similar in terms of dietary composition and specialization. The four species shown are a) *C. lunulatus*, from the Pacific, b) *C. trifasciatus*, from the Indian Ocean, c) *C. austriacus*, found predominantly within the Red Sea, and d) *C. melapterus*, from the north-west Inidian Ocean, including the Red Sea and Persian Gulf. All photos by Luiz Rocha.

Chapter 10

A

B

Fig. 10.1 Good and bad candidates for indicator species. A) *Chaetodon trifascialis* is an obligate coral feeding butterflyfishes that exhibits strong association with specific coral types, mainly *Acropora*, Photo by Luiz Rocha; B) *Chaetodon auriga* is a generalist feeder, which sometimes feeds on corals, but may also be abundant on reefs dominated by macro-algae. Photo by Michael Berumen.

Chapter 11

Fig. 11.1 Wild harvesting of coral reef butterflyfishes is undertaken by artisanal fishers and marine aquarium collectors; A) *Chaetodon* butterflyfishes for sale at the food markets in Papete, Tahiti. Photo by Morgan Pratchett. B) *Heniochus* butterflyfishes in a public aquarium. Photo by Charles Delbeek.

Chapter 12

Plate 12.1 *Chaetodon ulientensis* in an aquarium full of scleractinian corals at the Waikiki Aquarium. This butterflyfish is not an obligate coral feeder, but like many *Chaetodon* butterflyfishes will normally eat small quantities of coral (Pratchett, 2005). If however, there is plenty of alternative prey, or there is a large quantity of coral, then any effects from feeding on display corals should be minimal. This specimen has been on display since 1986 and would only pick at soft corals and anemones. Photograph by author.

Plate 12.2 *Heniochus diphreutes* are easy to keep in aquaria, but do require large tanks. This exhibit is at the Monterey Bay Aquarium. They normally feed in mid-water (on plankton) and will readily take most foods offered in aquaria. Unlike some *Heniochus* sp. they rarely feed on benthic prey, and will leave coral alone. Photograph by author.

Chapter 13

Fig. 13.1 Pre-settlement larvae (tholichthys) of *Chaetodon plebeius*, showing the conspicuous head shield, which is characteristic of larval butterflyfishes. Photo by C. Wen.